Studies of Brain Function, Vol. 14

Studies of Brain Function

Volumes already published in the series:

1 *W. Heiligenberg*
Principles of Electrolocation and Jamming Avoidance in Electric Fish

2 *W. Precht*
Neuronal Operations in the Vestibular System

3 *J. T. Enright*
The Timing of Sleep and Wakefulness

4 *H. Braak*
Architectonics of the Human Telencephalic Cortex

5 *H. Collewijn*
The Oculomotor System of the Rabbit and Its Plasticity

6 *M. Abeles*
Local Cortical Circuits

7 *G. Palm*
Neural Assemblies

8 *J. Hyvärinen*
The Parietal Cortex of Monkey and Man

9 *E. Zrenner*
Neurophysiological Aspects of Color Vision in Primates

10 *U. Bässler*
Neural Basis of Elementary Behavior in Stick Insects

11 *G. A. Orban*
Neuronal Operations in the Visual Cortex

12 *M. Heisenberg, R. Wolf*
Vision in Drosophila

13 *Yu. I. Arshavsky, I. M. Gelfand, G. N. Orlovsky*
Cerebellum and Rhythmical Movements

14 *G. Roth*
Visual Behavior in Salamanders

Gerhard Roth

Visual Behavior in Salamanders

With 113 Figures

Springer-Verlag Berlin Heidelberg New York
London Paris Tokyo

Professor Dr. Dr. GERHARD ROTH
Universität Bremen
Fachbereich Biologie
Abteilung Verhaltensphysiologie
Bibliothekstraße
2800 Bremen, FRG

ISBN-13: 978-3-642-72715-3 e-ISB N-13: 978-3-642-72713-9
DOI: 10.1007/978-3-642-72713-9

Library of Congress Cataloging-in-Publication Data. Roth, Gerhard. Visual behavior in salamanders. (Studies of brain function; vol. 14) Includes index. 1. Salamanders – Behavior. 2. Salamanders – Physiology. 3. Salamanders – Anatomy. 4. Vision. 5. Amphibians – Behavior. 6. Amphibians – Physiology. 7. Amphibians – Anatomy. I. Title. II. Series: Studies of brain function; v. 14. [DNLM: 1. Wl ST937KF v.14/QL668.C2R845v] QL668.C2R68 1987 597.95 87-12134

Softcover reprint of the hardcover 1st edition 1987

Media conversion: Daten- und Lichtsatz-Service, Würzburg
Offsetprinting and binding: Konrad Triltsch, Graphischer Betrieb, Würzburg
2131/3130-543210

Preface

Salamanders are subject to misconceptions even among vertebrate zoologists and physiologists. They are often said to exist only in northern temperate zones, being bound to aquatic or very moist cool habitats. In reality, more than half of all salamander species live in subtropical and tropical zones, exclusively in the New World. Again, more than half of the salamander species have become totally independent of aquatic habitats following the loss of a free larval stage. Many of the subtropical and tropical salamanders have become adapted to rather high temperatures up to 26–28 °C.

The brain and the sensory systems of salamanders are often considered to be primitive, and their behavior is thought to be simple and uninfluenced by learning. However, careful studies show that the salamander brain possesses virtually all the anatomical and functional properties found in anurans, which are usually regarded as being much more evolved with respect to the guidance of comparable behavior. Most of the salamander species not only possess a highly efficient visual system, which is the topic of the present work, but can orient themselves almost as effectively by means of olfaction, vibration sense, and electroreception. Furthermore, it has recently been shown that at least part of their behavior, especially that concerned with feeding and prey preferences, is strongly influenced by individual experience.

Accordingly, in the present work the author wants to establish a better knowledge of the visually guided behavior of these animals and the related sensory, neural, and motor mechanisms. Special emphasis has been laid on the presentation of natural life conditions and behavior within the natural habitat as far as these are known. Ecology and life history are two disciplines that have been largely neglected by neuroethologists, and only very recently have vertebrate as well as invertebrate neuroethologists felt the necessity to "go to the field". The fruitful combination of behavioral, neuroanatomical, and neurophysiological approaches within the discipline "neuroethology", has to be complemented by the ecological aspect. In salamanders this is easily done because their ecology, especially of the European and North American species, has been well studied, and the respective studies of tropical species are undergoing rapid progress.

Another necessary expansion of neuroethology is in the direction of comparative work. Past neuroethology has concentrated too often on the study of one single species. Neuroethological studies of closely related species that have undergone adaptive radiation are the key to understanding the principles of evolution in the sensory and neural systems controlling behavior. The largest salamander family, the Plethodontidae, is one of the vertebrate groups about whose evolutionary history, ecology, life habits, morphology, and behavior we are best informed, making them the ideal group for neuroethological studies.

The first chapter of this monograph contains a brief survey of the classification of the order Caudata with comments on the main characteristics of the families and genera. Special emphasis is laid on those groups which have been preferred objects of neuroethological studies, the Salamandridae and the Plethodontidae. I considered it important to give at least some information about habitats, life habits, and evolutionary history of the salamanders as a necessary background for the discussion of visually guided behavior and its morphological and physiological basis.

In the second chapter data on visually guided behavior are presented. Information on natural diet is included, which gives an important basis for behavioral experiments in the laboratory as well as for electrophysiological studies. The third chapter deals with the morphology of the eye and the retina, but also includes functional aspects, especially in the discussion of the consequences of miniaturization, a process that has taken place several times within the order Caudata.

The fourth chapter contains a general description of the salamander brain and a detailed presentation of the neuroanatomy of the central visual system, mostly based on modern tracing techniques. The fifth chapter presents electrophysiological data on the response properties of neurons in the retina, in the diencephalon and especially in the tectum opticum, as well as results from electric stimulation and lesion experiments.

The last chapter gives a summary of the behavioral, anatomical, and physiological data and a discussion of the question whether these data, combined with relevant data from anurans, are sufficient to establish at least the outlines of a model for neural guidance of visual behavior in salamanders and in amphibians in general.

A large number of colleagues have made very helpful comments on this book. I am especially indebted to D. B. Wake, R. Jaeger, K. Nishikawa, W. Himstedt, and G. Manteuffel for reading large parts of the manuscript and making numerous, valuable suggestions, R. Linke has made substantial contributions to Chapter 3, Chr. Manteuffel, G. Rettig, B. Fritzsch and H. Wicht to Chapter 4, and D. Ingle, G. Manteuffel and U. an der Heiden to Chapter 6, for which I am very grateful.

GERHARD ROTH

Contents

Chapter 1 Taxonomy, Geography and Ecology of Salamanders 1

Chapter 2 Visually Guided Behavior 14

2.1 Feeding Behavior 14
2.1.1 Functional Morphology of Feeding in Salamanders 14
2.1.2 Prey Preferences and Natural Diets of Salamanders 27
2.1.3 Experimental Analysis of Visually Guided Prey-Capture Behavior 34
2.1.4 Interaction Between Visual and Olfactory Guidance of Feeding Behavior 59
2.1.5 Ontogeny of Feeding Behavior and the Role of Learning 65
2.2 Optomotor Behavior. 82
2.3 Visual Mate Recognition in Newts. 85

Chapter 3 Morphology and Function of the Salamander Eye. 89

3.1 General Morphology. 89
3.2 Morphology of the Retina 95
3.2.1 Light Microscope Overview of the Structure . 95
3.2.2 Classification of Retina Ganglion Cells . . . 106
3.2.3 Neuronal Connections in the Retina 111
3.2.4 Size, Number, and Density of Photoreceptors and Retina Ganglion Cells 114
3.3 Functional Aspects 119
3.3.1 Dioptrics of the Salamander Eye 119
3.3.2 Accommodation 121
3.3.3 Visual Acuity 124
3.4 Eye Degeneration in Troglobitic Salamanders . . . 127

Chapter 4 Anatomy of the Visual System 129

4.1 General Anatomy of the Salamander Brain 129
4.1.1 Forebrain 132
4.1.2 Diencephalon 136
4.1.3 Praetectum 139
4.1.4 Midbrain 140
4.1.5 Cerebellum 143
4.1.6 Medulla Oblongata and Cervical Spinal Cord 144
4.2 Visual and Visuomotor Projections and Pathways 152
4.2.1 Visual Afferents to the Brain 152
4.2.2 Comparison with Anuran Retinofugal Projections 161
4.2.3 Topic Organization of Retinal Projections to the Diencephalon and the Tectum 162
4.2.4 Development of Retinofugal Projections . . . 166
4.3 Intrinsic Organization and Cytoarchitecture of the Optic Tectum and the Diencephalon 170
4.3.1 Tectum Opticum 170
4.3.2 Diencephalon 180
4.4 Central Visual and Visuomotor Pathways 182
4.4.1 Tectal Afferents and Efferents 182
4.4.2 Afferent and Efferent Pathways of the Praetectum 190
4.4.3 Afferents and Efferents of the Thalamus . . 192
4.4.4 Summary and Comparison of Tectal and Diencephalic Afferents and Efferents 192

Chapter 5 Neurophysiology of Visually Guided Behavior 199

5.1 Electrophysiological Methods 199
5.2 Response Properties of Retina Ganglion Cells to Preylike Stimuli 201
5.3 Color Coding in Retina Ganglion Cells 210
5.4 Response Characteristics of Tectal Neurons . . . 214
5.4.1 *Salamandra salamandra* 215
5.4.2 *Hydromantes italicus* 222
5.5 Recordings from Diencephalic Visual Areas . . . 232
5.5.1 Rostral Thalamus 232
5.5.2 Caudal Doral Thalamus 233
5.5.3 Praetectum 234
5.6 Electric Stimulation Experiments 239

5.7 Brain Lesion Experiments 240
5.8 Responses of Tectal Neurons to Monochromatic Light 241

Chapter 6 Conclusions and Speculations on the Neural Guidance of Visual Behavior in Salamanders 243

6.1 Neural Guidance of Feeding Behavior 243
6.1.1 Prey Object Identification 243
6.1.2 Neural Mechanisms of Prey Recognition . . 244
6.1.3 Classification of Tectal Cells Involved in Prey Recognition 248
6.1.4 The Constitution of Tectal Response Properties 250
6.1.5 Tecto-Pretectal Interaction 252
6.1.6 The "Recognition Module" Concept 255
6.1.7 The Problem of Stimulus Size Recognition . . 257
6.1.8 Distance Estimation Mechanisms 263
6.1.9 Neural Guidance of the Feeding Sequence and the Problem of Sensori-Motor Integration 269
6.2 Enemy and Barrier Avoidance. 275
6.3 Optomotor Behavior 275
6.4 Conclusion: What is Known and What is not yet Known About Neural Guidance of Visual Behavior in Salamanders? 277

Abbreviations Used in Figures 279

References 285

Subject Index 299

1 Taxonomy, Geography and Ecology of Salamanders

The following overview on taxonomy, geography, and ecology of salamanders is based mainly on the most recent checklist of the amphibian species of the world, edited by Frost (1985), the classical study on plethodontid salamanders by Wake (1966), Thorn's work on European salamanders (Thorn 1968), *Grzimeks Tierleben* Vol. 5, Part 2, (Freytag 1970), as well as various field guides (Behler and King 1979, Smith 1978, Conant 1975, Stebbins 1951). More detailed literature on the ecology of salamanders is cited within the text of this chapter.

The amphibian order Caudata (Urodela), comprising salamanders and newts, consists of nine families with about 360 species. Thus, they are larger than the second amphibian order Gymnophiona (Apoda or Caecilia) which consists of five families with 164 species restricted to the tropics and subtropics, and much smaller than the third amphibian order Anura (Salientia), comprising frogs and toads which have a worldwide distribution with 23 families with about 3,400 species (Frost 1985).

The nine families of the order Caudata are the following (Edwards 1976, Hecht and Edwards 1977, Frost 1985): (1) Cryptobranchidae; (2) Hynobiidae; (3) Sirenidae; (4) Amphiumidae; (5) Proteidae; (6) Salamandridae; (7) Dicamptodontidae; (8) Ambystomatidae; (9) Plethodontidae. The Cryptobranchidae and Hynobiidae together form the suborder Cryptobranchoidea, the Amphiumidae, Proteidae, and Salamandridae form the suborder Salamandroidea; the Dicamptodontidae, Ambystomatidae, and Plethodontidae form the suborder Ambystomatoidea; the Sirenidae form by themselves the suborder Sirenoidea.

Suborder Cryptobranchoidea. The suborder Cryptobranchoidea, comprising the families Hynobiidae and Cryptobranchidae, represents a primitive form of salamander. They are distinguished from all others in having strictly external fertilization. The hyobranchial apparatus of adults still shows two pairs of epibranchials, the first one being fused with the respective ceratobranchial.

Family Hynobiidae. The family Hynobiidae is represented by the genera *Hynobius* (+ *Salamandrella*), *Pachyhynobius*, *Paradactylodon*, *Pachypalaminus*, *Onychodactylus*, *Ranodon*, *Batrachuperus*, and *Liuia*. *Hynobius* (18 species) occurs in Japan, Korea, eastern China, and in extreme northeastern Europe (the species *H. keyserlingii* is placed in a separate genus, *Salamandrella*, with *S. keyserlingii* as its only species), *Pachypalaminus* (one sp.) occurs in Japan, *Onychodactylus* (two species) occurs in Japan, Korea, Far Eastern part of the USSR, and in north-eastern China, *Ranodon* (two sp.) is found in Central Asian USSR and western China. All these genera live on land during the non-reproductive period near mountain streams. *Batrachuperus* (six sp.) is found in western China, Iran and Afghanistan, *Paradactylodon* (one sp.) occurs only in Iran, *Liuia* (one sp.) and *Pachyhynobius* (one sp.) occur only in China. These latter genera are permanently aquatic, living in mountain streams. Many species of the hynobiid genera show signs of adaptation to periodic or permanent aquatic life, such as reduction or loss of lungs (in order to reduce or abolish buoyancy), as is the case in *Ranodon* and *Onychodactylus*.

Family Cryptobranchidae. The family Cryptobranchidae or "giant salamanders" comprises the genera *Cryptobranchus* (one sp.) and *Andrias* (two sp.). All giant salamanders are strictly aquatic. They undergo only partial metamorphosis: the larval gills are reduced and lungs are formed, but in *Cryptobranchus* (Hellbender, total length up to 70 cm) the branchial openings persist. This genus occurs in clear, fast-flowing streams and rivers of eastern and southeastern North America. The genus *Andrias* occurring in Japan and China and inhabiting similar habitats as *Cryptobranchus*, represents the largest living salamanders with a total length sometimes beyond 150 cm.

Suborder Sirenoidea. The suborder Sirenoidea is represented by only one family, Sirenidae. It comprises two genera: *Pseudobranchus* (one sp.) and *Siren* (two sp.). They are restricted to the southeastern USA. They are permanently aquatic larvae with external gills and no hind legs. They occur in swamps, shallow warm water, and ponds of the coastal plains. The species of the family Sirenidae are able to survive droughts by secreting a moisture-sealing cocoon over the body.

Suborder Salamandroidea, Family Salamandridae. Among the suborder Salamandroidea, comprising the families Salamandridae, Proteidae, and Amphiumidae, the family Salamandridae is the most derived and by far the largest one. It occurs with 14 genera and 53 species in Europe, North Africa, East Asia, Asia Minor, and North America. Almost all of them undergo complete metamorphosis. The genus *Pleurodeles* (two sp.) occurs in Western Europe and North Africa. It lives mostly in stagnant waters. *Tylototriton* (seven sp.) occurs in East Asia; these animals are terrestrial (with water breeding), inhabiting mostly mountain forests. The genus *Salamandra* (two sp.) is found in the whole western Palaearctic region with the species *S.*

salamandra, whereas the species *S. atra* occurs only in the alpine regions of Europe and in mountain habitats of the western Balkan peninsula.

S. salamandra (11 subspecies) is found mostly in deciduous woods of lowlands and at elevations up to 1,800 m, sometimes even higher, where it is found under logs and rocks, sometimes inside rotten logs, in rock crevices or burrows of small mammals, all near small streams. It is a very sedentary salamander. It usually produces aquatic larvae, but subspecies from the Iberian peninsula and from Corsica produce already metamorphosed offsprings. The species *S. atra* is strictly viviparous. It occurs in the European Alps and the mountain regions of the western Balkan peninsula, at elevations of 400–3,000 m (mostly 800–2,000 m). Due to its viviparity it is strictly terrestrial; it prefers dark and moist habitats under stones and logs and in crevices.

The genus *Mertensiella* (two species) is found in the Caucasus Mountains and in Turkey and the Greek island Karpathos. It lives at elevations of 0–2,800 m, mostly near mountain streams. The genus *Salamandrina* (one sp., *S. terdigitata*) has only four toes on the hind legs and reduced lungs. It lives on the lower western slopes of the Italian Appenines beside mountain streams. In a similar habitat the genus *Chioglossa* (one sp., *Ch. lusitanica*) is found; it also possesses reduced lungs and, like *Salamandrina*, a highly protrusible tongue. It occurs in northwestern Spain and Portugal.

The genus *Triturus*, the European newts, comprises 12 species and like the genus *Salamandra*, it has a distribution over almost all of Europe. The most common and widely distributed species is *Triturus vulgaris* (nine subspecies) which can be found from southern Italy and Asia Minor to Ireland, Scandinavia, and western Siberia. It shows very variable habitats, but prefers small to medium-sized waters with rich vegetation on open areas, from plains to elevations up to 1,400 m. On land it is found under stones, wooden planks, logs, and bark. It is often found together with *T. alpestris*. *Triturus italicus* is found in central and meridional Italy up to 830 m in habitats similar to *T. vulgaris*. *T. helveticus* is found in northwestern Europe including Great Britain and northern Spain at elevations up to 800 m. It prefers small, cool, clear waters, slowly running streamlets and ditches, which are not inhabited by other newts.

Triturus cristatus, whose four subspecies *T. c. cristatus, T. c. carnifex, T. c. dobrogicus* and *T. c. karelinii* are now recognized as species (Bucci-Innocenti et al. 1983), has a large distribution from France to the Russian Ural Mountains, and from Great Britain and southern Scandinavia to Italy and western Yugoslavia. Its habitat is stagnant waters and litter in littoral zones up to 1,000 m elevation. *T. marmoratus* is found on the Iberian Peninsula and in southern France. *T. alpestris* can be found in northern Spain, central Europe, northern Italy, the Balkan peninsula, and southern Russia. Its optimum habitats are small to medium-sized sunny or half-shaded shallow waters near forests. In the Alps this species occurs at elevations up to

3,000 m. The remaining species of the genus *Triturus* are *T. vittatus* (Caucasus and western Asia Minor), *T. montandoni* (Carpathian mountains), and *T. boscai* (western Iberian peninsula).

The newt genera *Paramesotriton* (five sp., including the former genus *Hypselotriton*) and *Pachytriton* (one sp.) are found in Vietnam and China. They are mostly aquatic and inhabit mountain streams, but can sometimes be found on land outside the breeding season. The genus *Cynops* (seven sp.) is found in Japan and China. It occurs in habitats very similar to the previous ones. The genus *Neurergus* (four sp.) occurs in the Middle East.

The genus *Euproctus* (three sp.) occurs in European mountain regions of the Pyrenees, Corsica, and Sardinia. It lives in cold streams and lakes, and its lungs are reduced.

The newts of the New World are separated in an eastern genus, *Notophthalmus*, and a western one, *Taricha*. *Notophthalmus* (three sp.) from eastern North America is primarily aquatic. It lives in quiet stretches of streams, swamps, ditches, and neighboring damp woodlands. The transforming animals leave the water for 1–3 years as brightly colored "efts", then return to water assuming adult characters.

Taricha (three sp.), occurring in western North America, is primarily terrestrial in its adult form. It lacks an eft stage. Its habitats are ponds, lakes, and slow-moving streams and adjacent humid forest or grasslands or cool mountain streams and surrounding coastal redwood forests.

Wake and Özeti (1969) recognized four groups within the family Salamandridae: Group A: *Salamandra* (+ *Mertensiella*), *Chioglossa, Salamandrina*; Group B: *Triturus, Euproctus, Neurergus, Paramesotriton, Cynops, Hypselotriton, Pachytriton, Taricha, Notophthalmus*; Group C: *Pleurodeles*; Group D: *Tylototriton*. Groups C and D are considered as the most primitive forms, group A as the most derived one. *Chioglossa* and *Salamandrina* are the most specialized species of the whole family. Among the "newt group" B, *Pachytriton* is regarded as the most developed species and *Cynops* as the most primitive one (see also Estes 1981).

Family Proteidae. The family Proteidae comprises two genera, *Necturus* and *Proteus*. Both genera are aquatic, permanent larvae, having external gills, but also possessing lungs. *Necturus* (five sp.) occurs in eastern North America. It inhabits medium-sized to large streams, lakes, and rivers from muddy shallows to greater depths (up to 27 m). *Proteus* (one sp., *P. anguinus*) is a permanent larva living in subterranean waters of the Yugoslavian karst. There it occurs at water temperatures of 6–9 °C. It has an elongated head and rudimentary, pointlike eyes, which in the adult animal are covered by skin.

Family Amphiumidae. The family Amphiumidae comprises only one genus with three species. Its range is restricted to southeastern North America.

This eellike salamander does not undergo complete transformation. Adults lose their gills but retain one pair of gill slits. The species live in swamps, bayous, drainage ditches, and streams.

Suborder Ambystomatoidea. The suborder Ambystomatoidea comprises the three families Ambystomatidae, Dicamptodontidae, and Plethodontidae. With the exception of one genus, all genera of this suborder are found in the New World.

Family Dicamptodontidae. The family Dicamptodontidae, comprising the genera *Dicamptodon* (three sp.) and *Rhyacotriton* (one sp.), is found only on the Pacific coast of the USA. *Dicamptodon* inhabits damp, coastal forests. The species includes rather large salamanders (*D. ensatus* up to 30 cm, *D. copei* to 17 cm total length). They either transform into adults facultatively (*D. ensatus*) or remain larvae permanently. The genus *Rhyacotriton* (*R. olympicus*) lives in forests in or beside streams or springs. Their larvae all transform.

Family Ambystomatidae. The family Ambystomatidae or mole salamanders comprises two genera: *Ambystoma* and *Rhyacosiredon*. The Ambystomatidae commonly burrow in the ground; for that behavior they have a robust body and stout limbs. They are externally distinguished from the lungless salamanders of the family Plethodontidae by lack of the nasolabial groove. The whole family occurs in North America, from southeastern Alaska and Labrador to the southern edge of the Mexican plateau.

The members of the genus *Ambystoma* or true mole salamanders (27 species) are typically terrestrial, but all lay their eggs in water with the exception of *A. opacum* (marbled salamander) which lays its eggs on land but has aquatic larvae. Larvae of the species *A. tigrinum, A. talpoideum*, and *A. gracile* may transform to land-dwelling adults or may remain permanent aquatic larvae. The famous "axolotl" (*A. mexicanum*) is a permanent larva which, however, transforms in captivity after treatment with thyroxin. All other species undergo complete metamorphosis. The terrestrial adult members of the genus typically inhabit deciduous or pine forests, live under debris near swamps and ponds, and along stream banks from sea level to 3,500 m. Some are also found in damp meadows and in subterranean habitats. The most widespread species, *A. tigrinum*, is frequently found in cellars and mammal burrows. In the west, *A. tigrinum* is often perennibranchiate. Some specimens become large and reach more than 33 cm total length.

The genus *Rhyacosiredon* (four sp.) occurs in Mexico and lives in clear and cold mountain streams.

Family Plethodontidae. The family Plethodontidae, or lungless salamanders, is by far the largest group of salamanders with 27 genera and more than 230 species, some of which have been described only recently (for a more complete survey cf. Wake 1966, Wake and Lynch 1976, Wake and Elias 1983). The Plethodontidae are believed to have originated in eastern North America, in the Appalachian highlands, where they inhabited mountain brooks (Wilder and Dunn 1920, Dunn 1926, Wake 1966). Apparently as an adaptation to this habitat they underwent reduction and finally complete loss of lungs, a process which occurred several times among other salamanders living in similar habitats, such as *Salamandrina, Chioglossa*, and *Rhyacotriton*. Plethodontids compensate their lunglessness by oxygen uptake through a highly vascularized skin and mouth cavity.

Despite deriving from aquatic ancestors, most plethodontids have become strictly terrestrial, i. e., they do not produce free aquatic larvae, but lay eggs on land from which already metamorphosed salamanders hatch (direct development). The members of the family occupy a large variety of terrestrial habitats, from cold and barren highlands to heavily forested lowlands. Some of them have become arboreal, and a few of them are semifossorial. More primitive species of plethodontids live in eastern North America, primarily in the Appalachian Mountains. Derived species have spread to western North America, Central and South America, and Europe.

Terrestrial plethodontids can be relatively abundant and in some regions may exceed birds and small mammals in biomass density (Merchant 1972, Burton and Likens 1975).

The family Plethodontidae is divided into two subfamilies, Desmognathinae and Plethodontinae.

Subfamily Desmognathinae. The subfamily Desmognathinae comprises the genera *Desmognathus, Leurognathus*, and *Phaeognathus*. Salamanders of the genus Desmognathus (dusky salamanders, 11 sp.) are found in the eastern part of North America from Quebec and Nova Scotia to the coastal plains of the Mexican Gulf. The genus is found mostly in aquatic to semiaquatic habitats; the largest species (*D. quadramaculatus*) is the most aquatic one, living in mountain streams throughout life; the smallest species (*D. imitator, D. ochrophaeus, D. aeneus, D. wrighti*) are the most terrestrial ones living under logs, rocks, and other moist debris. *D. wrighti* and *D. aeneus* have no aquatic larval stage. The larger species are heavy feeders on smaller salamanders from the same species or genus.

The aquatic genus *Leurognathus* (one sp.) occurs in mountain brooks in the southern Appalachian Mountains. *Phaeognathus* (one sp.) is primarily terrestrial and restricted to a small area in Alabama, USA. It has an elongate form with very short legs and lives in burrows.

Subfamily Plethodontinae. The subfamily Plethodontinae comprises all other Plethodontid species. It is divided into three major groups (tribes): Hemidactyliini, Plethodontini, and Bolitoglossini.

Tribe Hemidactyliini. The tribe Hemidactyliini comprises eight genera: *Gyrinophilus, Pseudotriton, Stereochilus, Eurycea, Typhlotriton, Typhlomolge, Haideotriton*, and *Hemidactylium*.

The members of the genus *Gyrinophilus* (two sp.) are large, stout salamanders which occur in springs, seepages, rivulets, and cave waters from Quebec to Alabama and Georgia. One species (*G. palleucus*, Tennessee cave salamander) is permanently larval; it can be transformed in the laboratory when treated with thyroxin.

The genus *Pseudotriton* (two sp.) occurs in springs, pools, and seepages in eastern North America. These salamanders are strikingly colored and very stout and short-legged. *P. ruber* is more terrestrial than *P. montanus*.

The one species of the genus *Stereochilus* is nearly aquatic. It lives in swampy, slow-moving waters of the Atlantic coastal plains from Virginia to Georgia.

The 13 species of the large genus *Eurycea* range over most of the eastern United States with isolated populations on the Edwards Plateau of Texas. They are all slender, active animals which inhabit a large variety of habitats, from stream sides and swamps to springs, caves, and subterranean waters. Of all species, *E. bislineata* has the largest distribution, ranging from the mouth of the St. Lawrence River, Quebec, to northern Florida and west to Ontario, Illinois, and the Mississippi River. It lives in rock-bottomed brooks, springs, seepages, river swamps, and flood plain bottoms in coastal plains to damp forest floors at high elevations (near 2,000 m). The species *E. junaluska, E. longicauda, E. quadridigitata, E. aquatica*, and *E. guttolineata* which together with *E. bislineata* form the eastern group of the genus, occur in similar habitats and all undergo full metamorphosis. The species of the central Texas groups, *E. neotenes, E. nana, E. latitans, E. troglodytes*, and *E. tridentifera* inhabit waters of caves, sinkholes, springs, and underground channels of the eroded limestone of the Edwards Plateau in Texas. All these species are permanently larval showing an uncolored belly and external gills. The eyes are mostly small or covered by skin and the snout flattened.

The remaining species *E. lucifuga* (Virginia, West Virginia, Alabama, Oklahoma) and *E. multiplicata* (Missouri, Arkansas, Oklahoma, Kansas) also show a strong tendency towards cavernicolous life. Whereas *E. lucifuga* regularly transforms, *E. multiplicata* is facultatively perennibranchiate. Permanent larvae are found most often among cave-dwelling populations.

Three more genera of the Hemidactyliini, *Typhlomolge, Haideotriton* and *Typhlotriton*, are permanent cave dwellers. *Typhlomolge* (two sp.) and *Haideotrition* (one sp.) are blind, unpigmented, and all neotenic. *Typhlomolge*, showing long, spindly legs and a flattened snout, occurs in the Edwards Plateau, Texas; *Haideotriton* has no flattened head. It occurs in the limestone area of southwestern Georgia and adjacent Florida. *Typhlotriton* (one sp.) is the only cave salamander which completely transforms to a terrestrial stage.

Larvae may develop in outside waters but return to caves before transformation and spend the rest of their lives there.

The members of the genus *Hemidactylium* (one sp., *H. scutatum*) are diminutive inhabitants of swamps, sphagnum bogs, and neighboring woods. They range from southern Canada to the southern Appalachian Mountains and to Florida and Louisiana, with several disjunct southern and western populations.

Tribe Plethodontini. The tribe Plethodontini comprises the genera *Plethodon, Aneides*, and *Ensatina*.

The genus *Plethodon* with more than 30 species (several cryptic species will soon be described) represents one of the largest genera of plethodontid salamanders. It occurs in many regions of eastern and northwestern North America and in restricted parts of the Rocky Mountains. These salamanders are all strictly terrestrial with direct development of offspring. One of the most common and dispersed species is *P. cinereus* (red-backed salamander) which occurs from western Ontario and Newfoundland to North Carolina and Indiana. It is found in cool moist coniferous, mixed, and hardwood forests up to 1,700 m. Another widespread and more southern species is *P. glutinosus* (slimy salamander). It occurs from New York State to Florida and west to Texas and Arkansas. It is a nocturnal salamander living in shaded ravine slopes, shale banks, wooded floodplains, and near cave entrances from sea level up to 1,700 m. Its skin glands secrete a thick, gluey substance. Most of the remaining species have rather restricted ranges, inhabiting woodlands and rocky areas. *P. neomexicanus*, which is found isolated from other *Plethodon* species in the Jemez Mountains, New Mexico, occurs at elevations between 2,000–2,800 m.

The genus *Aneides* (five sp.) is found in the Appalachian Mountains, the high mountains of New Mexico, and along the west coast of North America. *A. aeneus* is scansorial and lives on vertical surfaces, *A. hardii* is a log dweller. *A. ferreus* and *A. lugubris*, the two western species, are arboreal, sometimes found in trees up to 15 m high in low elevation forests.

Ensatina (one sp., *E. eschscholtzii*) is a stout-bodied, long-legged terrestrial salamander that is found only on the Pacific coast of North America, ranging from southern British Columbia to southern California.

Tribe Bolitoglossini. The tribe Bolitoglossini is by far the largest group of plethodontid salamanders and comprises, with about 165 species, nearly half of all salamanders in the world. This group comprises the genera *Hydromantes* and *Batrachoseps* and 11 other genera with about 150 species in the neotropics. All species of these genera are strictly terrestrial and show direct development. The Bolitoglossini are the most derived group of plethodontids and of the urodeles in general. They have highly specialized feeding mechanisms (cf. Chap. 2) and have invaded the most diverse habitats.

The genus *Hydromantes* (five species or more) occurs in terrestrial habitats in limestone or granite areas in the mountains of California (three sp., *H. shastae*, *H. brunus*, *H. platycephalus*), in southern France, and northern Italy (one sp., *H. italicus*, which includes probably at least two separate species) and on Sardinia (one sp., *H. genei*, a diverse taxon which probably should be divided into two to three separate species). It is the most geographically disjunct genus among amphibians although the American and European species remain very similar in ecology, morphology, and genetics (Wake et al. 1978). During the dry season, *H. shastae* and the Italian and Sardinian species *H. italicus* and *H. genei* inhabit caves. The European species most probably reached Europe during early Tertiary across the Bering land bridge between northwestern North America and northeastern Asia (Wake et al. 1978).

The genus *Batrachoseps* (worm salamander, eight to nine sp.) comprises small, attenuate salamanders with four toes and very long tails. All are terrestrial and live in underground burrows (e.g. earthworm holes). They are found along the western coast of North America from northern Oregon to northern Baja California (Mexico). Two species (*B. aridus* and *B. campi*) are found in the Californian desert in moist microhabitats. *B. attenuatus* (Californian slender salamander) is found from southwestern Oregon along coastal ranges to the region of Monterey Bay. It inhabits coastal mountains and interior foothills, redwood forests, and grasslands with scattered trees. It can be very common in parks and backyards, too.

The diverse group of tropical salamanders comprises nearly half of the known living species of salamanders. They are found throughout the breadth of the neotropics, from northern Mexico to southern Brazil, Bolivia, and Peru. The success of these salamanders in tropical environments is largely related to their terrestrial life history mode: eggs are deposited in nonaquatic sites, and hatchlings are terrestrial. These salamanders are thereby freed from a biphasic life cycle and from dependence on free surface water, features which constrain most other amphibians.

Ancestors of the tropical salamanders moved southward into America during early Tertiary times, when tropical forests extended far north of their present limits. Subsequent drying and warming of the climate has segregated the ancestral stock in disjunct mesic uplands. These geographic isolates formed the basis of the subsequent radiation of tropical plethodontids. Foci of evolutionary activity were: (1) southeastern Sierra Madre Oriental of Mexico; (2) northern nuclear Central America (= land between the Isthmus of Tehuantepec and the Nicaraguan depression); (3) the Talamancan region of Costa Rica and western Panama (Wake and Lynch 1976).

The presence of extensive highlands was the most important factor for the evolution of tropical species: high mountain masses increase precipitation by trapping rain clouds. Plentiful moisture distributed over a long elevational gradient favors the development of complex vertical zonation. "The presence of suitable arboreal microhabitats, especially bromeliads, appears to be the single most important factor which increases the potential for a local tropical habitat to support additional salamanders" (Wake and Lynch 1976). Tropical salamander species are segregated by elevation and microhabitat. Most tropical salamanders occur at moderate to high elevations and have small geographic ranges, but there are many lowland species, some of which have relatively extensive distribution.

Tropical plethodontids are presently divided among 11 genera: *Bolitoglossa* (70 sp. or more); *Chiropterotriton* (9 sp.); *Pseudoeurycea* (25 sp.); *Oedipina* (16 sp); *Thorius* (11 sp. or more); *Parvimolge* (1 sp.); *Lineatritition* (1 sp.); *Nyctanolis* (1 sp.); *Bradytriton* (1 sp.); *Dendrotriton* (5 sp.); and *Nototriton* (6 sp.).

The genus *Bolitoglossa* with over 70 species is by far the largest genus in the family Plethodontidae and in the whole order Caudata. It is by far the most diverse and broadly distributed group and probably will be divided into at least two separate genera. It ranges from northeastern Mexico through Central America to central Bolivia and the mouth of the Amazon River in Brazil and probably further south. Two groups, *Bolitoglossa*-alpha and *Bolitoglossa*-beta are recognized; the alpha group contains the more derived and the beta group the more primitive species, although relative to other salamander genera all species of *Bolitoglossa* are highly derived forms. Lowland species have evolved fully webbed hands and feet, often in parallel and convergently in different lineages (Wake and Brame 1968, Alberch 1981).

The *Bolitoglossa*-alpha group with 46 species extends from the Central Sierra Madre Oriental of Mexico to the Amazon Basin region and further south. The *Bolitoglossa*-beta group contains fewer species (27) than alpha. Its distribution is centered in Guatemala and Chiapas (Mexico); no species occurs south of Central Costa Rica.

In both groups, a wide range in adult body size is found. *B. dofleini* of the beta group exceeds 10 cm snout-vent length (SVL); *B. robusta* of the alpha group reaches 11-cm SVL. On the other hand, *B. rufescens* of the beta group hardly exceeds 4 cm and *B. minutula* of the alpha group never grows beyond 3.7-cm SVL. In both groups, lowland species are often either exceptionally large or unusually small. Both contain arboreal species with partially webbed hands and feet and "acrobatic" climbing abilities. Arboreal forms like *B. franklini* of Guatemala and *B. savagei* of Colombia, inhabit bromeliads in cloud forests at intermediate elevations. There are morphologically generalized species like *B. rostrata* of Guatemala and *B. subpalmata* of Costa Rica, which have maintained the ancestral way of life in terrestrial, upland microhabitats (cf. Vial 1968). Within both groups occupation of lower elevations has been paralleled by an increase in arboreality. Some of the species occur at high elevations up to 4,000 m, such as *B. hypacta* of Colombia. Several groups, however, reach sea level, such as the two closely related species *B. rufescens* and *B. occidentalis* from the beta group. These small salamanders with fully webbed hands and feet live in the leaf axils of bananas, bromeliads, and other large-leaved plants.

Lowland species of the alpha group are found in the "mexicana" subgroup. Its members are moderate to large in size (e.g., *B. platydactyla*, SVL 10 cm) again with extensively to fully webbed feet. They extend from Veracruz, Mexico, to central Honduras. They are often found in bromeliads, in the leaf axils of bananas, on the surfaces of leaves and in sugarcane. Sometimes, however, they can be found in terrestrial habitats, e.g., under logs, in leaf litter, in burrows, etc.

The genus *Chiropterotriton* (nine sp.) extends from northeastern Mexico to the Isthmus of Tehuantepec. Several of these species have terrestrial habits, but others are adapted to scansorial or arboreal ways of life. They are generalized in skull morphology.

The five species of the new genus *Dendrotriton*, which in part comprises species of the former beta group of the genus *Chiropterotriton* (Elias and Wake 1983), are locally very abundant. The best known species is *D. bromeliacia*, a species which is found only in a small section on the lower slopes of Volcan Tajumulco in southwestern Guatemala, at elevations of 1,900–2,700 m (Wake and Lynch 1976, Wake 1986). They are abundant in arboreal bromeliads. It and its related species all have long tails, fairly long limbs, a slender body, and large, very frontally oriented eyes. Many of the *Dendrotriton* species have large nostrils, especially as juveniles.

The genus *Pseudoeurycea*, with 25 species, is the second largest among tropical salamanders. It is found from northeastern Mexico to southern Guatemala. The species range in size from the relatively small *P. mystax*, which does not exceed 5-cm SVL, to the largest living terrestrial salamander, *P. bellii*, which reaches 13-cm SVL or more. Most species are generalized in morphology and occupy unspecialized microhabitats in forested montane areas. A few species inhabit arboreal bromeliads. In northern Central America the genus is restricted to humid habitats at high elevations. The species *P. rex* reaches extremely high densities above 3,400 m and can still be found beyond 4,000 m. *P. gadovii* seems to live at the highest altitude of all salamanders at nearly 5,000 m (Wake and Lynch 1976).

The species *P. brunnata* and *P. goebeli* are species of similar size occurring in sympatry. They live in western Guatemala where they inhabit terrestrial microhabitats between 2,400 and 3,000 m elevation. These two species may partially divide the food resources on the basis of prey size; *P. goebeli* possesses much larger teeth and stronger jaw muscles than *P. brunnata* (Wake and Lynch 1976).

The genus *Thorius* comprises at least 15 species that are among the smallest tetrapods. They occur in southern Mexico, north of the Isthmus of

Tehuantepec. Most species are terrestrial, some (e.g., *Th. narisovalis*) are log-dwellers and at least one (*T. dubitus*) is occasionally found in bromeliads. Due to their small size, the species of this genus are able to inhabit the leaf litter microhabitat to an extent not possible for larger salamanders. In this respect and in several morphological aspects, *Thorius* is comparable to the Californian plethodontid genus *Batrachoseps*. The morphology of *Thorius* is highly derived and unique among plethodontid salamanders (Wake 1966, Hanken 1980, 1982, 1983) as it relates to size reduction, or miniaturization. The smallest of all salamanders is *T. pennatulus* in which males may achieve sexual maturity at sizes of 16-mm SVL. It occurs in several localities in Veracruz in the vicinity of Volcan Orizaba at elevations of 800–1,200 m which is the lowest elevation of all species of the genus.

T. pulmonaris and *T. narisovalis* are two terrestrial species from the montane forests of Cerro San Felipe, Oaxaca (Mexico) living at different elevations (*T. pulmonaris* 2,000 m, *T. narisovalis* 2,600–3,000 m) and different microhabitats: *T. pulmonaris* is typically found in leaf litter, *T. narisovalis* under the bark of fallen logs. *T. dubitus* and *T. troglodytes* are found in the Veracruz mountains at elevations of about 3,000 m. *T. dubitus* occurs under moss and other plants, whereas *T. troglodytes* is found under rocks. The remaining species of the genus occur normally under fallen logs.

Oedipina is a highly specialized genus with 16 species which ranges from southern Mexico through Central America to eastern Colombia and northern Ecuador. All species have strikingly elongated forms, mostly with greatly reduced limbs, small heads, and very long tails. They are characteristically lowland to lower montane forms. The highest elevation of 2,500 m is reached by *O. altura* and *O. paucidentata* in the Cordillera de Talamanca. The genus is primarily found in fossorial habitats, deep surface under well-set cover, and in burrow systems, with the exception of *O. elongata* and *O. carablanca* which live within rotten logs or under slabs of bark of fallen logs.

The genus *Parvimolge* is found in Veracruz, Mexico. Its single species, *P. townsendi*, resembles the species of *Thorius* in size and habits.

The single species of *Lineatrition*, *L. lineola*, is a highly specialized, elongate, semifossorial form similar to some *Oedipina* species, with regard to extremely shortened limbs and a very long tail. It occurs in central Veracruz, Mexico.

The recently described genus *Nyctanolis*, with one species *N. pernix*, combines many ancestral morphological traits in an ecological specialized form (Elias and Wake 1983). Its relatively large-sized single species occurs at the Caribbean slope of nuclear Central America. It is arboreal, has long limbs, and a long whiplike tail.

The new genus *Bradytriton* (one sp., *B. silus*) lives in an area of very humid cloud forest of the northeastern slopes of the Sierra de los Cuchumatanes in northwestern Guatemala (Wake and Elias 1983). It is a short-snouted, stout terrestrial salamander. It has a vertically compressed tail, small limbs, and diminutive syndactylous hands and feet.

The new genus *Nototriton* comprises in part the beta group of the former genus *Chiropterotriton* (Wake and Elias 1983). Its six species range from Guatemala to Costa Rica. They are all diminutive bromeliad and moss-bank specialists and represent, together with the genera *Oedipina* and *Thorius*, the most derived plethodontids.

2 Visually Guided Behavior

2.1 Feeding Behavior

Salamanders possess an unusual variety of sensory mechanisms for the detection of prey. These include vision, olfaction, vibration sense, mechanoreception, and electroreception by means of the lateral line system. The first three senses are all used during terrestrial feeding, whereas vision, olfaction, lateral line system, and electroreception are used during aquatic feeding.

A number of aquatic salamanders are totally blind as adults, so that they have to rely on nonvisual mechanisms for the detection of prey (i. e. olfaction, mechanoreception, or electroreception). Others live under environmental conditions such as in muddy water where vision is of limited value. For all other salamanders vision is by far the most important sense involved in the guidance of feeding. Visual cues are used even under conditions of very dim light, and the use of olfaction and vibration sense seems to be restricted to feeding on very dark nights or underground (e. g., leaf litter, crevices, or periodically frequented caves). Under sufficient light conditions vision seems to dominate strongly over the other senses. This phenomenon will be discussed in detail later in this chapter.

2.1.1 Functional Morphology of Feeding in Salamanders

There is a close connection between the sensory guidance of feeding behavior, the behavioral feeding strategy, and the functional morphology of the feeding apparatus, including the jaws. This is especially true for visual guidance of feeding behavior. As will be demonstrated later, a very specialized, greatly extendable and high-speed tongue makes feeding on fast moving objects through the use of a low-energy "ambush strategy" possible, but this system requires more highly evolved capabilities of the visual system, for example with regard to prey recognition and depth perception, than is necessary for a salamander with an unspecialized, massive tongue of short range. The functional morphology of the feeding apparatus will, therefore, be discussed here in detail.

As shown in Fig. 2.1, in salamanders, as well as in amphibians in general, two distinct types of feeding motor responses are found (Wake 1982): one is

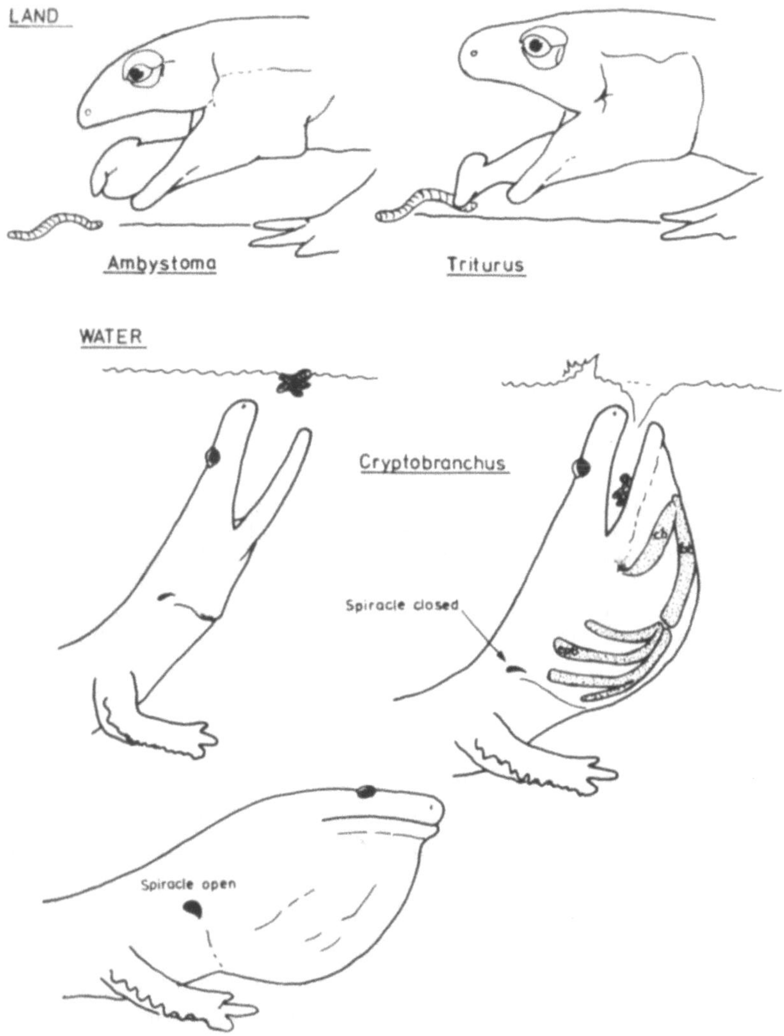

Fig. 2.1. Examples of terrestrial and aquatic feeding in salamanders. Hyobranchial apparatus in *Cryptobranchus* is *stippled.* (After Wake and Sweet unpublished)

present in permanently or temporarily aquatic salamanders including larval stages, the "Saugschnappen" or "gape and suck" type which consists of a rapid opening of the mouth at the same moment at which the throat is expanded. This results in a rapid inflow of water including the prey item. The tongue is rather immobile and plays only a minor role in this feeding sequence. The other type is found in terrestrial salamanders and includes a movement of the tongue out of the mouth such that the prey is caught primarily by gluing it onto the tongue pad. In many salamanders this is combined with the use of the jaws and teeth, especially if prey items are large and possess defense reactions. Other groups of salamanders, for example, the

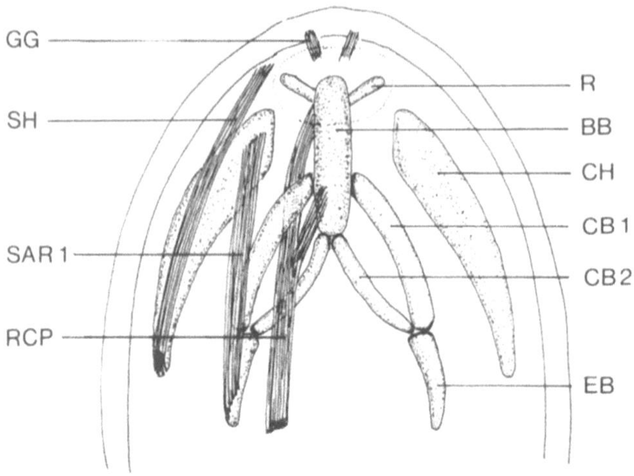

Fig. 2.2. Generalized hyobranchial apparatus and associated muscles in lung-breathing salamanders. During breathing, simultaneous contraction of *SH*, *SAR1*, and *RCP* expands the skeleton and the buccal cavity to fill the lungs with air. During feeding, *SAR1* pulls the skeleton forward, GG pulls the tongue pad (*stippled outlines*) over the front of the lower jaw, and *RCP* retracts it. Due to the stiffness which is related to its dual function, the motility of the skeleton is strongly restricted. For abbreviations, see list. (After Wake unpublished)

Bolitoglossini group of plethodontid salamanders, never make use of their jaws during feeding. The terrestrial feeding responses of salamanders differ primarily with respect to the mechanisms underlying tongue protrusion and the reach of the tongue.

In most aquatic and terrestrial feeding mechanisms the hyobranchial apparatus plays a decisive role (Figs. 2.2 and 2.3). It develops from the skeleton of the hyoid and branchial arches of the larvae. In salamandrids, ambystomatids, and plethodontids this apparatus consists of an unpaired median basibranchial (BB) which lies in the floor of the mouth and is located far anteriorly, a short distance behind the mandibular symphysis. One to two pairs of radial elements, or radii, are attached to the anterior end of the basibranchial. Two pairs of ceratobranchials (CB) articulate with the posterior part of the BB. The first CB articulate with the BB near its midpoint and the second CB with the BB at its posterior end. The first and second CB on either side extend posteriorly, approaching each other, and together articulate with the epibranchial (EB). The length of this element varies greatly among salamanders.

At each side of this apparatus a pair of ceratohyals (CH) lie in the floor of the mouth. They do not make contact with each other or with elements of the remaining hyobranchial apparatus; in plethodontids their posterior end is cylindrical and hooked, while the anterior portion forms a flattened blade

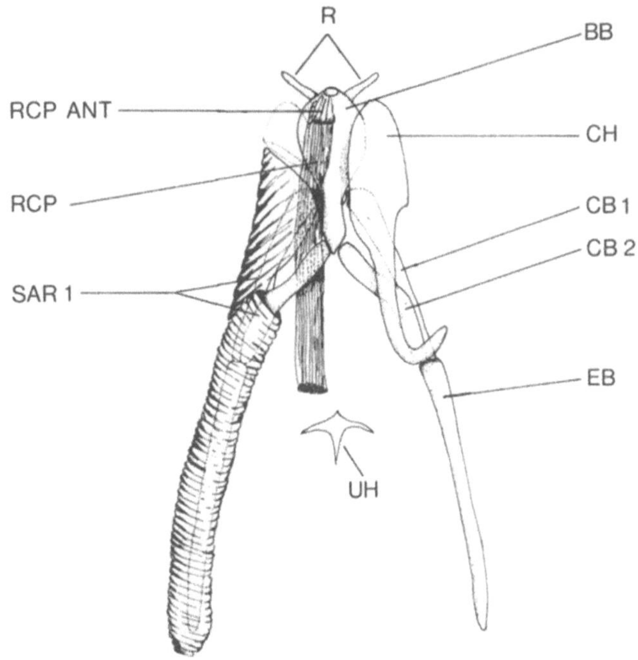

Fig. 2.3. Hyobranchial apparatus and associated muscles in a free-tongued plethodontid salamander, *Eurycea bislineata*. Here, the tongue skeleton is elongated and foldable. The *SAR1* originates from the ventral surface of the *CH*, then sweeps posteriorly to envelop the posterior parts of the two *CB* and all of the *EB*. The *RCP* passes anteriorly between the *CH* to insert on a common myocomma with the *RCP* anterior. For abbreviations, see list. (After Lombard and Wake 1977)

(Fig. 2.3). The CH is attached posteriorly by the hyoquadrate ligaments to the suspensoria. The final element of the apparatus, if present at all, is the urohyal which is a remnant of the larval second basibranchial and lies at the juncture of the rectus cervicis superficialis and geniohyoideus muscles.

The tongue pad, or tongue muscle proper, is situated at the anterior end of the BB. The BB lies in its base and the radii extend into the pad. The pad surface is covered by a large number of mucous glands.

The main muscles associated with the hyobranchial skeleton are:

1. Subarcualis rectus I (SAR1, Fig. 2.3) This large muscle encircles the caudal end of the EB (or first CB in those species which lack EB; see below) and extends rostrally along this element, forming a muscular sheath. More anteromedially it attaches broadly to the ventral surface of the flattened anterior part of the ceratohyals.
2. Rectus cervicis profundus (RCP, Fig. 2.3). This muscle, a direct continuation of the rectus abdominis muscle, originates from the puboischium. The muscle extends forward mediolaterally along the body axis and passes below

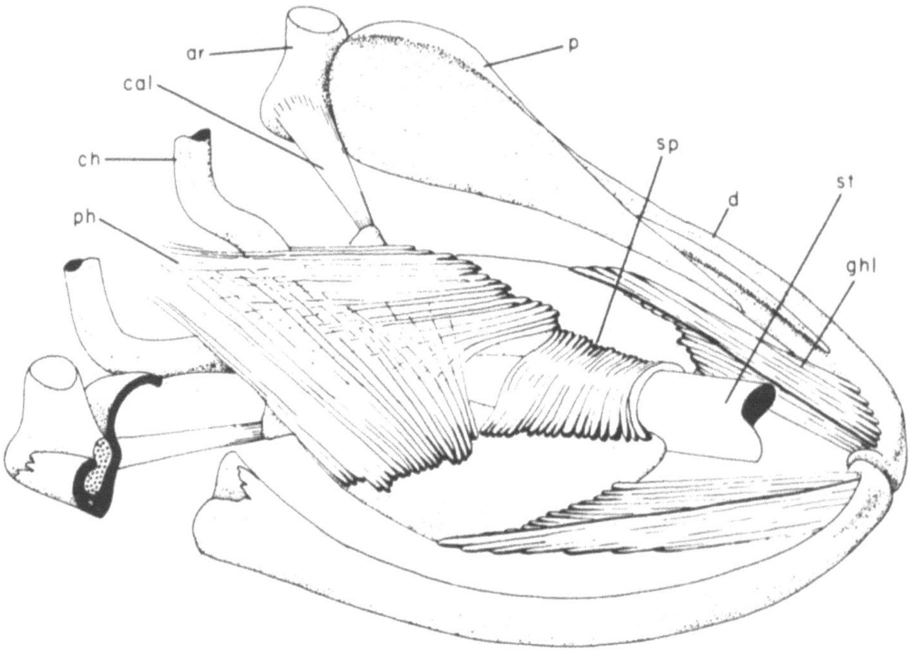

Fig. 2.4. Generalized view of the mandibular region in a free-tongued plethodontid salamander. View from above and in front of the lower jaw. The stalk of the tongue (*st*) passes through a cylinder formed by the two ceratohyals (*ch*), the floor of the mouth, and geniohyoideus medialis and subarcualis rectus muscles (not shown). The entire cylinder is directed by the geniohyoideus lateralis muscles (*ghl*) working antagonistically with the ligaments. The tongue stalk emerges from a hole in the floor of the mouth, and the anterior margin of the suprapeduncularis muscle (*sp*). *ar* Articular; *cal* cerato-articular ligament; *ph* pharyngeus; *p* prearticular. (From Lombard and Wake 1977)

the second CB and above the first. It enters the tongue pad inserting in various ways dorsal to the apex of the BB.

3. Subhyoideus (Fig. 2.2). This muscle originates on the posterior tip of the CH and extends anteriorly, parallel and ventral to the SAR1. It inserts near the mandible on the fascia of the m. intermandibularis posterior.

4. Geniohyoideus (GH). The geniohyoideus medialis (GHM) arises from the ventral surface of the mandible, just lateral to the symphysis, and extends posteriorly, parallel to the body axis, on the urohyal where it inserts. The geniohyoideus lateralis (GHL) arises lateral to the GH medialis and extends to the anterior part of the ceratohyal (Fig. 2.4).

5. Genioglossus (GG, Figs. 2.3, 2.5a, and 2.6). It extends posteriorly from the ventral surface of the mandible on each side of the mandibular symphysis, above the geniohyoideus muscles, and inserts dorsally in the substance of the rostral part of the tongue.

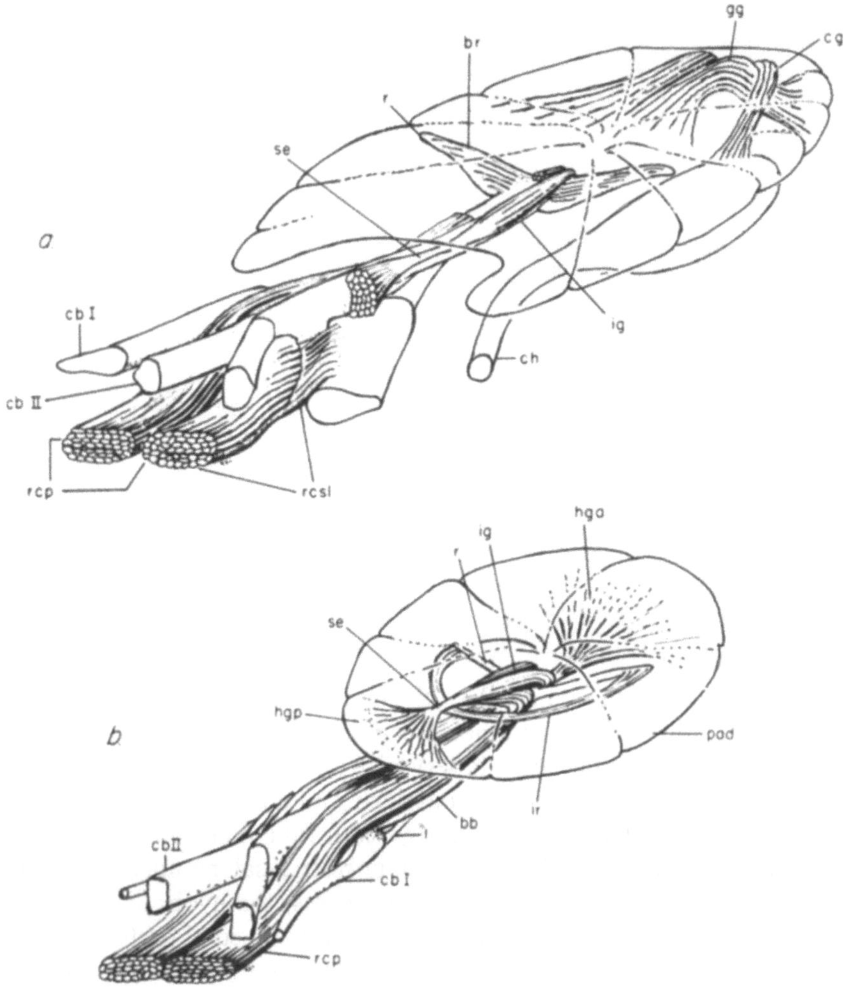

Fig. 2.5 a, b. Tongue tips and associated structures in *Plethodon* **a** and *Bolitoglossa* **b**. In **a** the hyoglossus muscle, and in **b** the ceratohyals are not shown. Note the absence of the genioglossus muscle in *Bolitoglossa*. For abbreviations, see list. (From Lombard and Wake 1977)

6. Tongue pad muscles involved in shaping the tongue pad during feeding (Fig. 2.5):

a) Hyoglossus (HG). A small muscle which orginates on the anterior tip of the BB and extends posterodorsally to insert into the substance of the tongue pad.
b) Basiradialis (BR). A small- to moderate-sized muscle extending from the dorsal side of the apex of the BB to the ventrolateral margin of the radii, near their tip.
c) Interradialis (IR). This small unpaired muscle extends between the distal tips of the radii above the BB.

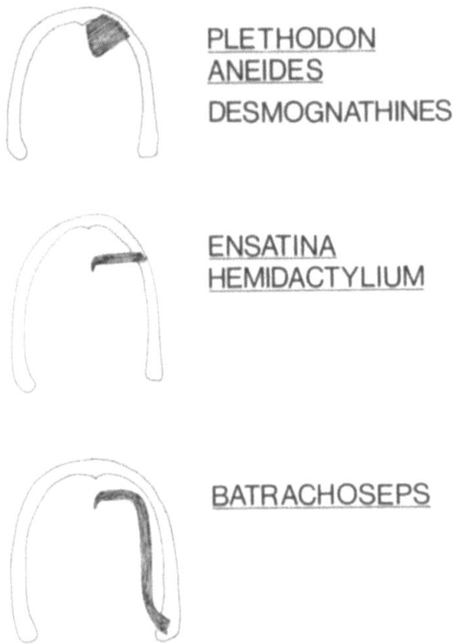

Fig. 2.6. Diagrammatic ventral views of the lower jaw of plethodontid salamanders, showing the course of the genioglossus muscle on one side. (From Lombard and Wake 1977)

In all generalized salamanders, the hyobranchial apparatus is involved in a dual function in larval and metamorphosed animals: respiration and feeding. During the larval stage it serves to move the gill arches for aquatic respiration and to expand the throat during aquatic feeding or "Saugschnappen". After metamorphosis it retains both functions in those salamanders which remain permanently aquatic. In terrestrial salamanders, after substantial reorganization and loss of some elements, the hyobranchial apparatus develops into a buccal pump for respiration: the medial parts of the apparatus are drawn posteriorly and ventrally to expand the buccal cavity. Then the nostrils are closed, and the buccal cavity is constricted to force air into the lungs.

The presence of both an aquatic larval stage and lung-breathing in metamorphosed animals seems to be a major constraint on the evolution of highly specialized feeding mechanisms by means of the tongue. The abolishment of one of these two constraints alone does not necessarily lead to such mechanisms, as is shown by some hynobiids, ambystomatids, and salamandrids that are lungless or have no aquatic larvae but not both. Only the combination of reduction of both constraints has led, as will be shown, to highly specialized feeding mechanisms (Wake 1982, Roth and Wake 1985a).

In Hynobiidae, Dicamptodontidae, and Ambystomatidae we find distinct biphasic life cycles which represent important constraints on the evolution of the hyobranchial apparatus. This results in very conservative feeding mechanisms. During feeding, *Ambystoma tigrinum* relies mostly on the use of its

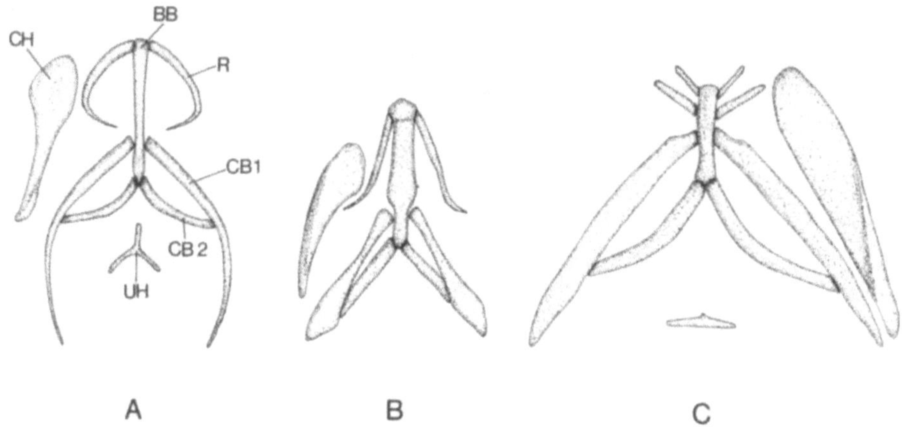

Fig. 2.7 A–C. Hyobranchial apparatus of **A** *Chioglossa lusitanica*, **B** *Salamandrina terdigitata*, **C** *S. salamandra*. Note the absence of epibranchials and the elongation of the first ceratobranchials. This is most prominent in **A**. For abbreviations, see list. (After Özeti and Wake 1969)

jaws. The hyobranchial apparatus is moved forward to propel the tongue, but there is only a minor degree of projection of the tongue pad beyond the snout (Larsen and Guthrie 1975; cf. Fig. 2.1)

I will discuss here the morphology of the hyobranchial apparatus of the two most evolved families of urodeles, the Salamandridae and the Plethodontidae, which also show the most evolved feeding mechanisms. The presentation is based mostly on the studies of Wake (1966, 1982), Özeti and Wake (1969), and Lombard and Wake (1976, 1977).

Within the family Salamandridae, the three genera *Salamandra* (including *Mertensiella*), *Chioglossa*, and *Salamandrina* have tongues specialized for feeding in terrestrial situations; all other genera have rather poorly developed tongues which are exclusively or mainly adapted for aquatic feeding. In all newts which rely upon "Saugschnappen" the hyobranchial apparatus is extensively ossified and is thus a relatively rigid structure. The contraction of the subarcualis rectus and subhyoideus muscles pulls the apparatus forward. A more or less simultaneous contraction of the rectus cervicis muscles holds the medial and anterior parts of the apparatus. This counteraction of the protractor and retractor muscles results in a forward movement of only the distal parts of the apparatus and in a great expansion of the mouth. At the same moment the mouth is opened by contraction of the depressor mandibulae, and water, together with the prey, is sucked in. Relaxation of the subarcualis rectus and the subhyoideus and continued contraction of the rectus cervicis restores the tongue and its skeleton.

The hyobranchial apparatus of *Salamandra, Chioglossa* and *Salamandrina* differs from that of all other salamandrid genera by lack of separate epibranchial elements (Fig. 2.7). In *Chioglossa* and *Salamandra* the first ceratobranchials extend caudally beyond the posterior end of the second CB, thus functionally replacing the epibranchials. *Chioglossa* has by far the the longest first CB, which curves dorsally around the margin of the throat.

In *Salamandra* the basibranchial (BB) bears two pairs of radii which both extend in an anterolateral direction. *Chioglossa* and *Salamandrina*, like most species of Salamandridae, possess only one pair of radii. These, however, are exceptional because they extend caudally and are extremely long, in both cases almost as long as the BB. In *Chioglossa* the radii curve backwards and then inwards, their tips being situated at the level of the rostral end of the first CB; in *Salamandrina* the radii are relatively straight, extending caudolaterally with laterally curving tips. In both genera, the radii have highly mobile joints with the BB. They are capable of extensive rotation, flipping the tongue pad out of the mouth.

Of the three salamandrid genera, *Salamandra* has the least developed tongue both with regard to the distance of its protrusion and its effectiveness in hitting the prey. In all three genera, the tongue pad is fixed to the lower jaw only at its anterior end, being free at the anterolateral, lateral, and posterior margins. In *Salamandra*, the posterior, free part of the tongue is small, whereas the other two genera have relatively large, free posterior flaps which can be projected out of the mouth over a rather great distance.

In all three genera, tongue protrusion is started by a forward and upward movement of the hyobranchial skeleton. The contracting SAR pulls the skeleton forward against the fixed ceratohyal (CH), serving as an anchor. The subhyoideus, by contraction, pulls the anterodorsal end of the CH anteroventrally, causing their elements to tip and elevating the tongue. By the forward movement of the BB and the radii the tongue is pushed over the mandible out of the mouth. The genioglossus (GG) muscle seems to be at least partly involved by drawing the tongue pad towards and over the front of the mouth. Furthermore, the tongue pad is squeezed forward by contraction of the well-developed basiradialis muscles, extending below the second radii. Retraction of the tongue pad and the tongue skeleton is achieved by contraction of both RCP and GG.

In *Chioglossa*, the reach of the tongue is greatly extended by the very long radii and the large tongue with a free posterior flap, and additionally by the long, first CB. During forward movement of the whole hyobranchial apparatus, through contraction of the SAR and subhyoideus muscles, the relatively massive basiradialis muscles situated inside the tongue also contract swinging the elongate radii in 180° arcs and flipping the tongue pad out of the mouth.

In contrast to *Chioglossa*, the hyobranchial apparatus of *Salamandrina* is rather short and stubby and does not seem to be as much involved in tongue movement. However, its tongue pad is very large and has a large, free posterior flap. Again, the single pair of radii are greatly elongated; they articulate with the anterior tip of the BB, similar to the situation found in

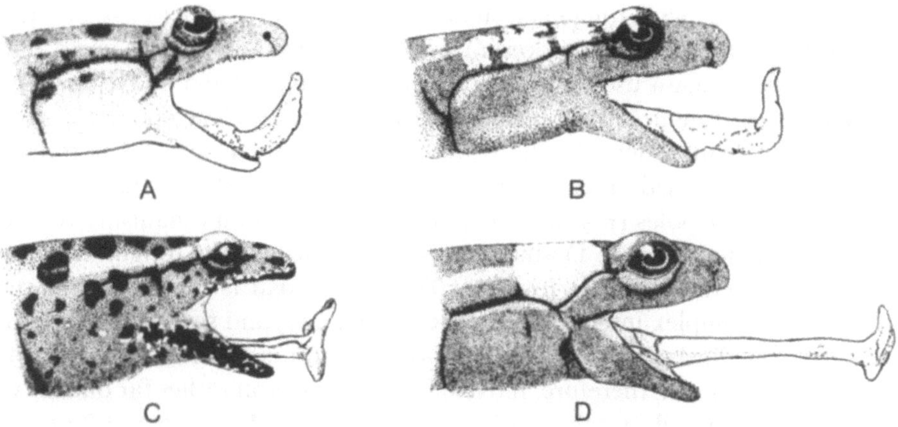

Fig. 2.8 A–D. Heads of plethodontid salamanders during tongue projection (tongue protraction phase). **A** *Desmognathus monticola*; **B** *Ensatina eschscholtzii*; **C** *Pseudotriton ruber*; **D** *Pseudoeurycea bellii*. (From Lombard and Wake 1977)

Chioglossa, and extend posteriorly into the free flap portion of the tongue. Tongue protrusion is mainly achieved by rotation of the radii at the anterior end of the BB, similar to *Chioglossa*. In both genera, the RCP is the main tongue retractor.

Within the family Plethodontidae, the tongue apparatus has undergone a dramatic evolution. We find a large variety of different tongues, from rather undifferentiated ones similar to the tongues of most salamandrids, to largely protrusible, but still attached tongues to finally free tongues in which the whole hyobranchial apparatus is projected out of the mouth (Fig. 2.8). The development of a highly protrusible tongue appears to be intimately connected to reduction or loss of lungs (Roth and Wake, 1985a).

The tongue of the genera *Desmognathus*, *Phaeognathus*, and *Leurognathus* (subfamily Desmognathinae), is the simplest tongue among plethodontids (Figs. 2.6 and 2.8 A). Feeding is accomplished by a combined use of the jaws and the tongue during rapid forward or sideward movement of the head; the tongue, however, is protruded out of the mouth for a short distance. The tongue pad of these genera is relatively large and narrowly attached at its anterior margin to the floor of the mouth. Posteriorly it has a large, free flap. The BB, first and second CB, and the EB are of about the same length. There are well-developed genioglossus, circumglossus and basiradialis muscles. During tongue projection the hyobranchial apparatus moves forward due to the contraction of the SAR, such that the tip of the BB overrides the insertion of the GG. Contraction of the GG pulls the anterior mass of the tongue pad to a vertical position, while the tongue pad muscles pull the posterior mass of the pad forward. Retraction of the tongue is achieved by contraction of the RCP.

A more specialized tongue mechanism is found in the genera *Aneides* and *Plethodon* of the tribe Plethodontini whose members are strictly terrestrial. These animals again use an extensive forward lunge of the body, but they protrude their tongues much farther outside the mouth. They possess a very large tongue pad which fills the whole intermandibular space. The tongue pad is again attached anteriorly but has a large free posterior flap showing paired posterior poles (Fig. 2.5). The tongue projection mechanisms are basically the same as in the Desmognathinae, except that the tongue skeleton moves considerably farther forward. The tongue pad is flipped out of the mouth by a complex interaction between SAR, GG, and several tongue pad muscles. The third genus of the tribe, *Ensatina*, possesses a rather extendable GG muscle and can, therefore, move its tongue skeleton rather far outside of the mouth, although it does not have a free tongue (Figs. 2.6 and 2.8 B).

The genera *Eurycea*, *Gyrinophilus*, *Pseudotriton*, *Stereochilus*, and *Typhlotriton* rely heavily on tongue projection to capture prey and in most cases seem to make little use of their jaws, although they are able to seize rather large prey such as other salamanders (Nishikawa, in litt.). They all have tongues with small pads. *Typhlotriton* has a very reduced GG, in the other genera the tongue is totally disconnected from the mouth, the genioglossus being completely absent. They, therefore, have a "free" tongue (Fig. 2.8 C).

In these genera, the EB is much longer than in the previously described plethodontids so that the tongue can be moved much further forward. During forward movement, the tongue skeleton, which at rest lies in the mouth, is compressed toward the midline by a three-dimensional folding resulting in the formation of a slender, compact projectile which travels out of the mouth. The RCP is slightly folded behind the tongue pad which increases the length of this muscle and allows the tongue to protrude out of the mouth before it is retracted. In addition to the forward movement of the skeleton, the tongue pad is flipped further forward by rotation of the radii.

The most specialized tongue of all salamanders with regard both to velocity and distance is found in the species of the tribe Bolitoglossini all of which have very fast, highly protrusible tongues (Fig. 2.8 D), with the exception of the genus *Batrachoseps* which has a tongue still attached to the lower jaw by the genioglossus. However, here the GG is greatly elongate, highly extendable, and barely restricts tongue protrusion (Fig. 2.6).

The hyobranchial apparatus of the Bolitoglossini differs from that of all other plethodontids in the size reduction of the first ceratobranchials, which in the genus *Thorius* results in a disconnection of this element from the basibranchial. The second ceratobranchials are short, and the epibranchials become very long. The reduction of the first ceratobranchials in Bolitoglossini is intimately connected with the loss of an aquatic larval stage in this group. In aquatic larvae the first CB is always larger than the second one. Larval feeding as well as respiration depends heavily on the role of the first

CB. All newly metamorphosed plethodontids, therefore, have a relatively large first and a relatively small second CB. In the Bolitoglossini, which all lack aquatic stages, the hyobranchial apparatus appears relatively late in ontogeny, and all parts develop almost simultaneously (Wake 1982). Therefore, there is no need for a major remodeling process from the larval to metamorphosed hyobranchial apparatus. Accordingly, in bolitoglossines, the size of the first CB can be reduced without negative consequences because there is no larval stage. The reduction of the first CB in mass and the shortening of the second CB result in much more efficient force transmission during tongue projection (Lombard and Wake 1976, 1977). Furthermore, the folding process is more rapid, and the additional increase of the epibranchials increases the reach of the tongue.

In the Bolitoglossini, the protractor muscle of the apparatus, the SAR, shows a specialized form, wrapping around the long EB and forming a complex sheath. The rectus cervicis, as the retractor muscle, is folded in the gular region. This muscle here runs uninterrupted by myocommata from the pelvic region to the anterior tip of the basibranchial, thus being greatly extendable.

As the SAR contracts, it pulls the skeleton forward and simultaneously "squeezes" the EB out of the muscular sheath. The skeleton folds in three dimensions to a very slender projectile which is fully projected out of the mouth. Retraction is achieved by contraction of the greatly extended rectus cervicis traveling out of the mouth during tongue protraction (Figs. 2.9 and 2.10).

In the Bolitoglossini the feeding apparatus clearly evolved toward an increase in velocity, feeding distance, and versatility of the tongue. Increase of velocity of tongue projection was achieved (1) by reduction of mass of the tongue skeleton; (2) by reduction of the length of the two pairs of CB which considerably shortens the time necessary for folding the skeleton; (3) by shifting the main line of force transmission from the first to the second pair of CB which optimizes the track of the movable tongue during protraction (Lombard and Wake 1976, 1977); and (4) by development of a complete muscular sheath around the EB by the SAR muscle which wraps around the EB in a spiral way, thus being able to both pull the EB forward and "squeeze" it out.

In such a way, very high tongue protrusion velocities are reached which, as feeding reactions, may be unique among vertebrates. In some species of the genus *Bolitoglossa*, the tongue is protruded out of the mouth to its full length of 20–25 mm in 2–4 ms (Thexton et al. 1977). In *Hydromantes*, which possesses by far the longest tongue with a reach of 45–50 mm, the protrusion lasts 6–8 ms (Roth 1974, 1976).

Tongue-retraction velocity is also increased by a unique interaction between the protractor and retractor system. In Bolitoglossini the main retrac-

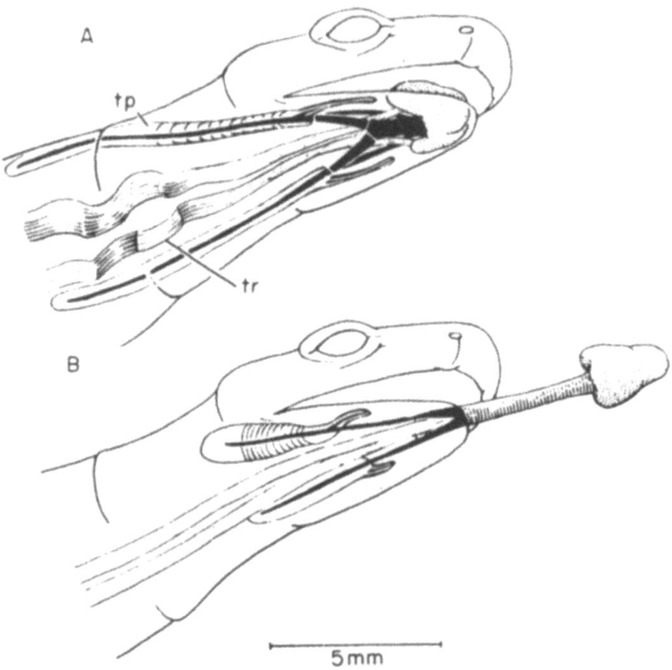

Fig. 2.9 A, B. Tongue projection in *Bolitoglossa occidentalis*. **A** Tongue at rest; note the coiling of RCP in the gular region. **B** Tongue at protraction. Tongue skeleton is shown in *black*. *tr* Tongue retractor muscle (RCP); *tp* tongue protractor muscle (SAR1). (From Thexton et al. 1977)

tor muscle, the rectus cervicis profundus, is folded in the gular region when at rest. Thexton et al. (1977) showed by means of electromyographic studies in *Bolitoglossa occidentalis* that the protractor SAR and the retractor RCP are activated more or less simultaneously. When both muscles contract, the SAR has an advantage over RCP due to the differing length-tension curves of the two muscles, such that the SAR can shoot the tongue out maximally until the RCP is under full tension and draws the tongue back. After Thexton et al. (1977) the whole tongue reaction takes place in about 10 ms in *Bolitoglossa occidentalis*. In *Hydromantes* the tongue reaction is considerably slower; it lasts 80–100 ms (Roth 1976).

Projection distance is increased mainly by an increase in the length of the epibranchials. In *Hydromantes*, the basibranchial is additionally increased in length. The BB has no radii and reaches far forward. The EB are very long and extend back to the middle of the body or beyond (Wake 1966).

Bolitoglossini are able to shoot their tongues out laterally, sometimes at an angle of 45° or more with respect to the head axis, as documented by high-speed motion pictures (Larsen and Wake unpublished). The shooting direction is controlled by the interaction of two types of muscles, the suprapeduncularis (SP) and the geniohyoideus lateralis (GHL) (Fig. 2.4). In the Bolito-

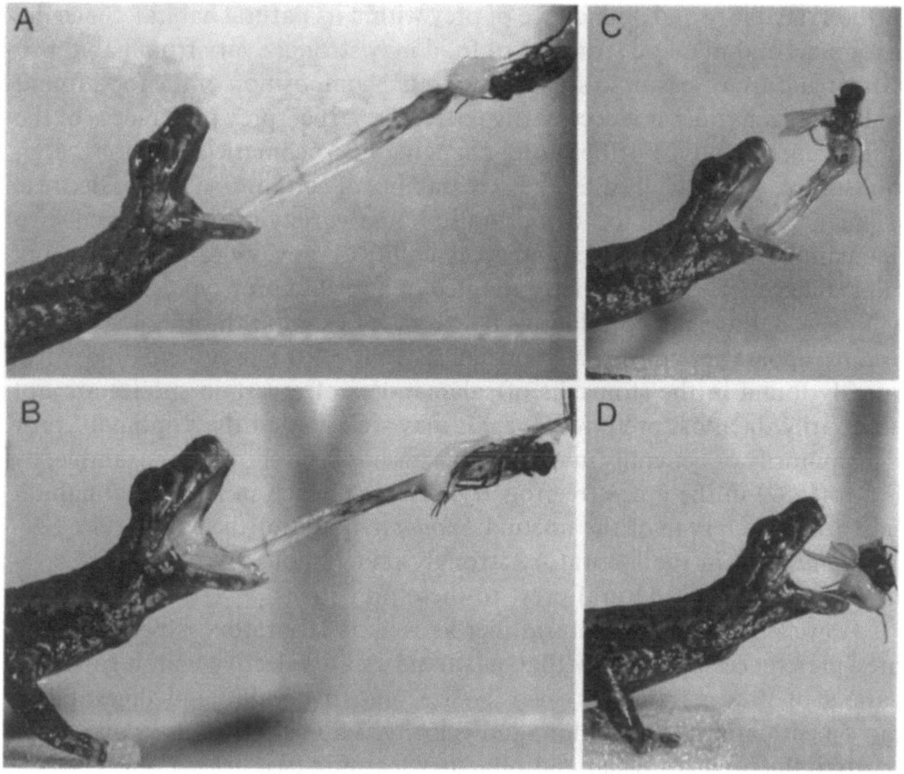

Fig. 2.10 A–D. Tongue projection in the plethodontid salamander *Hydromantes italicus*. **A** Early stage of retraction. The tongue shows about 3/4 of its full length. **B** Later stage of retraction. The tongue pad has just slightly contracted. **C** Late stage of retraction. Here, the tongue is turned 90°. Compression of the elements of the tongue skeleton by the suprapeduncularis muscle can be seen clearly. **D** End of retraction. The prey is glued to the contracted tongue pad. (From Roth 1976)

glossini, the suprapeduncularis extends dorsally over the BB between the flattened anterior parts of the ceratohyals. During tongue protraction this muscle forms a tight cylinder which presses the moving skeleton into a slender projectile. This cylinder is now directed by the pair of geniohyoideus lateralis muscles which run from the mandible near the symphysis to the anterior edge of the ceratohyal. Unequal contraction of these muscles leads to lateral deviation of the cylinder and, therefore, lateral projection of the tongue.

2.1.2 Prey Preferences and Natural Diets of Salamanders

Because salamanders eat such diverse sizes and species of prey, they are commonly believed to indiscriminately ingest whatever prey they encounter, limited only by the morphological constraints of the feeding apparatus (Hairston 1949, 1980). What a salamander eats is in the first place determined by the availability and abundance of prey within its natural habitat. Since the taxonomic and numeric structure of food may strongly vary from habitat to habitat and from season to season, the actual composition of the food found in the stomachs does not necessarily reflect the "true" prey preferences of the animal under study. Furthermore, changing microclimatic conditions, e.g., moisture, can drastically alter the accessability of certain prey types. Because salamanders may have to "compromise" on a variety of prey types, due to fluctuations in the abundance and accessability of prey, we cannot expect any salamander species to be strictly adapted to a certain prey type, although, as will be seen, there are correlations between, for example, morphology of the feeding apparatus and prey spectrum. Therefore, the prey items most frequently found in the stomachs of salamanders of a certain species are not necessarily the most preferred ones; it may well be that the "optimal" prey is temporarily or generally inaccessible. This has been shown, for example, by Freed (1982) in the green treefrog *Hyla cinerea* which in its natural habitat eats primarily larvae of the noctuid *Spodoptera* and of the cantharid *Chauliognathus*, but in the laboratory strongly prefers houseflies (*Musca domestica*), which are not natural prey, to these prey items.

However, studies on natural diet as well as laboratory experiments on prey preferences often show that salamanders are selective with regard to a variety of prey parameters such as size, nutritive value and digestibility, movement pattern, orientation, and velocity. This selectivity is known to be partly influenced by visual and olfactory learning (see below).

Laboratory experiments, which have the advantage of a systematic variation of several stimulus parameters, usually suffer from limitations with regard to the environmental conditions during feeding as well as to the simplicity of the stimuli offered to the animal. For example, using a homogeneous light background and high-contrast stimuli, one can obtain unreliable results with regard to size preferences: in such a test situation salamanders often snap at stimuli that are far too large to be swallowed. It can be shown that under such artificial conditions the subjects respond to the edges of the stimuli rather than to the stimuli themselves. Most reliable results are therefore achieved when data from natural diet and from laboratory experiments are compared.

A large amount of detailed knowledge exists concerning the natural diet of salamanders. Well-studied genera under this aspect are *Salamandra* (Szabó 1962, Thiesmeier 1982), *Triturus* (Kühlhorn 1959), *Desmognathus* (Hairston 1949, Keen 1975, Sites 1978), *Eurycea* (Sites 1978), *Plethodon* (Jaeger 1972, Fraser 1976), *Aneides* (Lynch 1985), *Ensatina* (Lynch 1985), *Batrachoseps* (Maiorana 1978a, b, Lynch 1985), *Bolitoglossa*, *Dendrotriton* (Wake et al. in prep.), and *Ambystoma* (Bishop 1941, Pope 1947, Anderson 1968, Freda 1983).

Salamanders are exclusively carnivorous animals. In contrast to most anurans, which in the tadpole stage usually feed on plants, salamanders and newts take animal food both before and after metamorphosis. Salamander larvae seem to be rather unspecific with regard to their diet. Szabó (1962) noted that according to the habitat the larval stomach of *Salamandra salamandra* contained Crustacea (42%), mainly from Amphipoda, and insect larvae (31%), mostly Diptera and Ephemeroptera. Additionally, *Salamandra* larvae swallowed Oligochaeta, Gastropoda, and tadpoles. Thiesmeier (1982) stated that the composition of the stomach contents in *Salamandra* larvae corresponds well to the composition of benthic invertebrates of the habitat. The amphipod crustacean *Gammarus*, the most abundant species in the brook investigated, amounted to 42% benthic animals and it had a proportion of 37% in the salamander stomach. In the stomach of the terrestrial adult *Salamandra salamandra* in Hungary, Szabó (1962) found that 31% of the prey objects belonged to Gastropoda (different kinds of slugs) and 61% Arthropoda including a great variety of insects such as Coleoptera larvae as well as adults, Dermaptera, and Diptera. Diplopoda, Chilopoda, and Isopoda were also taken. The remaining 8% of prey animals in the *Salamandra* gut were Oligochaeta.

Both seasonal and geographic variation in food have been observed in *Salamandra*. Bas Lopez et al. (1979) found that *Salamandra* in northwestern Spain fed less on gastropods than the Hungarian salamanders (19% instead of 31%) but consumed more coleopterans (21% instead of 7%). Generally, the prey consisted of rather slowly moving animals.

Larvae of the salamander *Ambystoma maculatum* eat mostly Cladocera, followed by Copepoda and Isopoda (Freda 1983). During the month of June, larvae of the newt *Notophthalmus viridescens* are found in their stomachs, especially at the end of the month (Freda 1983). Collins and Holomuzki (1984) showed that larvae of *Ambystoma tigrinum nebulosum* are dimorphic. While the typical larval morph ate primarily macroinvertebrates (aquatic insects, mollusks and crustaceans), the "broad-headed" morph was cannibalistic and most of their diet (84%) consisted in other salamanders.

Sites (1978) carried out a detailed study on the influence of season, environment and sympatric species in the prey spectrum of *Desmognathus fuscus*. In a permanent stream habitat where this species is sympatric with *Eurycea longicauda*, the most abundant prey items were larvae of terrestrial insects (32%), entomobryid collembolans (12%), asellid isopods (10%), and pselaphid coleopterans (10%) during fall. In spring insect larvae were again the most abundant prey (24%), but in contrast to fall gammarid amphipods (10%), lumbricid oligochaeta (10%), and carabid coleopterans (10%) were most abundant; collembolans were not found in the stomachs. In a different habitat, a seepage area, the fall diet of *Desmognathus fuscus* is different from that in the first one. Here, lumbricid oligochaeta make the greatest contribution to the diet (27%), followed by terrestrial insect larvae (18%), collembolans, isopods, and chilopods (10% each). In spring, the diet is again different: callipodid diplopods (13%), followed by a large variety of prey taxa constituting 4–5% of the diet such as lumbricids, mites, chilopods, collembolans, and ants.

Apart from the variation of the prey spectrum with regard to habitat and season the most interesting result of this study is the comparison between the two sympatric species. In the first habitat (permanent stream) the fall diet of *Eurycea longicauda* consisted mostly of sminthurid collembolans (45%), followed by galumnid and trombidiid mites (10%), ants and slugs (10% each). In spring, mites (38%) and entomobryid collembolans (25%) were now the preferred diet, followed by callipodid diplopods and insect larvae (13% each). This means that the two species eat rather different types of prey: whereas *Desmognathus fuscus* eats mainly earthworms and millipedes, *Eurycea longicauda* eats collembolans and mites.

This is an interesting result because the two sympatric species differ mostly with regard to the functional morphology of their feeding apparatus: whereas *Desmognathus* has a rather "primitive" flipping tongue combined with a specialized jaw opening mechanism, *Eurycea* possesses a fast, free, projectile tongue of medium reach.

Sites did not find a correlation between salamander size and prey size, as found by Lynch (1985). What he found was that larger salamanders ate larger prey, but continued to eat smaller prey. *Desmognathus* proved to be a generalist consuming most prey taxa falling within the proper size class. Similarly, Keen (1975) found that both *Desmognathus* and *Eurycea* are generalists with regard to their diet, but that *Eurycea* used a "sit-and-wait" or "ambush" strategy more frequently than *Desmognathus*, consuming the most abundant prey items present in the litter, i.e., mites and collembolans.

Jaeger (1972) has compared the natural diet of *Plethodon cinereus* and *P. richmondi* (now called *P. shenandoah*). *P. shenandoah* lives mainly in areas of talus and is there slightly larger (37.8-mm snout-vent length, SVL, 5.6-mm head width, HW) than *P. cinereus* (35.3-mm SVL, 5.0-mm HW) which lives in surrounding areas of deeper soil. However, some large *P. shenandoah* (43.7-mm SVL, 6.3-mm HW) are sympatric with *P. cinereus*. The diet of *P. shenandoah* consisted mainly of insect larvae (34%), followed by mites (22%), coleopterans (8%), and hymenopterans (7%), whereas *P. cinereus* ate mostly mites (39%), followed by insect larvae (28%) and homopterans (12%). Jaeger found a significant correlation between length and head width of the salamanders on the one hand, and length and width of the prey taken on the other. The prey items found in the stomachs of *P. shenandoah* from the talus had an average length and width of 3.4 and 1.0 mm, respectively, those sympatric with *P. cinereus* 4.0 and 1.1 mm, whereas the prey items found in the stomachs of *P. cinereus* had a mean length and width of 1.5 and 0.6 mm. Jaeger excluded species-specific differences, because when he compared comparably large adults of both species, there were no significant preferences for different sizes of prey.

In comparing sympatric *P. cinereus* and *P. hoffmani*, Fraser (1976) also found a significant correlation between body length and head width of the salamanders and the size of the largest prey item taken. *P. hoffmani*, being larger (mostly 50–55 mm SVL) than *P. cinereus* (here mostly 43–45 mm) ate significantly wider and longer prey. Fraser also found fluctuations in the utilization of prey taxa in spring and fall. In *P. hoffmani*, in both spring and fall, ants made the greatest contribution to the diet by number. By volume Hymenoptera, Coleoptera, and Pulmonata made the greatest contribution in fall and Araneidae, Diptera, Oligochaeta, and Coleoptera in spring. Interestingly, the smaller species, *P. cinereus*, ate significantly more collembolans than the larger species. It is surprising, however, that the length of most prey eaten by both species fell within the range of 0.5 and 3.5 mm, which is fairly small.

In an extensive study, Lynch (1985) compared four sympatric Californian plethodontid salamanders: *Aneides lugubris*, *A. flavipunctatus*, *Ensatina eschscholtzii*, and *Batrachoseps attenuatus*. He found that the available prey spectrum was partitioned according to size by this group of salamanders. The absolute lower size limit of acceptable prey changed relatively little with increasing body size, but both the relative and absolute number of small prey declined with increasing size of the salamanders. In *A. flavipunctatus* Diplopoda (cambalids), Coleoptera (carabids, curculionids, staphylinids), Hymenoptera (ants), Isoptera (termite workers of the family Kalothermitidae), Diptera and Collembola combined constituted about 80% of the food volume. In juvenile *A. flavipunctatus* the same diet was found plus dipterans and entomobryid, podurid and sminthurid collembolans, which were important prey in spring. There was a sharp ontogenetic decrease in the number of collembolans and mites eaten.

Maiorana (1978b) showed that the plethodontid salamander *Batrachoseps attenuatus* at times selectively ignores smaller prey. There was also evidence that *Batrachoseps* prefers sminthurid collembolans to oribatid mites, a taxon of equal size, but having an exoskeleton that is harder to digest.

Maiorana (1978a) also studied the prey spectra of sympatric *Aneides lugubris* and *Batrachoseps attenuatus*. She found that adult *Batrachoseps* and juvenile *Aneides* strictly overlap with regard to their diet, but that adult *Aneides*, while continuing to eat small prey, also ate larger prey in correlation with head width. In an island habitat, however, where large prey were scarce, even large *Aneides* ate the available small prey. Here, for both species 75% of prey items eaten were only 3-mm-long or smaller, and a substantial part of the prey was smaller than 0.5 mm.

Wake, Roth and Staub (in prep.) studied the natural diet in three tropical plethodontids, *Bolitoglossa rostrata*, *B. subpalmata*, and *Dendrotriton*. *B. rostrata*, the largest species within this series (SVL 28.0–64.3 mm) is a strict ground dweller, whereas *Dendrotriton bromeliacia*, the smallest species (SVL 17.4–33.8 mm), is strictly arboreal. *B. subpalmata*, in contrast, occurs both on the ground and on trees in bromeliads (terrestrial: SVL 26.6–55.2 mm, arboreal: 26.6–50.1 mm).

In this study, prey spectrum and prey size, based on stomach contents, were compared to head width and habitat of the salamanders. Furthermore, in *B. subpalmata*, seasonal changes in diet (dry vs rainy season) were studied.

In *B. rostrata* (salamanders n = 25, prey items n = 228) coleopterans were found in 17 stomachs, hymenopterans in 13 stomachs, followed by mites (11 stomachs), dipterans (9), collembolans (9), dermapterans (7), spiders (6) and hemipterans (6). The head widths of the salamanders ranged from 5.4 to 9.7 mm. The mean prey size was 3.9 mm in length and 1.4 mm in width. The maximum prey size was 26.7 mm in length and 10.0 mm in width, found in the stomach of a salamander of 9.0-mm head width, the minimum prey size was 0.33 × 0.33 mm.

During the rainy season, in terrestrial *B. subpalmata* (n = 31, prey items n = 152) coleopterans were found in 13 stomachs, followed by collembolans (11 stomachs), mites (11), dipterans (8), hemipterans (6), lepidopterans (5) and isopterans (5). The head width of the salamanders ranged from 4.7 to 7.9 mm. The mean prey size was 2.9 mm in length and 1.2 mm in width. Maximum prey size was 30 × 10 mm, found in a salamander of 6.9-mm head width, minimum prey size was 0.08 × 0.08 mm. During the dry season, in the same species (n = 124) mites were found in 33 stomachs, collembolans in 29 stomachs, and dipterans in 15 stomachs.

In arboreal *B. subpalmata* (n = 30, prey items n = 172) hymenopterans (ants) were found in 21 stomachs, collembolans in 13 stomachs, mites (9), and hemipterans (5). Head width ranged from 4.5 to 7.2 mm. Mean prey size was 1.8 × 0.5 mm. Maximum prey size was 13.3 × 2.2, found in a salamander of 6.4 mm head width, the minimum prey size was 0.17 × 0.08 mm.

Finally, in *Dendrotriton bromeliacia* (n = 50, prey items n = 172) dipterans were found in 26 stomachs, followed by mites (22 stomachs), ants (17), collembolans (8), coleopterans (8), and hemipterans (7). Head widths varied between 3.4 and 5.3 mm. Average prey size was 1.9 × 0.5. Maximum prey size was 15.8 × 2.5 mm, found in a salamander of 5.0-mm head width, minimum prey size was 0.2 × 0.05 mm.

The data show that the tongue-projecting plethodontids under study ate mostly small to very small prey which is considerably smaller than the prey size found in most nonbolitoglossines. Furthermore, terrestrial salamanders ate larger prey than arboreal ones even of the same species, as is the case in *Bolitoglossa subpalmata*. Here, in terrestrial specimens the average prey size was 2.9 × 1.2 mm (equal to a volume of 4.1 mm^3) and in arboreal ones 1.8 × 0.5 mm (= 0.5 mm^3). The largest average prey size (6.5 mm^3) is found in the largest species *B. rostrata*. But, very interestingly, the two arboreal groups, *B. subpalmata* from bromeliads, and *Dendrotriton bromeliacia*, despite their differences in size, show exactly the same average prey size (0.5 mm^3).

In the stomachs of arboreal Bolitoglossini mostly ants, dipterans, collembolans and mites were found, whereas the stomachs of terrestrial ones contained mostly coleopterans. Seasonal changes seem to have a great influence on the diet: during rainy season, the stomachs of *B. subpalmata* contained mostly coleopterans which are absent from the stomachs during the dry season.

The most important result of this study, however, is the fact that in contrast to all nonbolitoglossine salamanders studied so far, the Bolitoglossini studied did not show a correlation between head size and prey size. Maximum prey size did not increase with increase in head width. Although the few larger prey items were found in the stomachs of medium to large subjects, the largest salamanders usually ate prey of more or less the same size as the smallest ones.

One could assume that this noncorrelation between head width and prey size in tropical plethodontids was caused by the scarcity or absence of large prey in the natural prey spectrum, so that even large salamanders were forced to eat very small prey. This could be especially true for arboreal salamanders. However, as will be demonstrated later, experimental data also show that *B. subpalmata* prefer rather small prey stimuli of 2.5 × 2.5 mm when larger prey were also presented. In experimental studies with *Hydromantes italicus*, Roth (1974) also found no correlation between head width of the subjects and preferred prey size.

The main reason for such a preference for small to very small prey may lie in the fact that these salamanders exclusively use their tongues during feeding without use of their jaws, and they have difficulties in feeding on large and elongate prey. Thus, prey should be so small that it can be taken completely into the mouth when the tongue is returned.

Apart from prey abundance, microclimatic conditions, especially current ambient moisture, seems largely to determine the foraging success, at least in most plethodontid salamanders. As many ecological and physiological studies show, these animals suffer strongly from desiccation. Therefore, as Jaeger (1978) stated, prey items are periodically limited in availability, but not in abundance. Salamanders forage freely and consume large quantities of prey under favorable conditions of moisture. Prey items become increasingly unavailable as the soil and leaf litter dry and the salamanders are restricted to isolated pockets of moisture. Jaeger (1978) has shown that *Plethodon cinereus* leaves the forest floor during rainy and foggy nights and climbs plants where they forage on plant-dwelling homopterans (Aphididae, Cicadellidae, Delphacidae, Membracidae) and hemipterans (Miridae, Nabidae). Under those conditions salamanders eat significantly larger volumes of food than conspecifics do. Most of these plant-dwelling prey are considerably larger than mites and collembolans which compose a high percentage of the diet in the litter. Furthermore, it may be that search for prey is easier on plants than in the litter.

Thus, what a salamander eats in its habitat is influenced by a large variety of factors such as abundance of prey, availability of prey, ingestibility, width of the head and of the mouth of the salamander, feeding apparatus and feeding strategy. The correspondence between head/mouth width and prey size has been shown to be complex: larger animals tend to eat larger prey, but continue to eat small prey which might be very abundant or more readily accessible. However, this seems not to be true for Bolitoglossini; many of these salamanders seem to eat much smaller prey on average than they could do considering head and mouth width. They eat mostly small to very small prey items such as mites, fruit flies, and collembolans even when larger, but still edible prey are present. When a tongue-flipping and a tongue-projecting salamander are sympatric, as is the case with *Desmognathus fuscus* and *Eurycea longicauda*, then the former eats larger and more wormlike prey such as insect larvae, whereas the latter eats very small prey such as mites and collembolans. However, even for tongue-flipping salamanders like *Plethodon cinereus* it has been shown that these animals can switch from one type of prey to the other and from one feeding strategy to the other, depending on prey density or on prey type (Jaeger 1972; see also below).

The most interesting fact, for aspects of sensory physiology of the visual system, is the extremely small size of the prey eaten by many plethodontid salamanders. Many bolitoglossine salamanders eat mostly prey that are 0.5-mm-long or even smaller. If we consider that these animals feed mostly under conditions of dim light we have to assume that their visual acuity is very good even under these light conditions. This problem will be discussed later.

2.1.3 Experimental Analysis of Visually Guided Prey-Capture Behavior

Generally, prey-capture behavior in salamanders shows a sequence of the following reactions:

Orienting Movement. The first reaction to a prey object is a turning movement which permits the salamander to fixate the object binocularly. In contrast to frogs, salamanders perform the initial orientation with their heads alone, but not with the whole body. The orienting movement can be released by visual or mechanical stimuli.

Approaching. Since terrestrial salamanders do not jump like frogs, they approach their prey by walking toward the stimulus until it is within reach of the jaws or tongue. Approaching can be released by visual, chemical, mechanical, or electric stimuli. As will be discussed later, there are great differences among salamanders with regard to the feeding strategy and the distance at which a prey will elicit approach behavior. Whereas some salamanders will actively search for prey ("hunter strategy"), others will wait until the prey comes close ("ambush strategy"). However, some salamander species are known to switch from one strategy to another depending on prey density or prey type (Jaeger 1972, Wake et al. in prep.).

Olfactory test. When a prey is within reach, an olfactory test may occur. Some species show a characteristic posture during this reaction. *Triturus*, for instance, stiffens its forelegs such that the front part of the body is raised, and then points its head down at a steep angle until the tip of the snout is directly above the prey. In this position the frequency of ventilation movements is increased, which helps to suck the odor substances into the olfactory organ. Visual or mechanical stimuli may also release this reaction.

Fixation. After reaching the snapping distance, salamanders usually fixate the prey object binocularly for a shorter or longer period depending on the attributes (e.g., movement intensity) of the prey. There are, however, species-specific differences, too (see below).

Snapping. By snapping, the prey is finally seized and taken into the mouth. As described earlier, in water this is usually a brief sucking of water into the mouth which sweeps the prey in as well. On land, the tongue is protruded to catch the prey objects. Larger prey may be seized by the jaws, too. The snapping reaction is largely determined by tongue morphology, especially by the reach of the tongue, feeding strategy and the nature of the prey object. Whereas slowly moving prey items, such as earthworms, slugs, or insect larvae, can be pursued and caught even by salamanders with massive, short tongues, many arthropods such as collembolans, crickets, and flying insects possess highly effective escape systems and can often be caught only by means of an "ambush strategy" and a projectile tongue.

All amphibians with a short-range, unspecialized tongue snap only in a more or less frontal direction. For that reason they have to direct their heads directly toward the prey. In contrast, bolitoglossine salamanders can project their tongues laterally, due to a specialized firing cylinder described above. Angles of 45° and more between tongue tip and head axis have been observed.

In general, there are differences among salamander species with regard to the feeding sequence. Whereas most plethodontid salamanders and *Salamandra* show a rather "straightforward" sequence that consists mostly of orienting, approach, and snapping, or, as often in the case of bolitoglossines, of orienting and snapping with little approach, newts, especially during their aquatic stage, show extensive correcting movements while approaching, and long periods of fixation and olfactory tests. This is especially true for *Triturus cristatus* and *T. alpestris* which are guided mainly by olfactory cues even under daylight conditions, whereas in *T. vulgaris* and *T. helveticus* visual guidance dominates over olfactory guidance under daylight conditions (Himstedt 1967), as is the case in most other salamanders.

2.1.3.1 Analysis of Relevant Visual Parameters of Prey Objects

In amphibians feeding behavior can be elicited by relatively simple visual stimuli. Thus, one can test the relative contribution of certain visual parameters to the overall attractiveness of a prey object. Some of these parameters can be tested independently of each other, but others show a strong interdependence resulting in nonlinear effects. The most important visual parameters of prey stimuli are size, shape, orientation with regard to the direction of movement, velocity, movement pattern, and contrast to the background.

Stimulus Size. The data on size of natural prey given above showed that salamanders usually feed on prey of a wide size range. The lower size limit seems to be identical for most salamanders and is independent of their body size or head width. In contrast, in many salamanders (with the exception of Bolitoglossini) the upper limit, i.e., the maximum size of prey eaten, increases with body size and head width. The minimum length of natural prey is between 0.2 and 0.5 mm. The maximum length varies with regard to the capability of eating wormlike prey, but most plethodontid salamanders rarely ate prey items that were longer than 10 mm. In laboratory experiments with live prey, however, we observed that medium-sized plethodontid salamanders such as *Hydromantes italicus* or *Bolitoglossa subpalmata* ate crickets that were 10 mm or more long, and the animals had difficulty in swallowing the prey.

In the laboratory, experiments for determination of stimulus preference are usually carried out using pieces of black cardboard or plastic of different size and configuration moved at different velocities on a white background or a background consisting of fine-grained, random black and white structure. Using squares of different size, *Hydromantes italicus* responded to moving square stimuli within a size range of 0.5- and 10-mm edge length with a response maximum between 1 and 5 mm (Roth 1976; Fig. 2.11). Under similar test conditions, *Salamandra salamandra* showed a rather different range of accepted prey size: the minimum size was 2 mm and the maximum size was 32 mm with an optimum at 16 mm (Luthardt 1981; Fig. 2.12). Whereas stimuli of 2- and 4-mm edge length elicited only few responses, the

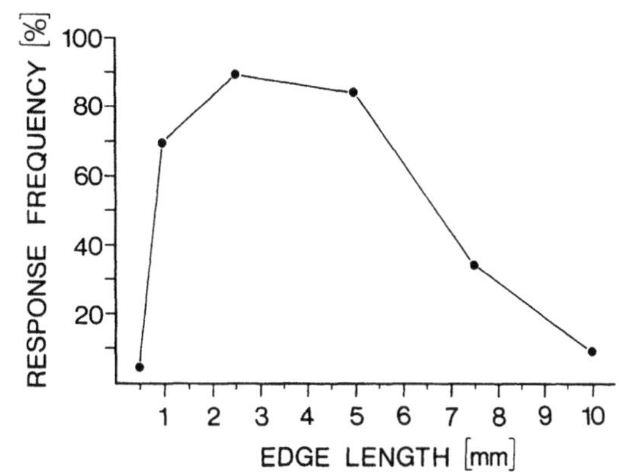

Fig. 2.11. Dependence of prey-catching responses on stimulus size in *Hydromantes italicus*. Black squares moving at a velocity of 3 cm/s in front of the animal were used as stimuli. *Abscissa* indicates edge length of the stimuli. (After Roth 1976)

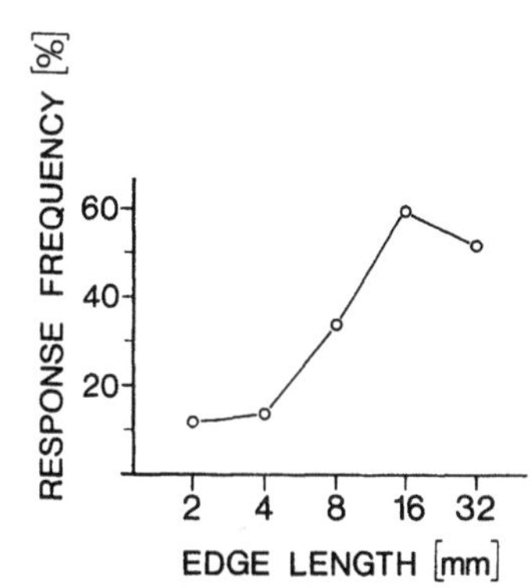

Fig. 2.12. Dependence of prey-catching responses on stimulus size in *S. salamandra*. Black squares moving at a velocity of 1 cm/s in front of the animal were used as stimuli. *Abscissa* indicates edge length of the stimuli. (From Luthardt 1981)

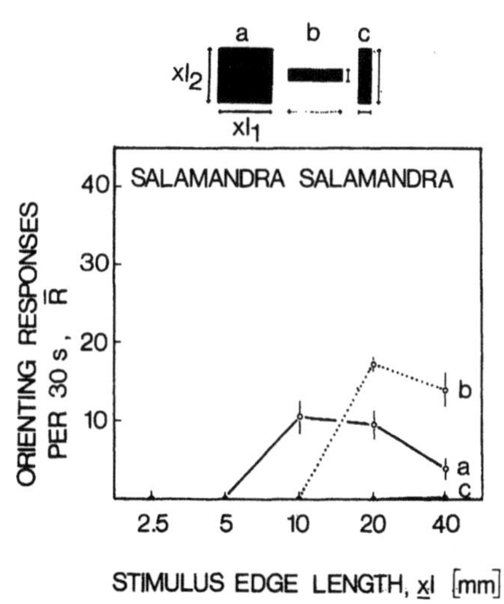

Fig. 2.13. Orienting behavior of *S. salamandra* in response to black configurational stimuli moving at a velocity of 10°/s in front of a white background. Starting with a 2.5 × 2.5 mm square, edge length of the stimuli was increased parallel (*b*), and perpendicular (*c*) to the direction of movement or in both directions (*a*). (After Finkenstädt and Ewert 1983)

32-mm stimulus elicited responses in about 50% of all tests. In a study where orienting responses were measured instead of snapping responses, Finkenstädt and Ewert (1983a) found that *Salamandra* did not respond to square stimuli with an edge length of 5 mm or smaller, but responded well to stimuli with edge lengths between 10 and 40 mm, thus, to stimulus sizes which are surprisingly large (Fig. 2.13).

These data show that *Salamandra* prefers larger prey than the plethodontid species tested even when the differences in head size are considered. Its optimum prey size ranges between 10- and 20-mm edge length, whereas the plethodontids preferred sizes of 2–5 mm, but still accepted prey of 0.5 mm and smaller. The fact that *Salamandra* snaps even at extremely large prey of 40-mm edge length, which as natural prey would be too large to be swallowed, indicates that here the subjects respond to the edges of the stimuli rather than to the actual size.

In a recent study *Plethodon cinereus*, *Plethodon jordani*, and *Bolitoglossa subpalmata* were compared with regard to stimulus size, shape, and orientation preferences (Roth and Kohlhase in prep.). When confronted with squares with edge lengths of 2.5 , 5, and 10 mm, all three species preferred the smallest stimulus to the larger ones, although they still responded to the largest one in a considerable number of tests. In tests with small stimuli moved irregularly on the ground, *Bolitoglossa subpalmata* snapped at objects 0.5-mm-long, although they preferred prey dummies of 1 mm and larger (Wiggers and Roth in prep.)

Stimulus Velocity. Stimulus movement is an almost necessary prerequisite for prey identification in amphibians: motionless prey objects are usually ignored, though at least in salamanders sufficient olfactory stimulation or individual experience with stationary prey can lead to feeding responses to stationary prey (see below). Systematic studies on minimum, optimum, and maximum prey stimulus velocity exist only for *Salamandra salamandra* (Himstedt 1967, Luthardt 1981), *Triturus vulgaris* (Himstedt 1967), and

Table 2.1. Stimulus velocity preferences in *Hydromantes italicus*, *Triturus vulgaris* and *Salamandra salamandra*[a]

	Minimum	Optimum	Maximum
H. italicus	0.05 cm/s 0.24°/s (D = 12 cm)	0.5–2.5 cm/s 4.8–72°/s (D = 6–2 cm)	6 cm/s 172°/s (D = 2 cm)
T. vulgaris	0.05 cm/s 1°/s	0.5–1.5 cm/s 5–20°/s	2 cm/s 45°/s
S. salamandra	0.5 cm/s 0.5°/s	0.5–1 cm/s 5–15°/s	2 cm/s 40°/s

[a] The data for *Triturus* and *Salamandra* are taken from Himstedt (1967), and for *Hydromantes* from Roth (1976). D = distance from subject to stimulus.

Hydromantes italicus (Roth 1976) (Table 2.1). In the study of Himstedt, *Salamandra* and *Triturus* showed almost identical data with regard to stimulus velocity: the minimum velocity that elicited feeding responses was 0.05 cm/s, and the maximum velocity was 2 cm/s. The range of optimal velocity was from 0.5 to 1 cm/s. In *Salamandra*, Luthardt found a minimum at 0.125 cm/s, a maximum at 6 cm/s and an optimum at 1 cm/s. At least this latter result is in good accordance with the data from Himstedt. In *Hydromantes* the minimum prey stimulus velocity was 0.05 cm/s and the maximum velocity was 6 cm/s. The optimum was situated between 0.5 and 2.5 cm/s, which is higher than in the salamandrids. These data are in good accordance with the overall differences in feeding behavior: *Hydromantes* is able to catch fast moving prey much better than the two other species, especially from short distance, where maximum angular velocities of about 180°/s occur (the maximum angular velocity for the other two species is about 40°/s). However, *Hydromantes* is a rather slow feeder compared to most tropical bolitoglossine salamanders. In behavioral experiments *Bolitoglossa* easily snapped at prey stimuli moving at a velocity of 10 cm/s or faster which at snapping distance results in an angular velocity of about 300°/s. This is in good accordance with the enormous speed of tongue protraction which enables these animals to feed on fast-moving or flying prey from a resting position.

Freed (1982) reported that *Hyla cinerea* orients to and attacks flying houseflies that have an average flight velocity of 94 cm/s. The angular velocitiy for *Musca* in flight 18 cm away from a tree frog would then be about 300°/s which is more or less the same as in the Bolitoglossini with the fastest feeding reactions.

Stimulus Configuration and Orientation. Amphibians are able to discriminate between different geometric configurations of prey stimuli. Ewert was the first to find that for the effectiveness of elongate stimuli, the orientation parallel and perpendicular to the direction of movement is important: toads (*Bufo bufo*) prefer elongate stimuli extended parallel to the direction of movement ("worms") to square stimuli and to elongate stimuli extended perpendicular to the direction of movement (Ewert 1968). In contrast, frogs prefer compact prey to elongate, wormlike prey.

Salamanders are also clearly able to discriminate between stimuli of different shape and orientation with regard to the direction of movement. Roth (unpublished data) studied the responses of *Hydromantes genei* to a square measuring 5 × 5 mm and rectangles measuring 2.5 × 10 mm oriented parallel and perpendicular to the direction of movement ("horizontal" and "vertical" rectangles, respectively). The stimuli were moved at 1 and 3 cm/s. At both velocities the animals showed a clear preference for the square, followed by the horizontal rectangle. The vertical rectangle, though being the least effective, elicited surprisingly many snapping responses.

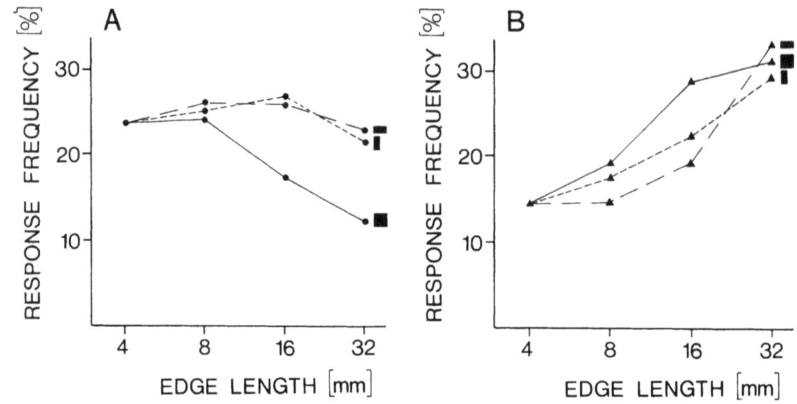

Fig. 2.14 A–B. Stimulus preference of *S. salamandra*. Square stimuli were increased in edge length either parallel (*long broken lines*), or perpendicular (*short broken lines*) to the direction of movement, or in both directions (*solid lines*) from 4° to 32°. **A** Orienting behavior at a stimulus distance of 3.3 cm. **B** Snapping behavior at a stimulus distance of 1.3 cm. (After Stenner 1976)

In a study with *S. salamandra*, Stenner (1976) studied stimulus configuration and orientation preference with regard to orienting and snapping behavior. In these experiments, squares and rectangles with horizontal and vertical orientation were used at a distance of 3.3 cm from the subject in the tests for orienting behavior and 1.3 cm in the tests for snapping behavior. The edge lengths of the square stimuli were geometrically increased from 4° to 32°, in the rectangles one edge was held constant at 4°, while the other edge was increased as in the squares. At a distance of 3.3 cm this is a range of absolute edge length from 2.31 to 20.62 mm, and at a distance of 1.3 cm this is a range from 0.91 to 8.12 mm. The stimuli were moved at 28°/s.

As shown in Fig. 2.14 A, in the test series for orienting behavior the animals responded equally well to the horizontal and vertical rectangles at all lengths. At an edge length of 16° (9.45 mm) the vertical rectangle was slightly superior to the horizontal one, whereas the opposite was the case at an edge length of 32°. The response activity decreased toward this longest edge length. At edge lengths of 16° and 32° the square proved to be significantly less effective than both rectangles. Here, the response frequency decreased rather sharply. The author ascribes this effect to the larger area of the square at these edge lengths compared to the rectangles.

In the tests for snapping behavior (Fig. 2.14 B) the responses to the three stimuli increased with increasing edge length and size. At 8° and 16° edge length the square was preferred to the vertical and the horizontal rectangle, whereas at 32° the horizontal rectangle was preferred. However, the differences were not significant except for the square preference at 16° edge length.

Stenner interpreted these results as evidence that different filter mechanisms exist for orienting and snapping behavior, and that the filter mecha-

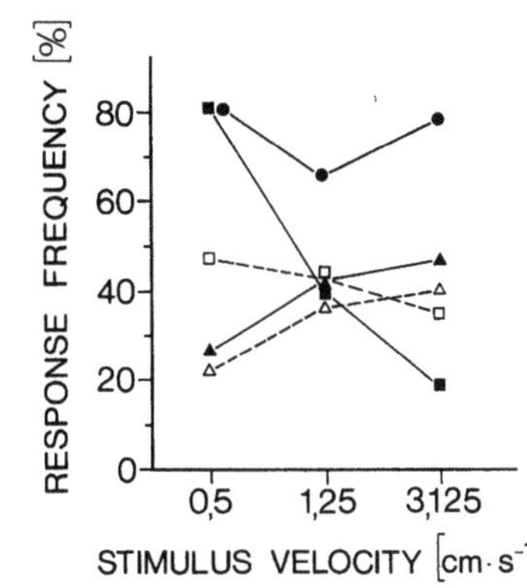

Fig. 2.15. Dependence of prey-catching behavior of *S. salamandra* on stimulus orientation and velocity. Stimuli were a square (16 × 16 mm, *filled circles*) and rectangles (4 × 16 mm) with orientation parallel (*filled squares*), diagonal with tip down (*open squares*), diagonal with tip up (*open triangles*), and perpendicular (*filled triangles*) to the direction of movement, and moved at 0.5, 1.25, and 3.125 cm/s. (From Luthardt and Roth 1979a and Luthardt 1981)

nism for orienting is more specific. However, the differences among the stimuli presented at the two distances are not large enough to corroborate such an assumption. The number of orienting responses to the test stimuli is higher than or equal to the snapping responses, except at an edge length of 32°.

In experiments with *S. salamandra*, Luthardt and Roth (1979a) and Luthardt (1981) used a square (16 × 16 mm) and horizontally, diagonally, and vertically oriented rectangular stimuli (4 × 16 mm) which were moved at three different velocities of 0.5, 1.25, and 3.125 cm/s. In these experiments, the number of snapping responses in percent of the total number of stimuli presented was taken as a measure for stimulus attractiveness. As shown in Fig. 2.15, the square stimulus proved to be most attractive at all velocities (low velocity: 80%, medium: 65%, high: 78%). At low velocity the horizontal rectangle proved to be far superior to the vertical one (80%:26%) whereas at high velocity the opposite was the case. Here, the vertical rectangle was superior to the horizontal one (46.5%:25%). At medium velocity both stimuli were almost equal in efficiency (horizontal: 38.5%, vertical: 41.5%). When diagonally oriented stimuli were tested (Luthardt 1981), a similar effect occurred, though with lesser strength. At low velocity a diagonally oriented rectangle inclining downwards in the direction of movement proved to be superior to a rectangle oriented in the opposite direction (46.5%:21.5%), whereas at high velocity the opposite was the case (upward: 39.5%, downward: 34.5%). The diagonal stimulus inclining downwards with regard to the direction of movement was apparently "interpreted" by the salamander as being near to horizontal direction, and the diagonal stimulus with the opposite orientation as being near to vertical orientation. These results show that the effiency of the parameter "orientation" varies with variation of velocity.

Finkenstädt and Ewert (1983a) obtained different results. They used a 15 × 15 mm square and 2.5 × 15 mm rectangles oriented horizontally and vertically and moved at 4°, 10° and 20°/s which equals 0.5, 1.25, and 2.5 cm/s.

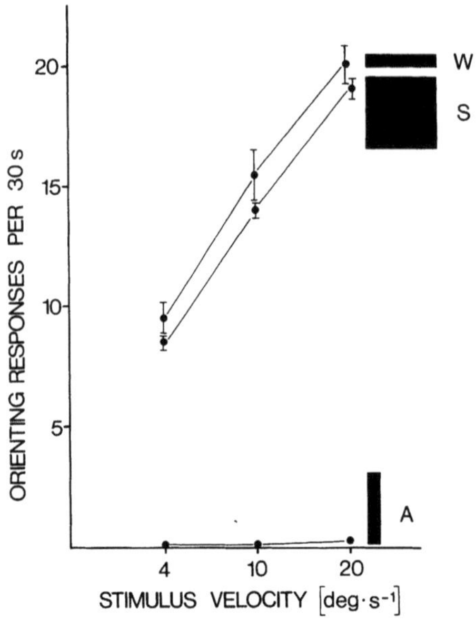

Fig. 2.16. Orienting behavior of *S. salamandra* in response to a black 2.5 × 15 mm wormlike (*W*) stripe, a 2.5 × 15 mm "antiwormlike" (*A*) stripe, and a 15 × 15 mm square (*S*) moving at 4°, 10°, and 20°/s in the horizontal direction. (After Finkenstädt and Ewert 1983)

The authors measured the orienting response of the salamanders to the moving stimuli, not snapping responses as Luthardt and Roth (1979a) did. As shown in Fig. 2.16, Finkenstädt and Ewert did not find any velocity dependence of the effectiveness of elongate stimuli of different orientation. At all velocities measured the horizontal rectangle proved to be far superior to the vertical rectangle, to which the salamanders never responded by orienting. The square stimulus had more or less the same effectiveness as the horizontal rectangle.

It is surprising that Finkenstädt and Ewert did not observe a response to the vertical rectangle in *S. salamandra*, because in all other studies known to the author a certain amount of such responses occurred. Again, the difference between the experiments of Finkenstädt and Ewert with regard to the effect of a velocity-dependent preference inversion is not easy to explain, especially because the same effect was observed when Luthardt and Roth repeated their experiments with the same animals (unpublished). Furthermore, a similar result was observed in experiments with bolitoglossine salamanders.

Here, *Plethodon cinereus*, *Plethodon jordani*, and *Bolitoglossa subpalmata* were tested with square stimuli of different size (2.5-, 5- and 10-mm edge length) and rectangular stimuli of different sizes (2.5 × 10, 2.5 × 20, and 2.5 × 40 mm), again oriented parallel and perpendicular to the direction of movement. The stimuli were moved at velocities of 0.3 and 3 cm/s (*P. cinereus*, *B. subpalmata*) or 1 and 3 cm/s (*P. jordani*) (Roth and Kohlhase in prep.).

At the velocities measured the smallest square was the preferred stimulus, in most cases followed by the shortest horizontal rectangle or the medium square. There was one exception: in *P. cinereus* the shortest horizontal rectangle was better than the smallest square at low velocity (0.3 cm/s). In *P. jordani*, after the smallest square, the shortest horizontal rectangle was the second best stimulus, and the medium square and the shortest vertical rectangle were significantly less effective at 1 cm/s. At 3 cm/s, however, these two latter stimuli were significantly better than the shortest horizontal rectangle, the smallest square being again the best stimulus. In *P. cinereus*, as already mentioned, the shortest horizontal rectangle was superior to the smallest square at low velocity (0.3 cm/s), but at high velocity (3 cm/s) the effectiveness of the horizontal rectangle strongly decreased, and the smallest vertical rectangle was most effective, followed by the medium square. In *Bolitoglossa subpalmata*, the smallest square was again always the best stimulus, followed by the smallest rectangle and the square. These animals did not or only rarely responded to the vertical rectangles.

Thus, in some salamanders there seems to be a tendency for elongate stimuli oriented *parallel* to the direction of movement to decrease in effectiveness with increasing velocity, and the opposite seems to be the case for elongate stimuli oriented *perpendicular* to the direction of movement. Luthardt observed that salamanders confronted with squares measuring 16 × 16 mm often snapped at the horizontal edge when the stimulus was moving slowly, and at the vertical edge when moving fast (Luthardt 1981). Thus, it may well be that the vertical bar is interpreted by the animal as the leading edge of a compact object.

Another possible explanation of the above mentioned discrepancies is the influence of experience with different types of prey. As will be discussed later in this chapter, stimulus preferences are at least partly influenced by experience with natural prey. The salamanders of Finkenstädt and Ewert (1983a) were fed exclusively mealworms, whereas those of Luthardt and Roth (1979a) were fed a mixed diet consisting of crickets, earthworms, and slugs. The Bolitoglossini were fed exclusively crickets and flies. However, the influence of natural diet on visual stimulus preference is most prominent at a juvenile stage, mostly immediately after metamorphosis, and is less effective at an adult stage.

In general, among the salamanders mentioned here, *Salamandra* seems to be the most unspecific one with regard to stimulus size, configuration, and orientation. This variability might also explain the differences between the experimental results from different laboratories. Plethodontid salamanders seem to be more restricted to small compact prey and elongate objects of short length. This is in accordance with the data on the natural diet.

Movement Pattern. The way in which prey stimuli move also influences selectivity of the feeding response. In *Hydromantes* the relative efficiency of stepwise moving stimuli was shown to vary as a function of both step frequency and velocity (Roth 1978). A stimulus (5 × 5 mm square) was moved

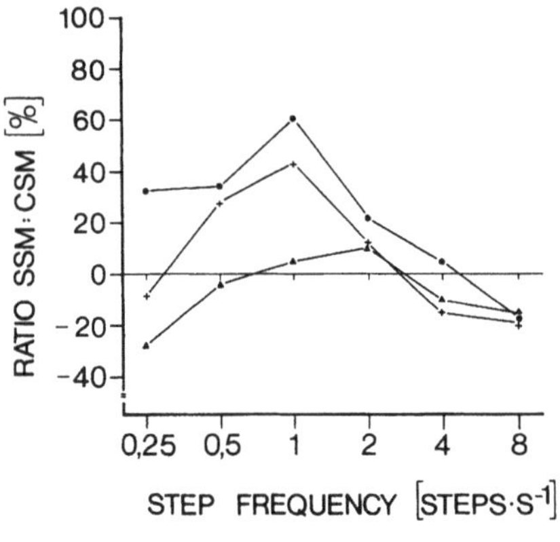

Fig. 2.17. Comparison between prey-catching responses of *Hydromantes italicus* to stepwise stimulus movement (*ssm*) and to continuous stimulus movement (*csm*) at different basic velocities and different step frequencies. The stimulus used was a black square measuring 4 × 4 mm. The *curves* show how much (in % of the respective smaller value) *ssm* is more efficient that *csm* (positive values) or vice versa (negative values). *Dots*: 0.5 cm/s; *crosses*: 1.25 cm/s; *triangles*: 3.1 cm/s. (From Roth 1978)

at three different velocities (0.5, 1.25, and 3.125 cm/s) and moved either continuously or stepwise at different step frequencies. As shown in Fig. 2.17, at low and medium velocity a step frequency of 1 step/s, and at high velocity 2 steps/s elicited most responses. The efficiency of both stepwise and continuous stimulus movement increased with an increase of the velocity, but the efficiency ratios of stepwise and continuous movement had an inverse relationship to each other: at low velocities, where continuous movement was less effective, stepwise movement of the same velocity was considerably more effective; at high velocity, where continuous movement was rather effective, stepwise movement was either not different or was less effective.

In similar experiments with *S. salamandra*, Luthardt and Roth (1979a) studied the interaction of orientation, velocity, and step frequency. Figure 2.18A shows that a bar measuring 4 × 16 mm and oriented *parallel* to the movement direction was most efficient at low velocity (0.5 cm/s) when moving continuously or when moving stepwise with 4 or 8 steps/s. It was very inefficient at low velocity when moving stepwise at 0.25–2 steps/s. This effect was less prominent at medium velocity. At high velocity (3.125 cm/s), there was no significant difference between continuous and stepwise movement, regardless of the step frequency.

With the same rectangular stimulus oriented *perpendicular* to the movement direction, there was again no significant difference between continuous and stepwise movement at high velocity. In contrast, at low and medium velocity, stepwise movement became much more efficient at 2 and 4 steps/s compared to continuous movement (Fig. 2.18B).

In many other experiments with salamanders, stepwise or irregular movement always proved to be more effective than smooth or continuous movement when the prey had a rather *compact* shape, and the opposite was the case with elongated, wormlike stimuli. Thus, movement pattern seems to be an important parameter for prey recognition, and a certain shape seems to

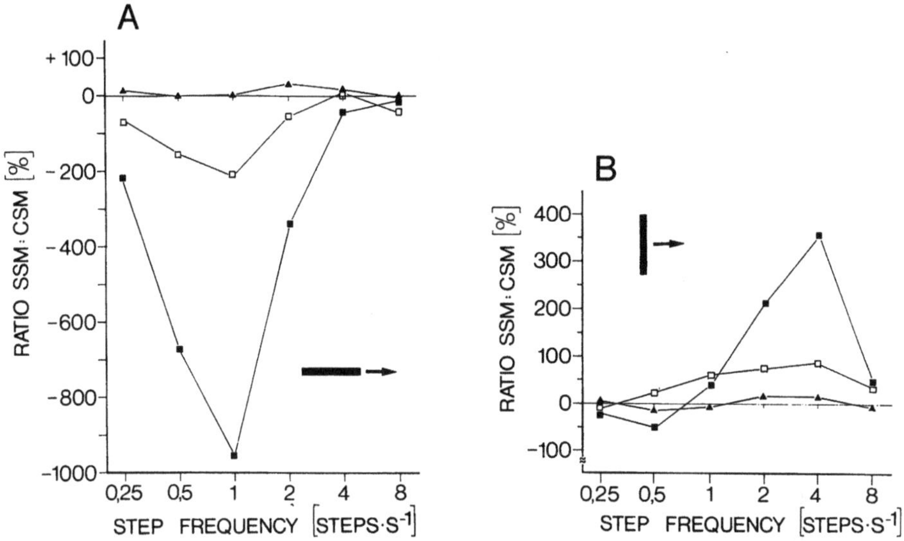

Fig. 2.18 A–B. Comparison between prey-catching responses of *S. salamandra* to stepwise and to continuous stimulus movement. Diagrammatic representation of data as in Fig. 2.17. Stimuli were rectangles measuring 4 × 16 mm and oriented parallel (**A**) and perpendicular (**B**) to the direction of movement. *Solid squares*: 0.5 cm/s; *open squares*: 1.25 cm/s; *triangles*: 3.125 cm/s. (From Luthardt and Roth 1979a)

be connected to a certain movement pattern. The experiments of Luthardt and Roth in which different movement patterns, velocities and configurations were used can be interpreted as indicating that salamanders associate an elongate, "wormlike" shape of a stimulus with slow and smooth movement, and that they associate a compact shape (plus vertical bars as leading edges of compact objects) with fast and/or jerky movement. In contrast, a jerky movement of "wormlike" objects are often disliked (as is the case in *Salamandra*). Such a movement pattern is more typical for fast moving insects of compact shape than for worms.

There is some correspondence between these laboratory results and the natural feeding habits. The most important is that those salamanders such as *Salamandra* that are known to feed preferentially on wormlike objects, respond best to elongate or large square stimuli moving slowly and continuously in laboratory experiments, whereas those salamanders that feed mostly on small compact arthropods like free-tongued plethodontids, prefer small compact objects moving fast and discontinuously to elongate ones in the laboratory.

Stimulus-Background Contrast and Illumination Level. Many salamanders are nocturnal animals: they are active during darkness, leaving their hiding places after sunset and returning before sunrise. Some other species (e.g., *P.*

cinereus) appear to feed under leaf litter both day and night, i.e., under continuous conditions of dim light. The reasons for such a behavior are complex. Salamanders do not avoid light per se, as is often believed: during daytime one can often observe them sitting in holes or caves with their heads turned toward light. Under laboratory conditions salamanders often expose themselves to light for many hours even when there are possibilities for hiding. The main reason for nocturnal behavior seems to be that temperature and moisture are more favorable at night. Another reason may be protection against predators, although we know very little about predation on salamanders.

Although salamanders have mechanisms to detect and capture prey in total darkness, e.g., by means of olfaction or vibration sense, they are able to use vision at astonishingly low illumination levels. Roth (1976) showed that *Hydromantes italicus* is well able to detect and capture odorless moving prey dummies at an illumination level of 0.003 cd/m^3 (0.001 lx) (Fig. 2.19). In 25% of the stimulus presentations under these conditions complete approach of the salamanders occurred, and in 13% they accurately snapped at the stimulus. In experiments with *S. salamandra*, Himstedt (1982) demonstrated feeding reactions at still lower illumination levels. As shown in Fig. 2.20, in these experiments both the illumination level and the stimulus-background contrast changed. The moving prey stimuli had a low ($C = 0.15$), medium ($C = 0.3$), or high ($C = 0.6$) contrast to the background. The behavioral responses measured were orientation toward the stimulus, approach, and snapping. The salamanders were observed by means of an infrared viewing apparatus. As in humans, infrared light is not visible for salamanders.

At an illumination level of 10^{-5} cd/m^2 the subjects showed a high percentage of reactions, which did not change much with a change in visual contrast. At an illumination level of 10^{-6} cd/m^2 there was still a relatively high percentage of orienting and approach responses, but only toward the medium and high-contrast stimuli, and snapping occurred only at the high contrast stimulus. At 10^{-7} cd/m^2 no feeding responses occurred. In nature, the illumination level on a moonless night is 10^{-4} to 10^{-5} cd/m^2. When the stars are covered by clouds, the illumination level is reduced by another logarithmic step. Since most prey objects do not have high contrast to the background, the absolute illumination level for visually guided feeding behavior seems to be around 10^{-5}. These conditions correspond to an open area on a clouded or rainy night. As can indeed often be observed, on rainy nights salamanders are frequently found in open space, e.g., on dirt banks or large logs or rocks which may reflect some remaining light.

The data, thus, show that salamanders are able to use vision at very low illumination levels. However, when feeding at night under covering objects or underground, they are probably forced to use either olfaction or vibration sense for prey detection.

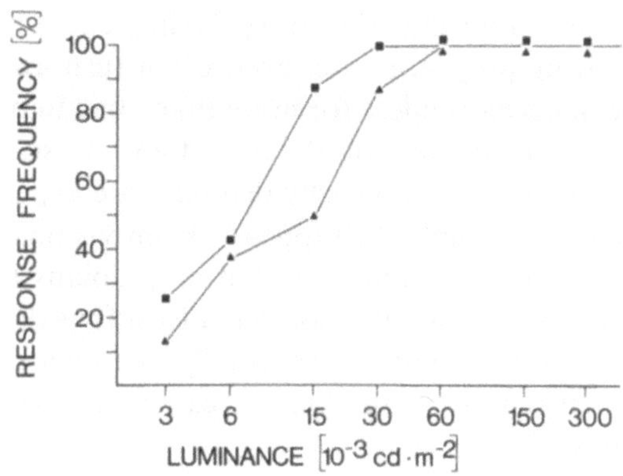

Fig. 2.19. Dependence of prey-catching responses on luminance in *Hydromantes italicus*. *Squares*: complete approach; *triangles*: tongue projection. (From Roth 1976)

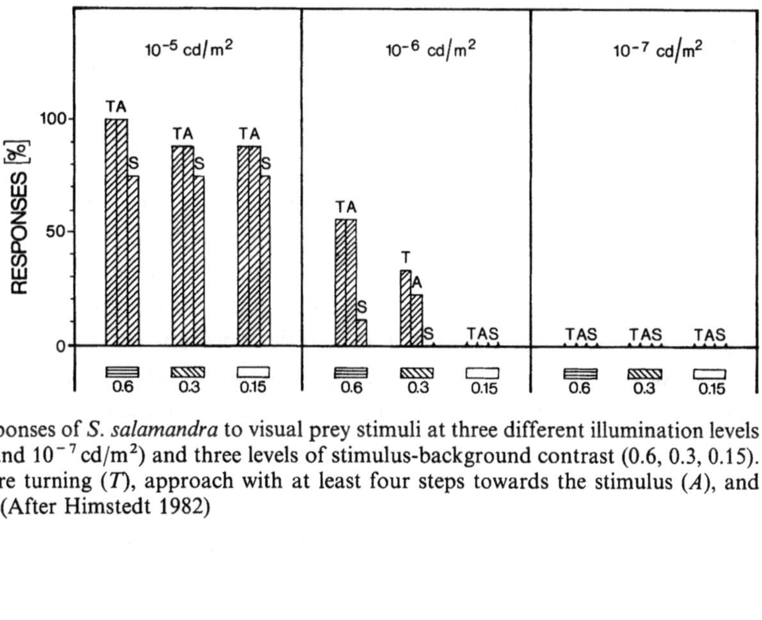

Fig. 2.20. Responses of *S. salamandra* to visual prey stimuli at three different illumination levels (10^{-5}, 10^{-6}, and 10^{-7} cd/m^2) and three levels of stimulus-background contrast (0.6, 0.3, 0.15). Responses were turning (*T*), approach with at least four steps towards the stimulus (*A*), and snapping (*S*). (After Himstedt 1982)

2.1.3.2 Detection of Stationary Prey Objects

Amphibians are usually believed to perceive only moving objects. In their famous article "What the frog's eye tells the frog's brain", Lettvin et al. (1959) stated that a frog will starve to death sitting in a heap of dead flies. At least for toads (Roth and Wiggers 1983) and for salamanders this statement is not correct. What is true is that amphibians, salamanders included, usually do not pay attention to stationary prey objects. As can often be observed during laboratory feeding experiments, salamanders that pursue a moving prey, e.g., a cricket, often turn away from this object as soon as it remains motionless for more than 30 s. In other cases, when the salamanders continue to fixate on the prey, they will snap at it only when it moves again. However, one can easily demonstrate experimentally that an immobile prey does not simply "disappear" from vision, as has often been argued on the basis of the assumption that amphibians lack involuntary eye movements. This assumption is incorrect at least for salamanders (Manteuffel et al. 1977). Nonmoving objects are usually not regarded as being prey, but, as will be discussed later, amphibians can be trained to accept stationary objects as prey.

Roth (1976) showed that *Hydromantes italicus* responds with feeding behavior to stationary objects. Himstedt et al. (1978) studied responses to stationary preylike objects in *Salamandra salamandra*. As in the experiments of Roth, stimuli of identical or different size and shape were presented pairwise at a vertical surface, 2 cm apart from each other. The animals were lured toward the stationary stimuli by means of a moving stimulus until they had reached a position symmetric to the stationary stimuli. Then the luring stimulus was withdrawn and it was tested whether the subjects approached one or the other stimulus and snapped at it (Fig. 2.21).

In the first series of experiments two black circles 2.5 mm in diameter were used. The subjects responded in 25 of 40 experiments with approaching the right or left stimulus and with 8 snapping reactions. In the next series of experiments two dissimilar stationary objects were offered: circles of different diameter and rectangles of different size and orientation, as shown in Fig. 2.21. When a circle and a rectangle were shown the subjects always preferred the circles to the rectangles of equal area, independent of the orientation of the latter. When two rectangles of equal size but of different orientation were shown the subjects responded with equal frequency to either stimulus.

The stimulus preferences measured differed strongly from those obtained in experiments with moving stimuli of the same size and shape. Here, the horizontal rectangle was preferred to the circle of equal area and this stimulus was preferred to the vertical rectangle. The same difference between the preference for moving and stationary prey stimuli were observed in the same kind of experiments with toads (Roth and Wiggers 1983). Horizontal and vertical edges seem to have different qualities for amphibians when moving and when at rest. The most pronounced change induced by movement takes place for vertical edges (at least at low velocity). At rest they are as attractive as horizontal edges, but when moving they strongly decrease in effectiveness in comparison to the horizontal edge.

It is important to note that neither in the experiments with stationary stimuli carried out by Roth (1976) nor in those of Himstedt et al. (1978) a

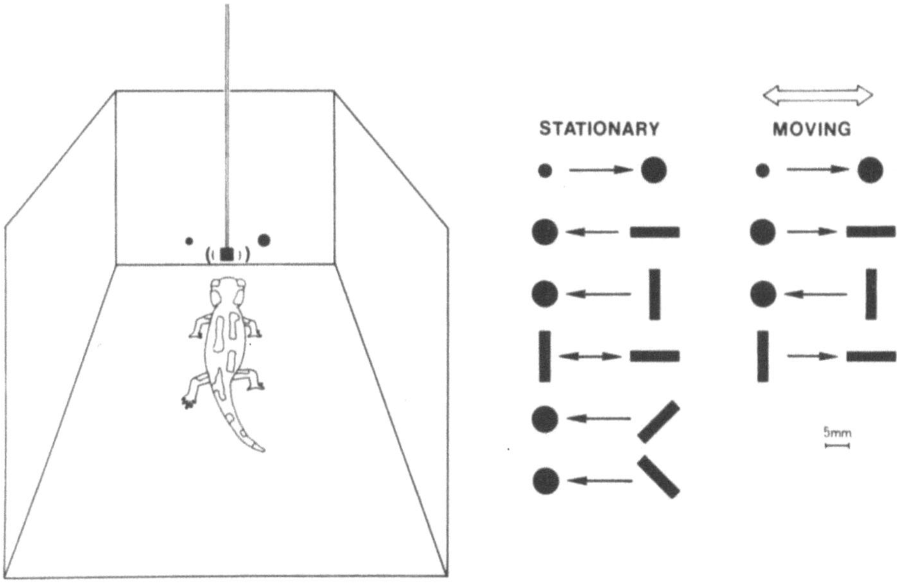

Fig. 2.21. Responses of *S. salamandra* to stationary patterns. *Left*: A salamander follows a moving prey dummy to the wall where stationary patterns are fixed. *Right*: Comparison of prey-catching reactions to stationary and moving patterns. *Arrows* point toward the pattern preferred in a choice experiment. (From Himstedt et al. 1978)

"vacuum activity" response, as described for the toad by Hinsche (1935) took place, that is a snapping in the direction of the luring stimulus after its disappearance. The subjects either turned away from both stimuli or approached one of them. However, the times elapsed between removal of the luring stimulus and the first approach movements were relatively long with a range between 12 and 87 s. Furthermore, when standing in front of one stimulus, the subjects waited another 8 to 85 s until they either snapped at it or turned away.

In these experiments with stationary stimuli the animals were motivated for feeding by a moving prey stimulus. In another series of experiments with *Hydromantes*, Roth (1976) used olfactory stimuli for motivation. Here, an invisible olfactory and a stationary visual stimulus were presented on a wall of the experimental box. The olfactory stimulus consisted of a black mass of dead and slightly homogenized houseflies in a hole on a black background such that this olfactory stimulus was more or less invisible. The stationary visual stimulus was a black circle on white background. The question was whether the subjects would prefer the odorous but invisible prey stimulus or the clearly visible, but stationary, visual stimulus. The experiments showed that after complete approach and exact olfactory localization, the animals responded in 70% of cases to visible odorless prey and in 23% to the invisible

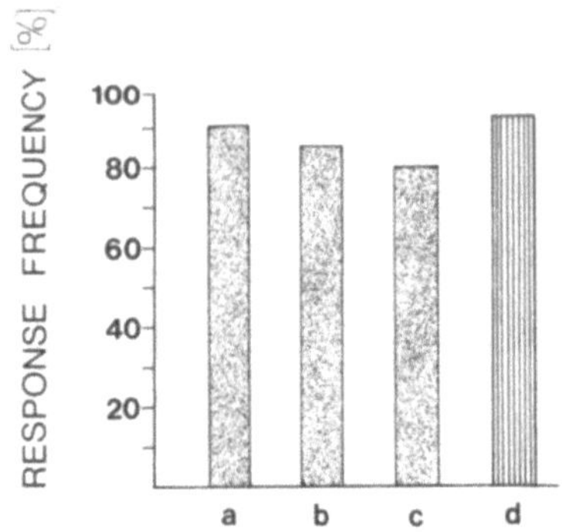

Fig. 2.22. Interaction of the visual and olfactory system during prey-catching behavior (expressed in percent of snapping responses) in *Hydromantes italicus*. Prey-catching responses to smelling prey. *a–c*: In darkness; *d*: in light. *a* Live houseflies; *b* dead flies; *c*, *d* squashed and homogenized flies. (From Roth 1976)

odorous prey. Very remarkably, the animals often remained in front of the invisible odorous prey for a long time, sometimes for 20 min or more without shooting. They often approached the prey until they touched it with their snouts, then moved back, showed olfactory test position, moved their heads from side to side, before finally turning away.

Interestingly, as shown in Fig. 2.22, in darkness the same olfactory stimulus elicited snapping in 80% of the tests, and under light and on white background more than 90%.

These results show that salamanders are able to detect stationary objects even when they have not seen moving prey objects before, i.e., without any possibility of "induction" of virtual movement to the stationary objects. Furthermore, body or head movements are not necessary for recognition of stationary stimuli. Before snapping both *Hydromantes* and *Salamandra* often stared at the stationary visual stimulus for more than 1 min without any apparent head movement.

Possible mechanisms for the detection of stationary prey objects are the following:

1. Sustained reactions of retinal neurons. In the retina of frogs, class-1 and class-2 ganglion cells show long-lasting reactions when a small stimulus moves into the receptive field and stops there. As the subject approaches the stationary stimulus, its image shifts upon the retina. When the subject stops moving, such neurons could continue discharging for a certain time. However, in *Salamandra* the longest sustained reaction of retina ganglion cells observed following stimulus movement lasted only 4–6 s (Grüsser-Cornehls and Himstedt 1973). Neurons with sustained reactions in the range of the "waiting period" mentioned above have not yet been found (though one cannot exclude finding them in unrestrained animals).
2. Minute body or head movements. However, this possibility is ruled out by motion pictures. These pictures show that, as already mentioned, often the subjects remain in front of the stationary stimulus without any movement

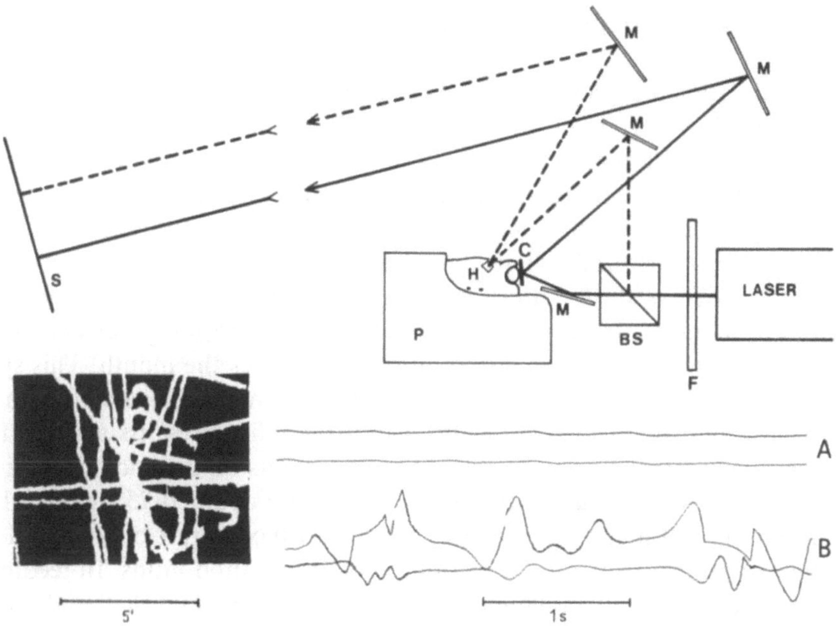

Fig. 2.23. Measurement of eye movements in *S. salamandra* by means of a reflected laser beam. *BS* Beam splitter; *C* mirror on cornea; *F* neutral-density filter; *H* head of salamander; *M* mirror; *P* plaster block with fixed head; *S*: screen or position-sensitive photodetector. *Below left*: Photograph of eye movement tracks on the screen. *Below right*: Signals from the photodetector recorded on an oscilloscope: *upper beam* represents deviations in the x-axis, *lower beam* deviations in the y-axis. **A** Movements of head mirror; **B** movements of cornea mirror. (From Manteuffel et al. 1977)

with an amplitude larger than 100 μm before snapping. This is also true for the experiments with toads (Roth and Wiggers 1983).

3. Respiratory eye movements. In the frog, Schipperheyn (1965) described "respiratory" eye movements which resulted from pressure changes exerted on the eyeballs during ventilation movements of the throat. The author inferred that because of these movements frogs could perceive stationary contours. These eye movements are vertically oriented in the frog. If this were the case in salamanders, too, this should induce a preference for the vertical rectangle, because it makes them appear more wormlike. This, however, was not observed.

4. Undirected, tremorlike eye movements. Such movements, which were long denied to occur in amphibians (Autrum 1959), were demonstrated in *Salamandra salamandra* by Manteuffel et al. (1977). These authors used a laser beam reflected from a mirror onto the cornea. As shown in Fig. 2.23, these rapid eye movements have an amplitude of approximately 12 minutes of arc, sufficient to displace the retinal image over several receptors. This latter mechanism alone could be sufficient for the perception of stationary objects.

2.1.3.3 Depth Perception and Distance Estimation

Estimation of prey distance is a necessary prerequisite for successful feeding. However, there are great differences among different species of salamanders with regard to the precision of snapping at prey which are rather strictly correlated with tongue morphology and feeding strategy: species with a morphologically and functionally elaborate and far-reaching tongue, mostly show an "ambush strategy" and high precision during use of the tongue, whereas those species with clumsy tongues of short reach mostly show a "hunting strategy" and poor precision of snapping. *Salamandra* has a clumsy tongue that during snapping barely leaves the mouth. This species is a classical "hunter" which during twilight and night makes active searches for prey and for this reason walks astonishing distances. Its abilities to hit even slowly moving prey are poor. In experiments on distance estimation of *Salamandra salamandra* Luthardt-Laimer (1983) counted 75 successful prey captures in 200 snapping responses, which is a success rate of only 39%. The other extreme is represented by bolitoglossine plethodontids. In feeding experiments with freely moving prey such as houseflies, fruit flies, and crickets, these animals rarely miss, mostly in those cases in which a prey item quickly moves at the very moment when the salamander projects its tongue. Plethodontids with shorter tongues such as *Plethodon*, *Eurycea*, and *Batrachoseps*, are less efficient, but still have a success rate well above 50%.

Precision of distance estimation has been studied systematically in the bolitoglossine species *Hydromantes italicus* (Roth 1974), *Bolitoglossa occidentalis* (Thexton et al. 1977) and *Bolitoglossa subpalmata* (Wiggers and Roth in prep.). In these experiments a live prey object (housefly, fruit fly, small cricket) was fixed on the tip of a needle in such a way that body movement of the prey was still possible in order to attract the attention of the subjects. By means of a micromanipulator the prey was made to slowly approach the animals which were quietly sitting at the edge of a platform or within a tube of glass or foam rubber. Bolitoglossine salamanders are well suited for this type of experiment because, due to their "ambush" feeding strategy, they remain more or less motionless until a prey comes close. In all these experiments the different authors observed no case in which the animals extruded their tongue precociously, i.e., when the prey was still out of reach of the tongue. In *Hydromantes*, Roth observed a maximum snapping distance of 40 mm. As shown in Fig. 2.24, in *Bolitoglossa occidentalis*, a much smaller bolitoglossine, Thexton et al. (1977) measured a maximum snapping distance of 17 mm. The maximum snapping distance observed in *Bolitoglossa subpalmata* which lies between *Hydromantes italicus* and *B. occidentalis* in size, was 30 mm (Wiggers and Roth in prep.)

Several mechanisms for visual distance measurement are known: (1) eye convergence; (2) lens accomodation; (3) motion parallax; (4) stereoscopic

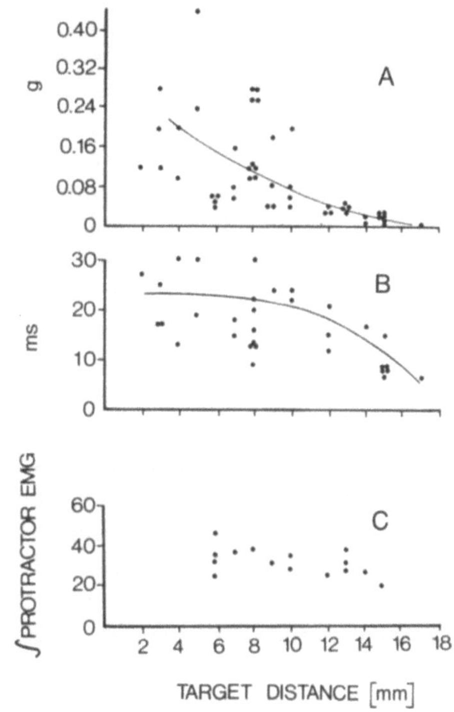

Fig. 2.24 A–C. Tongue projection in the free-tongued salamander *Bolitoglossa subpalmata*. **A** Relationship between the force exerted on prey (vertical axis, *g*) and target or tongue projection distance (horizontal axis, mm). The force exerted by the tongue was measured at peak value. The graph shows a general, and apparently nonlinear, decrease of force exerted with increasing distance to prey. **B** Relationship between duration of tongue contact with prey (vertical axis, ms) and tongue projection distance (horizontal axis, mm). With increase of projection distance, there is a general decrease of contact time, more marked at distances greater than 10 mm. **C** Relationship between integrated protractor (SAR) *EMG* activity (vertical axis, arbitrary units) and tongue projection distance (horizontal axis, mm). The graph shows not only that there is no indication of increased *EMG* activity with increasing projection distance, but, if anything, the reverse. (From Thexton et al. 1977)

vision based on lateral disparity; (5) binocular triangulation; (6) knowledge of absolute object size.

There is no evidence that eye convergence, i.e., an inward movement of the eye axes, takes place in salamanders or in amphibians in general, though salamanders, as other amphibians do, show stabilizing, eye-reflex movements (cf. Grüsser and Grüsser-Cornehls 1976). Careful observations of the eye position of *Hydromantes italicus* during the approach of a prey object did not reveal any convergence of the eyes (Roth 1974).

Motion parallax seems to play no or only a minor role in depth perception, because salamanders do not exert lateral head movement necessary for such a depth perception mechanism, when they approach or fixate a prey item. Furthermore, a motion parallax effect due to the salamander's own movement during approach does not occur when the animal is sitting quietly and "waiting" for an approaching prey item, as is typical for those salamanders with the highest depth perception abilities.

Furthermore, it has been often observed by the author that salamanders accurately feed on completely unfamiliar prey objects or without individual experience of any food item in general immediately after hatching (Roth unpublished observation). Therefore, individual knowledge about absolute prey sizes cannot be crucial for depth perception. Individuals respond to such a wide range of prey size (e.g., from 0.5 to 15 mm) that even "inborn knowledge" about prey size cannot play any major role.

As to the other mechanisms, viz., lens accomodation and stereopsis based on disparity cues and binocular triangulation, the first and the third are the most likely to play a major role in depth perception in salamanders. To date it is unknown whether salamanders exploit the disparity of the two retinal images of an object in binocular vision in order to create stereopsis as in higher vertebrates. Since amphibians have no fovea and no eye convergence to achieve foveal stereoscopic vision, retinal disparity is used for binocular triangulation. These topics will be further discussed in Chapter 6.

As to accomodation, it has been shown by Werner (1983) that *Salamandra salamandra* accomodates during fixation on a near prey object. The situation becomes complicated, however, by the fact that on the one hand binocular distance estimation is possible without accomodation and that on the other hand, binocularity, i.e., the presence of two functional eyes, is not necessary for good depth perception in amphibians (Schneider 1954, Ingle 1968, 1972, 1973, 1976). Collett (1977), Lock and Collett (1979), Jordan et al. (1980), and Douglas et al. (1986) could show by different procedures that monocular toads, for example, are almost as good in object distance estimation as binocular ones. These authors could also show that binocular toads have depth perception independent of accomodation and that they make use of binocular triangulation (cf. Collett 1977).

Luthardt-Laimer (1983) studied binocular and monocular depth perception in *Salamandra salamandra*. She compared three groups of salamanders: (1) binocular animals; (2) chronically monocular animals, in which one eye had been excised 1 year before the beginning of the experiments; (3) reversibly monocularized animals, in which one eye was covered with a tiny aluminum cap. The animals were compared with regard to their feeding behavior to live prey and to moving prey stimuli.

In the first series of experiments live crickets with a body length of 1 cm were used. Binocular salamanders showed a snapping success rate of 39%. The chronically monocular salamanders were only slightly inferior with 36.5% (difference not significant), but in the reversibly monocularized animals the snapping success rate was significantly lower with 26% (Fig. 2.25C).

In the second series of experiments, a black square measuring 8 × 8 mm served as prey stimulus. It was moved through a horizontal window of a screen at a velocity of 0.5 cm/s from the right to the left side. In all monocularized animals used in these experiments, the right eye was either covered or extirpated. In order to test distance estimation the author made use of the following behavior: when *Salamandra* approaches a prey object moving slowly in front of it from one side to the other, it does not approach the object directly, but the direction of its first steps are directed toward the prey with

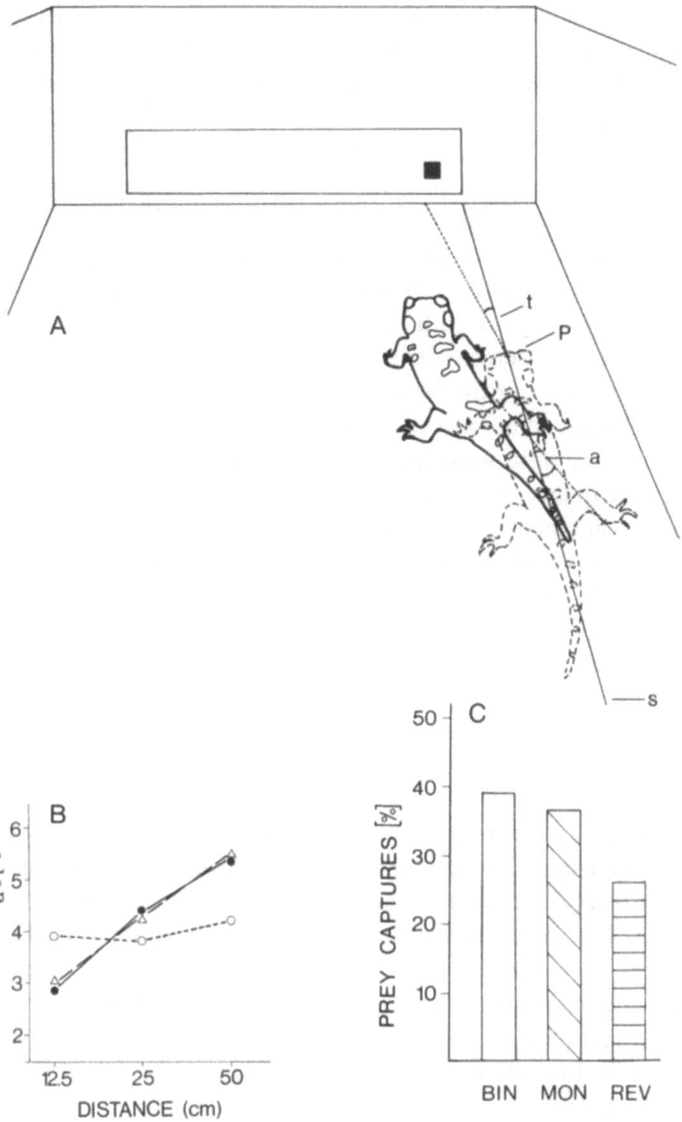

Fig. 2.25 A–C. Depth perception in *S. salamandra*. **A** Experimental arrangement for the investigation of the approach direction. The stimulus moved from right to left at 0.5 cm/s. *s* Line running perpendicular to the screen toward the right side of the window; *p* starting point, above which the central point between the eyes of the salamander was oriented at the beginning of the experiment; *t* the angle between the line *s* and the line joining *p* and the left edge of the stimulus at the moment of the first approach movements of the salamander; *a* angle between line *s* and line tracing the position of the hip joint of the right hind leg. *Broken lines*: Position of the salamander at the beginning of the experiment. **B** Increase in lead angle, expressed by the term $axt^{-0.5}$, in relation to the prey-object distance. *Black circles*: Binocular salamanders; *open circles*: reversibly monocularized salamanders; *triangles*: irreversibly monocularized salamanders. **C** Frequency of prey captures. Prey objects: live crickets. *BIN* binocular salamanders; *MON* irreversibly monocularized salamanders; *REV* reversibly monocularized salamanders. (From Luthardt-Laimer 1983)

a sort of lead angle that points ahead of the moving object (Fig. 2.25 A). This angle is strictly correlated with the distance between stimulus and animal: the more distant the stimulus is the larger is the lead angle.

A comparison between the three groups of salamanders showed that binocular and chronically monocular salamanders do not differ with respect to their lead angle; in both groups it increased with increasing distance between subject and stimulus (12.5, 25, and 50 cm). However, in reversibly monocularized animals the lead angle did not significantly increase with increasing distance. This means that these animals could not adapt their approach behavior with regard to the distance of the stimulus (Fig. 2.25 B).

In these experiments, Luthardt-Laimer could exclude the possibility that the loss of accurate distance estimation in reversibly monocularized salamanders was due to a disturbance by wearing a cap on one eye. Subjects were tested that were wearing an aluminum ring around the eye which did not prevent them from seeing. These subjects were equal in distance perception to the binocular salamanders.

In this study two important questions remained unanswered: (1) Was the "normal" behavior of chronically monocularized salamanders due to a sudden "switch" from one perception mechanism (binocular) to another one (monocular), or to a longer lasting, automatically proceeding compensatory process, especially because the experiments were carried out 1 year after the extirpation of the eye? (2) Would the loss of accurate distance estimation in the reversibly monocularized animals disappear after a certain time, as a function of learning?

These two questions could not be answered in this study, because (1) the operated salamanders did respond to both natural and artificial prey stimuli only several weeks after surgery, and (2) it proved to be impossible in the reversibly monocularized salamanders to fixate the aluminum cap for a longer period without major lesions to the eye. However, the author could exclude the possibility that the "normal" behavior of chronically monocularized salamanders was due to learning during the experiments, because these animals showed exact prey distance estimation in their first feeding responses after surgery.

In recent experiments with *Bolitoglossa subpalmata* and *Hydromantes* (Wiggers and Roth in prep.) the first question mentioned above was investigated. In contrast to *Salamandra*, most of the chronically monocularized Bolitoglossini showed feeding reactions almost immediately (i.e. 5–30 min) after awakening from anesthesia.

Preliminary experiments had demonstrated that reversible monocularization, i.e., covering of one eye, leads to a strong change in feeding behavior in bolitoglossine salamanders (Roth 1974). With one eye covered, *Hydromantes italicus* showed a conspicuous approach behavior in front of a live housefly fixed on a needle. It approached the fly up to snapping distance,

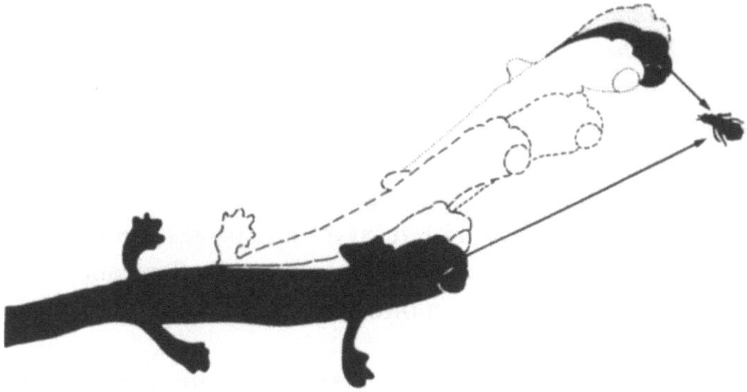

Fig. 2.26. Approach behavior of monocular *Bolitoglossa subpalmata* to a pinned-down, live cricket. The right eye of the salamander is covered. For further explanation, see text. (Wiggers and Roth unpublished)

then bent the body away from the prey toward the side of the seeing eye, but compensated this bending by turning the head back toward the prey such that head and body formed angles between 60° and 90°. In such a position the animals moved around the prey object on a circular line of 2–3 cm in diameter. In many cases an oblique head position declining toward the seeing eye was observed. In Fig. 2.26, this peculiar behavior is shown for *Bolitoglossa subpalmata*. In about half of the experiments the salamanders eventually snapped at the fly, in the other cases they turned away or the experiment had to be stopped after many minutes.

In detailed experiments with *Bolitoglossa subpalmata*, however, Wiggers and Roth found that there was no fundamental difference between chronically and reversibly monocularized salamanders. Under both experimental conditions both types of behavior could be observed in the same animal, i.e., a "normal" direct approach and body curvature. However, the authors found that in both chronically and reversibly monocularized animals there was a strict inverse relationship between approach velocity and the amount of curvature of the body of the salamander: the more readily and faster the salamander approached the prey, the less prominent was the bending of the body, and vice versa. Often, a highly motivated salamander approached a prey quickly without much bending during the first feeding reaction, but the curvature of the body as described above became more and more prominent during the subsequent reactions when the salamander, after having eaten a certain number of prey, approached the prey more slowly.

Sometimes, the bending of the head relative to the body was up to 90°; in addition, often the salamanders lowered the seeing eye down, until it almost touched the floor. At slow approach and strong curvature the snapping distance was mostly short (3–5 mm in *B. subpalmata* and about 10 mm in *Hydromantes italicus*). In the most remarkable case of monocular feeding, a *B. subpalmata*, with one eye covered, approached a pinned-down cricket and showed strong body curvature typical for slow approach. But when another cricket which had been "forgotten" in the experimental container suddenly moved along the salamander at the side of the covered eye in a caudorostral direction, the salamander instantly turned its head toward the cricket and caught it with lateral tongue projection. This again demonstrates that monocularized Bolitoglossini have excellent and fast depth perception with regard to the whole monocular visual field, even in the part beyond the frontal midline.

In the chronically monocularized salamanders the degree of body bending did *not* depend on the time elapsed after eye surgery. Even months after one eye had been excised and after hundreds of successful feeding reactions with one eye only, the salamanders showed strong body bending when slowly approaching the prey. Thus, feeding experience did not influence monocular feeding behavior, nor did any automatic behavioral recovery take place.

In some experiments, the authors tested bolitoglossine salamanders with one eye excised using freely moving crickets or flies. When these animals were highly motivated, they quickly pursued the prey and caught it, sometimes by means of lateral tongue projection both to the side of the seeing eye as well as to the side of the excised eye.

These experiments show, that at least in bolitoglossine salamanders, monocularization in general does not lead to any substantial impairment of depth perception. Under high motivation both chronically and reversibly monocularized animals behave more or less exactly as binocular ones. In nature, perfectly healthy, one-eyed bolitoglossine salamanders are not uncommon, especially of those species that show a high degree of aggressive behavior such as *Plethodon* or *Aneides* (Nishikawa pers. comm.).

This fact seems to contradict the findings of Luthardt-Laimer with *Salamandra* presented above which showed a loss of depth perception in salamanders with one eye covered. The reasons for these differences are unclear. They may be due to differences in the experimental conditions. Furthermore, since the two groups of salamanders represent very different feeding strategies, it may also well be that the differences between the experimental data are due to differences in the respective depth perception mechanisms.

In summary, despite these differences, all these experiments indicate that we have to assume *two* depth perception mechanisms, one for conditions of binocular vision where lens accomodation seems to play no or only a minor role and retinal disparity is used for triangulation, and one for conditions of monocular vision which seems to be based essentially on eye accomodation or on a sort of fixed-focus system. These questions will again be discussed in Chapter 6.

2.1.4 Interaction Between Visual and Olfactory Guidance of Feeding Behavior

Most salamanders possess a well-developed olfactory system which enables them to localize and capture prey without the help of the visual system, i.e., in darkness or with blinded or covered eyes (Nicholas 1922, Matthes 1924a, b, c, Roth 1974, 1976, Roth and Luthardt 1980, David and Jaeger 1981, Himstedt 1982, Lindquist and Bachmann 1982). Olfaction is also used in homing behavior (Madison, 1972), in intraspecific communication (Arnold 1976, Jaeger and Gergits 1979, Madison 1975, McGavin 1978, Tristram 1977), and in communication between congeners (Jaeger and Gergits 1979).

Since many salamanders are primarily active during twilight, at night, or in habitats where light is permanently scarce or absent, they seem to make extensive use of this ability. In some salamanders the olfactory system seems to work very precisely, not only in detecting and identifying prey but also in localizing it. In total darkness, *Hydromantes italicus* is able to localize prey so precisely from a distance of 2–3 cm that it can shoot its tongue towards it (Roth 1976). Other salamanders, like *S. salamandra*, usually have to come very close to the prey in order to localize it exactly by olfaction alone.

The rate of exact localization and successful feeding on prey in darkness by means of olfaction may vary greatly among different salamander species. Roth (1976) found that *Hydromantes italicus* was able to capture live houseflies in complete darkness in 90% of all experiments. In experiments with dead, but intact flies and with squashed and lightly homogenized flies, the percentage was 86% and 80%, respectively (cf. Fig. 2.22). Similarly, in experiments with *Plethodon cinereus* using dead fruit flies presented under conditions of light and of total darkness, David and Jaeger (1981) found only a slight decrease in feeding success (6.3 and 6.1 prey items eaten per 30 min per salamander under light and dark conditions, respectively). In contrast, in *Salamandra salamandra*, Roth and Luthardt (1980) found that the percentage of successful feeding responses to dead prey in darkness was only 40%. More experiments of this type comparing plethodontids and salamandrids are needed in order to determine whether there is a significant difference in the sensitivity to olfactory cues between the two salamander families.

Under sufficient light conditions both the visual and olfactory systems are active. One, therefore, would expect a cooperation of the two systems during feeding on visible odorous prey stimuli. As will be shown, this cooperation strongly depends on the visual properties of the stimulus. As already discussed earlier in this chapter, there is a strong difference in feeding responses to moving and stationary visual stimuli. Salamanders, as amphibians in general, will mostly disregard stationary visual objects. The visual system strongly "requires" motion as a decisive parameter of visual prey stimuli, and usually does not "interpret" stationary objects as prey, even if they are clearly visible.

The situation becomes complicated if the stationary visual stimulus emits odor. Then a certain conflicting situation is created which leads to a very conspicuous behavior of the salamanders. If one places dead intact or squashed and homogenized prey into an experimental container under sufficient light conditions such that it is clearly visible against the background, the prey odor will sooner or later elicit searching behavior. The salamander, for example *Salamandra* or *Hydromantes*, starts to walk around the container, mostly along the walls, but almost never directly toward the prey, even though the visual system clearly perceives the prey as an object in the environment. This searching behavior, therefore, seems to be guided exclusively by olfaction, and the animals seem to ignore the visual cues of the prey. Apparently, the animal does not correlate the olfactory information with the visual information.

One might argue that this behavior occurs because the olfactory information is still too diffuse to allow a precise localization as a basis for identification. But it often happens that subjects come very close to the prey or even pass over it without "recognizing" the visual object as the source of the prey odor. In many of these cases the animals then stop, exert olfactory investigation movements around the prey and show the olfactory test position typical for most urodeles (snout directed downward, neck raised), then move backward or sideward and finally fixate on the prey. But even then, most animals do not immediately snap at the prey but continue to search around although the prey is only a few millimeters distant and situated directly in front of their eyes and nose. This very conspicuous behavior may last from 10 s to 30 min until the animals finally snap at the prey, often with great vigor. But sometimes, the animals leave the prey after many minutes of searching, without snapping. In a few cases the animals did not direct their final snapping toward the prey but lateral to it, missing it entirely.

The interaction between visual and olfactory guidance of feeding behavior was studied systematically in *Hydromantes* (Roth 1976), *Salamandra* (Roth and Luthardt 1980, Luthardt and Roth 1983, Luthardt-Laimer and Roth 1983, Himstedt 1982), in *Plethodon cinereus* (David and Jaeger 1981) and *Ambystoma* (Lindquist and Bachmann 1982).

Both Roth (1976) and David and Jaeger (1981) showed that despite the large behavioral differences during feeding on live or dead prey under light, there are only minor differences with regard to feeding success. The reduction of feeding success from live to dead prey was 6.7% in *Hydromantes italicus* and only 3.2% in *Plethodon cinereus*. Thus, these animals will finally "accept" dead prey by learning despite the absence of movement.

In the experiments of Roth and Luthardt (1980), Luthardt and Roth (1983), and Luthardt-Laimer and Roth (1983) with *Salamandra*, dead (killed by freezing and then thawed) crickets were used as stationary visual prey. The authors measured (1) the "prey-catching time" elapsed from the beginning of the experiment to the feeding response (if there was any), as an indicator of the ability of the subjects to localize visually stationary and odorous prey; (2) the "fixation time", i.e., the time spent in front of the prey after complete approach and before snapping (if it occurred); (3) the percentage of snapping responses.

Under light conditions, the mean prey-catching time was 7.8 min and the fixation time was 76 s (Roth and Luthardt 1980). The percentage of snapping responses was 30%. Interestingly, with the same animals, but under dark conditions where the subjects were observed by means of infrared light, the prey-catching time was only 5 min, and the time between complete approach and snapping only 21 s. Furthermore, the total percentage of snapping responses in darkness was 40%. This surprising finding indicates that it takes longer for salamanders to localize and finally snap at stationary, odorous prey when they can smell *and* see it than when they smell it alone in darkness without the "help" of the visual system!

These results were corroborated by Himstedt (1982) in a series of experiments in which the responses to stationary, *odorless* prey dummies remained very low at illumination levels down to 10^{-6} cd/m^2, which represents the visual sensitivity threshold of *Salamandra*, whereas the responses to stationary, *odorous* prey strongly increased with decreasing ambient illumination level. This again demonstrates that the subjects responded better to stationary odorous prey the less visible it was.

It seems clear that this paradoxical behavior has to do with the important role played by motion in visual behavior. In the case of stationary, visible *and* odorous prey, the visual system does not immediately "accept" the olfactory information; rather the absence of motion seems to inhibit the release of the snapping response by the olfactory system alone. Under sufficient light conditions, the visual system seems to require prey motion to elicit snapping. If motion is not present, the snapping response is considerably delayed or even totally suppressed. This could be seen very nicely in those cases in which an animal had stared at a dead fly or cricket for many minutes without feeding, but snapped at the prey almost immediately as soon as the light was turned off. The inhibitory effect of lack of motion on the visual system disappears in the absence of light, and the olfactory system alone is able to elicit the feeding response.

The inhibitory influence of the visual system on the olfactory guidance of feeding toward stationary, odorous prey seems to concern mostly the final snapping response after reaching the prey. During the search period, the animal seems to be guided by olfaction alone and simply disregards visual cues. The strength of visual inhibition on the snapping response due to the absence of prey motion seems to be different in different groups of salamanders. In *Hydromantes italicus* the percentage of snapping after exact localization of dead prey under light was 90% in one series of experiments and 70% in another series (Roth 1976). In *Bolitoglossa rufescens* the percentage of exact localization of squashed *Drosophila* was about 60%, with 41% snapping (which is 62% of all cases of exact localization) (Roth unpublished data).

Juvenile *Salamandra* 10 months after metamorphosis, showed exact localization of stationary prey (dead crickets) in 60% of trials and 30% snapping responses; i.e., 50% of all animals which localized and fixated on the prey, finally snapped at it (Roth and Luthardt 1980). In 2-year old *Salamandra*, the percentage of prey localization was again about 60%, but now 52% (i.e. 87% of all cases of exact localization) snapped at the prey. The mean fixation time was almost the same in juveniles and 2-year-old ones (76 and 68 s, respectively) (Luthardt and Roth 1983). Since the two age groups were tested in different experimental series several years apart, it cannot be determined whether or not the increase in snapping responses after complete approach in the older ones is due to an age- or experience-dependent reduction of visual inhibition with regard to stationary odorous prey. Both groups were raised with live prey only; therefore, the older ones had no experience with stationary prey.

The interaction between the visual and the olfactory system was further studied by Roth (1976) in *Hydromantes italicus* and by Luthardt and Roth (1983) in *S. salamandra*. In the experiments with *Hydromantes*, pulp of squashed and lightly homogenized house flies was filled into small holes (4 mm in diameter) within one wall of the experimental container. One hole had a white and one a black background such that the pulp was clearly visible within one hole and invisible in the other. Additionally, a third hole between the two was filled with odorless black material (plastic) and had a white background. Therefore, three types of nonmoving prey were presented to the animals: (1) odorous prey with visual contrast (OC), (2) odorous invisible prey (O), and (3) odorless prey with visual contrast (C). After complete approach and exact localization (= 100%), the animals responded to OC in 70% of cases, and to O in 23%. One animal snapped at C after long fixation of the invisible odorous stimulus.

Figure 2.27 I gives the results of the experiments with *S. salamandra* exerted under the same conditions. Here, the animals localized and fixated OC in 59%, and snapped at it in 52% (= 88% of all cases of complete approach). The invisible, odorous prey (O) was localized in 49% with 21% (= 42.8%) snapping responses. Interestingly, in 35% the animals fixated the odorless, contrast prey (C) and snapped at it in 6% (= 17%). While the total prey-catching time did not differ significantly with respect to all three stimuli (OC: 7.8 min; O: 8.6 min; C: 7.5 min), there were significant differences between the fixation times: OC: 68 s; O: 92 s; C: 15 s. The distances from

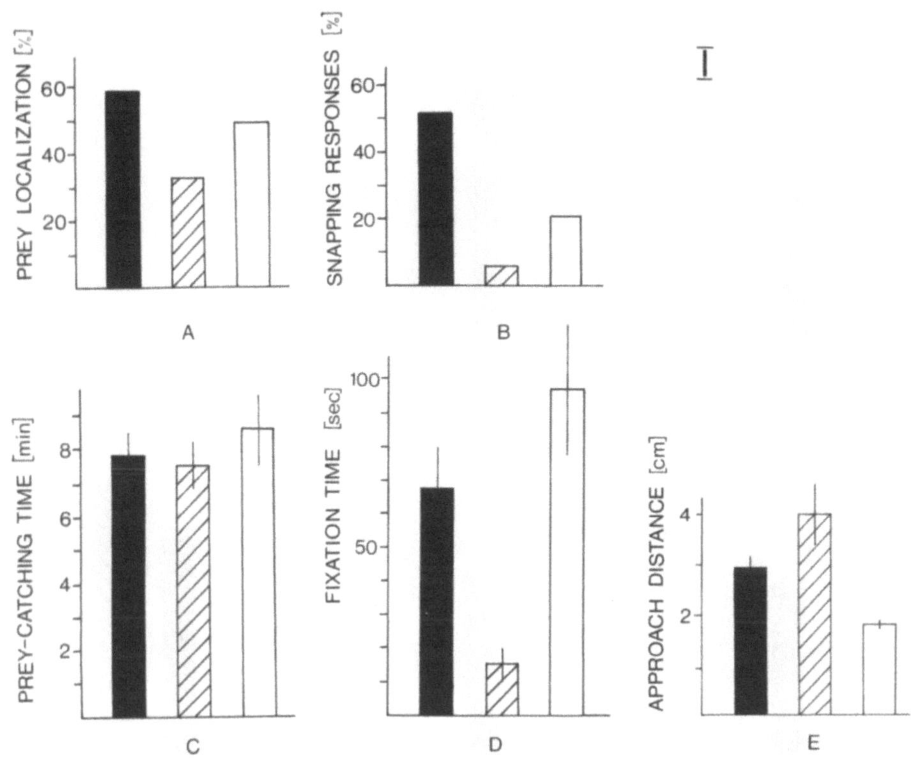

Fig. 2.27 I, II. Responses of *S. salamandra* to stationary prey. I Subjects raised with live prey. II Subjects raised with dead food. *Black columns*: Odorous contrast prey; *hatched columns*: contrast prey with neutral odor; *white columns*: invisible odorous prey. *A* Frequency of prey localization; *B* frequency of snapping at prey; *C* prey-catching time; *D* fixation time; *E*: approach distance. *Vertical bars* indicate standard error of the mean (From Luthard and Roth 1983) (Part II see page 64).

which direct approach occurred were rather different: OC: 2.9 cm, O: 1.8 cm, C: 4.0 cm (the differences between O, on one the hand, and OC and C, on the other, are significant).

These experiments show that both *Hydromantes* and *Salamandra* prefer stationary, visual stimuli with prey odor to both invisible, odorous prey and visible prey of neutral odor. It is surprising, however, that sometimes *Salamandra* snapped at odorless visible prey which *Hydromantes* did only in one case Furthermore, it is remarkable that in those cases where *Salamandra* responded to C, the fixation time was much shorter than in the experiments using OC and O, although the number of responses was much lower. It may be that those 6% of subjects that localized the prey and then snapped at it were the most motivated ones which, therefore, showed little snapping response "delay" after complete approach.

In their experiments with *Ambystoma tigrinum* Lindquist and Bachmann (1982) used moving, visual prey stimuli instead of stationary ones. The first stimulus type was a mechanically moved piece of earthworm (= moving,

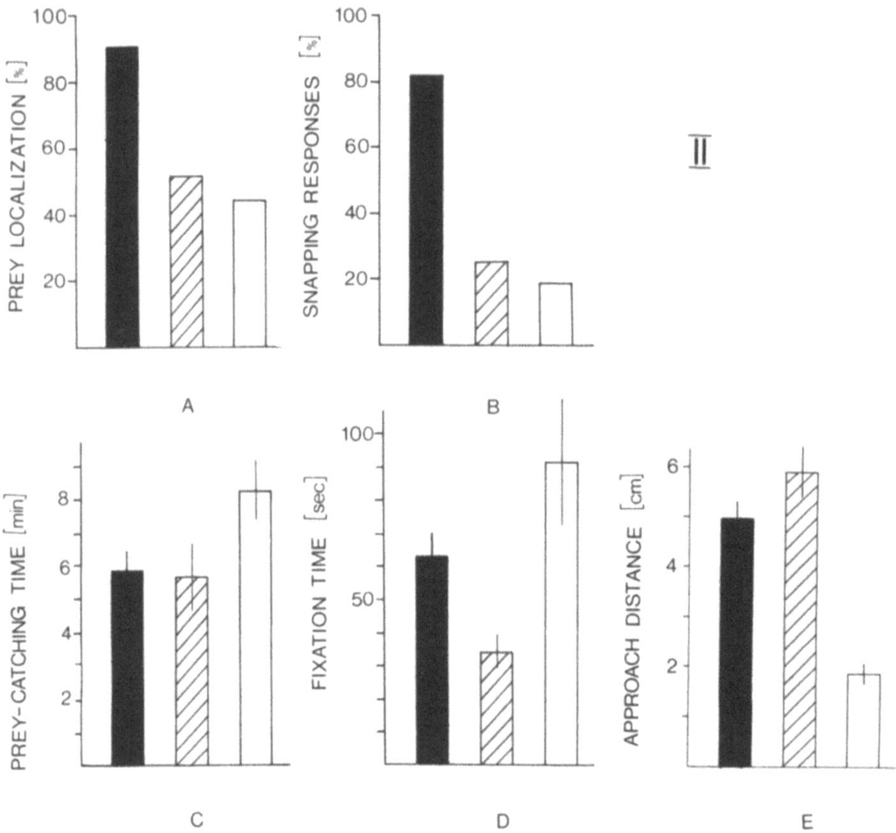

Fig. 2.27 II. (Legend see page 63)

visible, odorous stimulus, MOC), the second one a ground piece of earthworm, dissolved in distilled water (= invisible, odorous stimulus, O), and the third one a mechanically moved piece of an artificial earthworm (= moving, visible, odorless stimulus, MC). The experiments were carried out in light.

The percentage of exact prey localization was 97.9% under MOC conditions, 88.9% with O, and 97.2% with MC as stimulus. The percentage of snapping responses after prey localization was 100% with MOC, 28.1% with O, and 88.6% with MC. This shows that under sufficient light conditions *Ambystoma* is almost equally able to localize prey objects either by visual or olfactory cues alone, although the combination of both cues increases the localization rate. As in the experiments discussed before, major differences occurred with respect to the final snapping response: whereas all (MOC) or almost all (MC) subjects that had approached the visible stimuli snapped at it, only a small proportion of those that had localized the invisible, odorous stimulus (O) snapped at it. This shows that in *Ambystoma*, as in the other species tested, the inhibition on the final snapping response due to the absence of visual cues is rather strong.

It is interesting that there is a difference between the percentage of snapping to moving, visual prey with (MOC) and without odor (MC), although the percentages of exact localization were almost equal. This means that the release of snapping is enhanced by prey odor even in the case of moving visual stimuli, although the visual cues strongly dominate feeding behavior. This interaction is further shown by a comparison of the mean prey-catching times. When the three different prey types were presented at a distance of 10 cm away from the starting point of the subjects, the prey-catching time was 70 s in the experiments with MOC, 240 s in those with O and 164 s in those with MC. It is not surprising that it takes much longer for the subjects to localize the invisible, odorous prey, but it is remarkable that the MOC prey was approached in less than half the time in comparison to MC, apparently due to the presence of prey odor. Since the percentage of localization was almost equal in both cases, we have to assume that the prey odor had an influence on the feeding motivation, but plays practically no role in the localization of moving, visual stimuli.

In summary, the interaction between the two main sensory systems for the guidance of feeding behavior takes place in a complicated way. Without doubt, the presence of prey odor enhances feeding behavior under light conditions. In all experiments visible prey stimuli were more easily localized and more readily "accepted" as prey when they emitted prey odor. This is true for moving and as well as for stationary stimuli. In the latter case, prey odor significantly reduces the "visual inhibition" created by the absence of movement. This could be seen in those experiments where the percentage of localization of invisible odorous prey and visible, odorless prey did not differ very much (49% to 35% in *Salamandra*), but the differences between the snapping reactions were much larger (21% to 6%). But it remains remarkable that in the case of invisible odorous prey stimuli less than half of all subjects which had localized that prey exactly also snapped at it, whereas in darkness this percentage was much higher.

This corroborates the idea that salamanders possess at least two systems for sensory guidance of feeding behavior, one for dim light or darkness (olfaction) and one for sufficient light conditions (vision) which have evolved and are active rather independently. Vibration sense seems to play an important role, too, for feeding in darkness, although an experimental analysis of this ability is still lacking.

2.1.5 Ontogeny of Feeding Behavior and the Role of Learning

The feeding behavior of amphibians is generally considered to be highly instinctive. In most cases the sequence of feeding appears to be rather stereotyped (cf. Tinbergen 1951). As discussed above, feeding responses can be elicited easily by simple prey dummies. However, it is known from several studies with anurans that responses to prey objects can be modified by experience. According to Schaeffer (1911), Buytendijk (1918), Cott (1936), Eibl-Eibesfeldt (1951), Brower and van Zandt-Brower (1962), and Sternthal (1974), anurans are able to distinguish noxious or unpleasant from edible food, sometimes after one trial. Brzoska and Schneider (1978) could train toads to avoid unpalatable food or to snap at motionless food using an auditory conditioning stimulus.

Salamanders are usually regarded to be even more instinct-bound and stereotyped than are frogs and toads. Individual experience seems to play only a minor role, if any, in their feeding behavior and prey preferences. After metamorphosis, hatching or birth, salamanders show a behavior at their first feeding activity which is more or less identical with that of older juveniles or adults. This is the case not only in salamandrids, but also in bolitoglossine plethodontids with extensive tongue projection, in which appropriate use of the tongue, including precise depth estimation, is essential. *Hydromantes italicus* and *H. genei*, which were born and raised in our laboratory, showed very precise and successful use of their tiny projectile tongues at their first feeding activity about 10–14 days after hatching from eggs. Since these salamanders have no aquatic larvae but direct development, larval experience cannot play any role. We must assume that salamanders have an inborn feeding program with regard to moving prey. However, as will be shown later, this does not exclude a maturation process, on the one hand, and a modifying influence of postmetamorphic prey experience on the other.

Himstedt et al. (1976) investigated the ontogeny of configurational prey preferences in *S. salamandra*. As shown above, in many amphibians the configuration of prey stimuli, especially the orientation of the stimulus with respect to the direction of movement, is essential. Figure 2.28 shows that when aquatic larvae of *Salamandra* were tested using moving, visual stimuli of different size, shape, and orientation, they generally did not show any differences in their responses to squares and to horizontally and vertically oriented rectangles. They preferred an edge length of 16 mm in all three stimuli.

Salamanders of an age of 1–3 months after metamorphosis still responded more or less equally to all three stimuli, but they no longer preferred a specific edge length, except in the case of the horizontal rectangle, where the longest stimulus (32 mm) was slightly preferred to the shorter ones. The overall turning activity, however, was lower than that of the larvae.

At an age of 5–7 months after metamorphosis, juvenile salamanders either showed more or less the same preferences as the previous age group (B1), or demonstrated a strongly modified behavior such that they no longer responded to squares and vertical rectangles of any size, and responded to the horizontal rectangle only at a length of 16 mm and 32 mm (B2). At an age

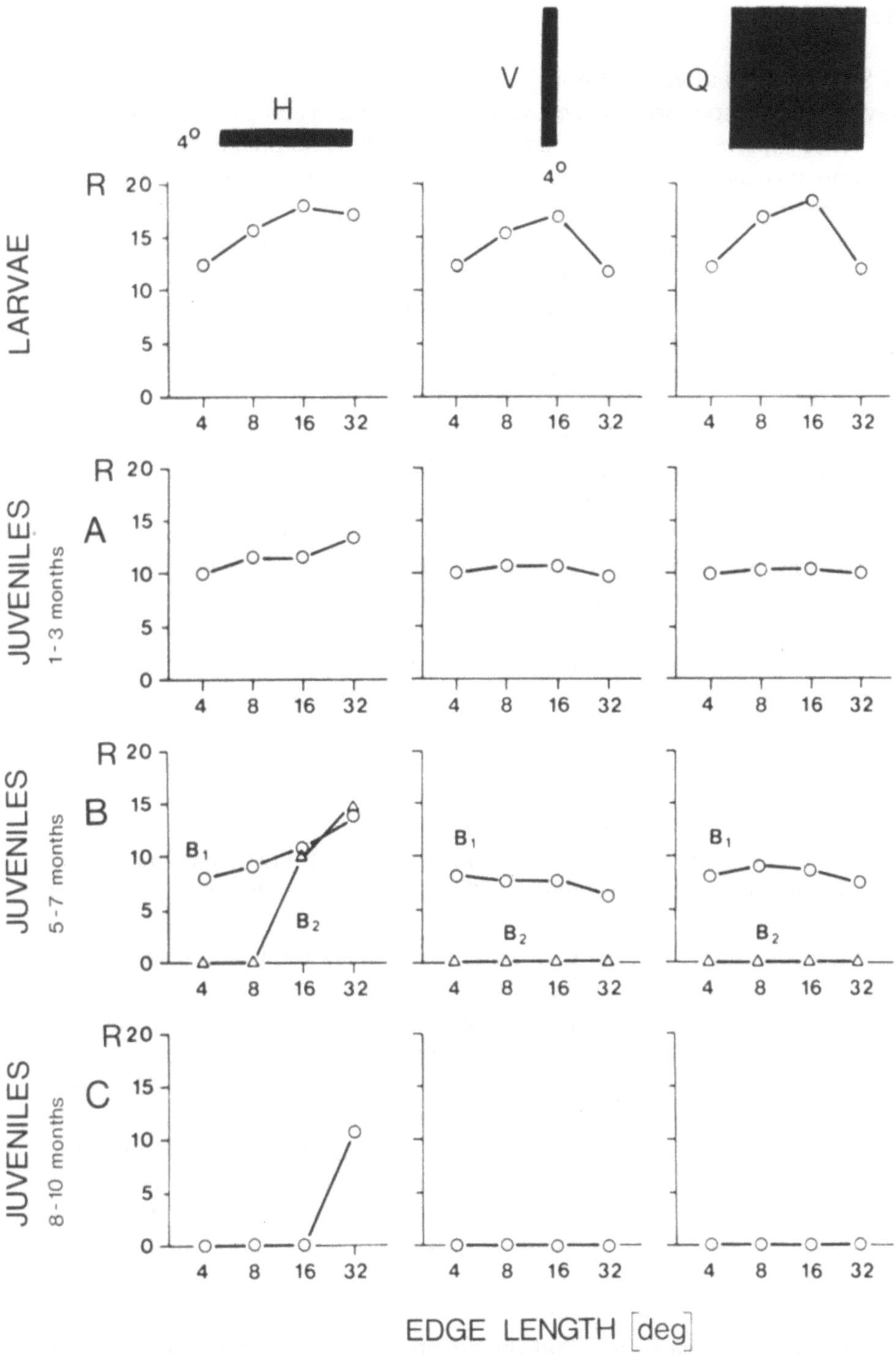

Fig. 2.28. Ontogenetic changes in prey stimulus preferences in *S. salamandra*. Reactions measured were orienting responses (*R* = orienting turns/min) of larvae, juveniles 1–3 months (*A*), 5–7 months (*B1*, *B2*) and 8–10 months (*C*) to moving stimuli in which either the horizontal edge (*H*), the vertical edge (*V*), or both edges (*Q*) were geometrically increased from 4° to 32°. (After Himstedt et al. 1976)

of 8–10 months after metamorphosis, all juveniles showed this type of behavior; they responded to the longest horizontal rectangle (32 mm) only.

These experiments seem to indicate that in these salamanders an ontogenetic process takes place between the larval stage and the age of 5–10 months after metamorphosis, such that a broad spectrum of prey-size and prey-configuration acceptance is "narrowed down" to a rather specific prey preference for elongate, "wormlike" stimuli. Since the larvae had been fed tubifex, a food that has no resemblance in shape to the wormlike dummies preferred later on, this process cannot be ascribed to larval learning. It is surprising, however, that in these series of experiments the salamanders of an age of 8–10 months responded exclusively to the longest "worm" dummy. As many other experiments mentioned above with older *Salamandra* show, these animals respond better to square stimuli than to wormlike ones, and even respond to vertically oriented rectangles (Luthardt and Roth 1979a).

Becker (1980, 1985) studied the influence of early sensory experience on visual prey preferences. Newly metamorphosed *S. salamandra*, which as larvae had been fed tubifex, were divided into two groups with different food regimens: group 1 was fed live crickets under normal light conditions (L-group) such that the subjects gained primarily visual experience with moving prey. Group2 was fed exclusively live earthworms or pieces of earthworms in darkness (D-group); in the time between the feeding sessions they were housed in semitranslucent containers such that they were stimulated by light but could not perceive any prey movement.

At an age of 3–4months after metamorphosis both groups were tested with regard to their responses to horizontal and vertical rectangles measuring 1 × 8 mm and moved at velocities of 0.25 and 1 cm/s. The results are given in Fig. 2.29.

At a stimulus velocity of 0.25 cm/s (Fig. 2.29A), the L-group snapped at the horizontal rectangle significantly more frequently than at the vertical one (46% vs 20%, $p = 0.005$). The D-group showed the same preference, but the overall activity level as well as the percentage of snapping reactions at the vertical rectangle was considerably higher (horizontal rectangle: 72%, vertical rectangle: 52%, $p = 0.025$). The situation was somewhat different at the higher velocity of 1 cm/s (Fig. 2.29B). Here, the L-group again significantly preferred the horizontal rectangle to the vertical one, though at a much lower activity level (26% vs 10%, $p = 0.025$); but the D-group now showed no significant differences between their responses to the two stimuli (vertical rectangle: 25%, horizontal rectangle: 23%), although their activity was higher than that of the L-group.

These experiments show that the preference of *Salamandra* for wormlike stimuli is at least partly inborn. Both tested groups had no visual experience with those stimuli: the L-group had visual experience, but not with wormlike prey; and the D-group was raised with worms, but since these animals had

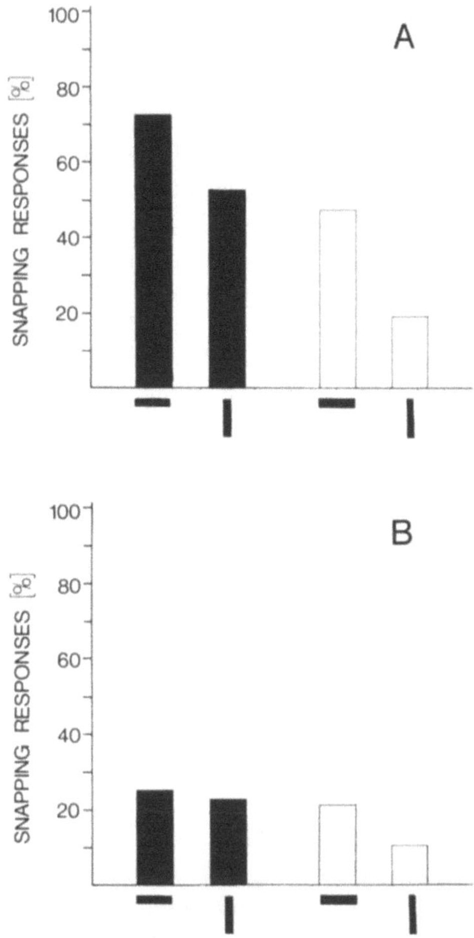

Fig. 2.29 A–B. Stimulus orientation preferences of juvenile *S. salamandra* raised in the dark (*black columns*) and under normal light conditions (*white columns*). Frequencies of snapping responses to vertically and horizontally oriented rectangles (1 × 8 mm) moved at 0.25 cm/s (**A**) and 1 cm/s (**B**) are shown. (From Becker 1980)

been fed in darkness, they possessed no visual experience with that type of prey. Nevertheless, both groups significantly preferred the horizontal to the vertical rectangle, at least at a low velocity of 0.25 cm/s. However, the fact that at higher velocity the D-group, which had no visual prey experience at all, showed no difference to the two stimuli, may indicate an influence of the conditions under which the animals were raised, in the sense that the *strict* preference for horizontal vs vertical rectangles is established only with sufficient *general* prey experience. Such an *unspecific* influence of environmental sensory conditions on the establishment of stimulus preferences is well known for higher vertebrates.

Most interesting is the fact that the D-group, which was naive with respect to visual prey, responded more frequently than the experienced ones to the prey dummies. This finding occurred several times in different experimental series with experientially deprived subjects. Visual experience with natural prey seems to reduce the readiness of salamanders to respond to artificial prey; in other words, naive salamanders do not yet "know" the difference between natural and artificial stimuli.

An interaction between "inborn" behavioral patterns and early experience was found by Himstedt with regard to the ontogeny of prey size preferences (cf. Grüsser-Cornehls and Himstedt 1976). Newborn larvae of *S. salamandra* were divided into two groups. The first group was kept in tap water without food. The second group received *Cyclops* and *Daphnia* as food for 14 days. After that period animals of each group were tested for prey size preferences in a choice test, in which white spots of either 2 mm or 6 mm in diameter were used as stimuli. The first one corresponded in size to the natural food objects, whereas the second one was far too large for a young salamander larva to swallow.

The experiments revealed that both groups significantly preferred the smaller stimulus of "natural" size. However, there was a clear difference between the two experimental groups: whereas the smaller stimulus was chosen by the naive larvae in 65% of the tests, in the experienced animals the percentage was 89%. This means that there is an innate preference for "natural" prey size which seems to be somewhat unstable: one third of the naive animals chose prey that were far too large. During ontogeny this innate preference is apparently "sharpened" by experience.

The influence of early visual experience on feeding preferences seems to be greater with regard to stimulus movement pattern than with regard to stimulus size or orientation. This was shown by Luthardt and Roth (1979b) in an ontogenetic study, again with *S. salamandra*. As mentioned earlier, the way prey moves seems to be a very important cue for prey recognition. As demonstrated by Luthardt and Roth (1979a), adult *S. salamandra* showed very different stimulus preferences depending on the movement pattern of the stimuli: at low velocities they significantly preferred continuous movement of wormlike objects, whereas at higher velocity jerky movements of compact objects or vertical bars (which might be interpreted as the vertical edges of the compact stimuli) were preferred.

The authors tested to what degree the preference for certain prey-movement patterns is experience-dependent. Five groups of newly metamorphosed salamanders were raised with prey types which differed characteristically with regard to the way they moved, namely:

(1) stationary prey (dead crickets, mealworms, tubifex, pieces of liver); (2) live crickets; (3) live mealworms; (4) dead food equal to (1), moved continuously at a velocity of 0.5 cm/s by means of an electric carrier onto which the food was placed; (5) mixed live food (crickets, tubifex, wingless *Drosophila*, slugs).

The different groups, therefore, were confronted either with no prey movement at all (1), with slow, slightly discontinuous movement (2), with faster and more jerky movement (3), with strictly continuous movement (4), and with a mixture of different types of slightly to clearly discontinuous movement (5). The subjects were tested at ages of 1, 3, and 6 months after metamorphosis with respect to their specific preferences to stimulus movement patterns. As test stimulus a horizontal rectangle was used being moved either continuously or stepwise at step frequencies of 0.25, 0.5, 1, 2, 4 and 8 steps/s. The speed was 0.5 cm/s.

For clarity only the results of the 6-month-old salamanders are presented in Fig. 2.30. At that age the salamanders of group 1 (stationary prey, *A*) preferred continuous stimulus movement (csm) to stepwise stimulus movement (ssm) at all frequencies except 2 steps/s where no difference occurred between ssm and csm. Group 2 (live crickets, *B*) significantly preferred csm to ssm at low frequencies (0.25–1 Hz) and ssm to csm at high frequency of 4 Hz. In group 3 (live mealworms, *C*) the preference was quite different: whereas the animals responded equally to ssm and csm when ssm moved with the two lowest step frequencies, they significantly preferred csm to ssm at medium or high step frequency. Group 4 (continuously moved dead prey, *D*) preferred csm to ssm significantly at 1 and 2 Hz, whereas at other step frequencies there was no significant difference between csm and ssm. Finally, the control group 5 (*E*) preferred csm to ssm at a frequency of 0.5 steps/s.

During the test period of 6 months all groups showed a general increase in preference for continuous movement, regardless of the step frequency preferences. This also occurred in the control group which had been fed mixed, live food. This may be considered as a maturational process similar to that described above for stimulus orientation preferences. However, the differences between the different groups not only persisted but became more prominent, especially the differences between groups 2 and 3 (crickets vs mealworms).

An interesting similarity evolved between group 1 (stationary food) and the control group: the genetically determined movement pattern preferences of all "naive" animals became almost identical with the preferences of the animals with experience of "mixed" movement pattern. Furthermore, the "naive" animals responded generally more frequently to the stimuli than all other groups which had some kind of prey movement experience. This difference is highly significant ($p < 0.001$). This phenomenon also occurred in Becker's experiments described above.

The experiments show that the preferences of *S. salamandra* for stimulus-movement patterns are significantly dependent on early prey experience. A learning process takes place which leads to a preference for a movement pattern which is similar to that of the food with which they have been raised. Salamanders that had been fed crickets preferred discontinuous movement of higher step frequencies to continuous movement, and they preferred continuous movement over discontinuous movement of low step frequencies. This

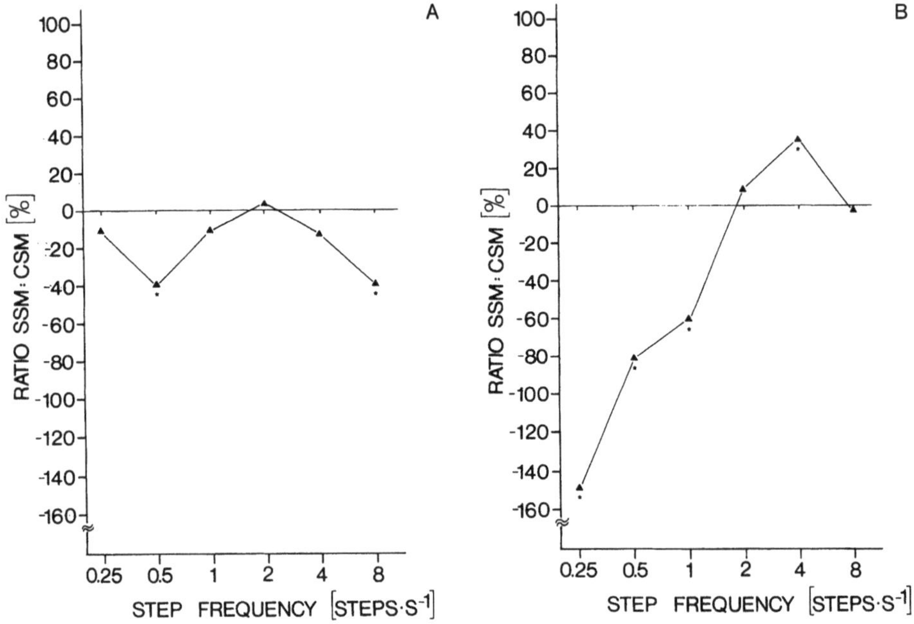

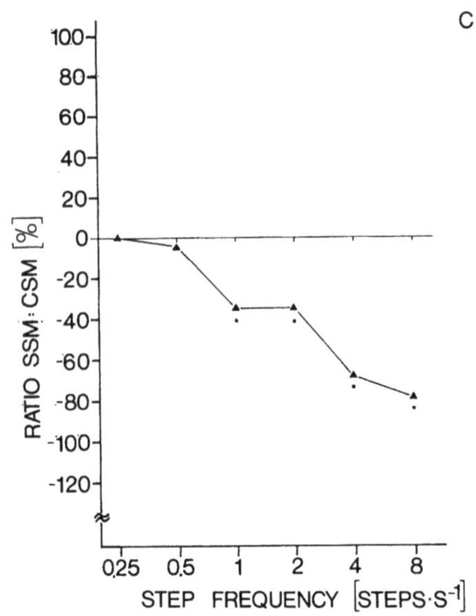

Fig. 2.30 A– . Movement pattern preference in juvenile *S. salamandra* raised with food that differed in movement pattern. Data (frequency of snapping responses) from 6-month-old subjects are shown. Same data presentation as in Figs. 2.17 and 2.18. **A** Salamanders raised with dead food; **B** salamanders raised with live crickets; **C** salamanders raised with live mealworms;

preferred movement pattern corresponds with the movement of young crickets and other small insects. On the other hand, subjects raised with mealworms generally respond equally well to continuous and discontinuous movement at low step frequencies, and they strongly "disliked" discontinu-

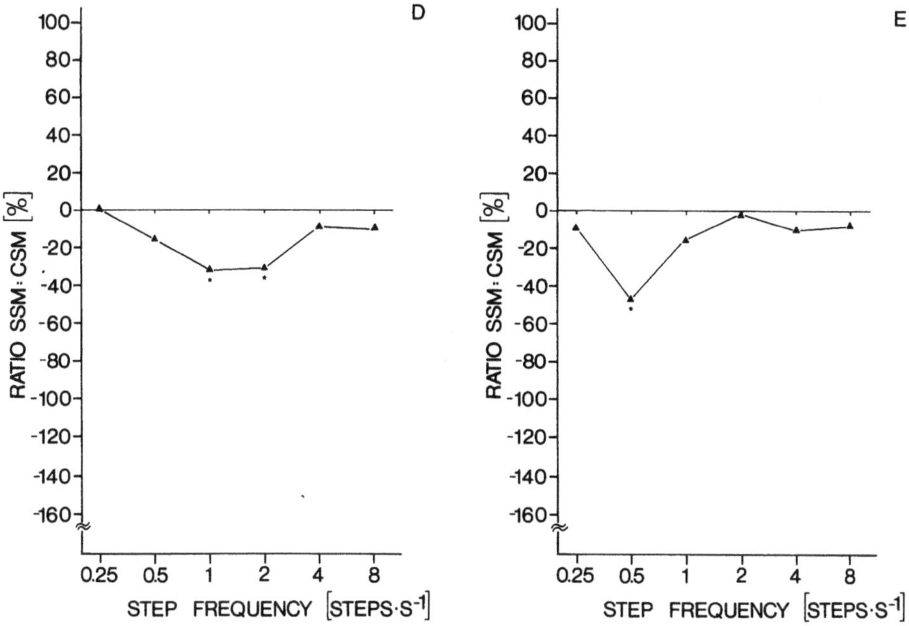

Fig. 2.30 D, E. D salamanders raised with dead food, moved continuously; **E** salamanders raised with mixed live food. *Asterisks* indicate significant differences in efficiency of *ssm* and *csm*. (From Luthard and Roth 1979 b)

ous movement at high step frequencies; this preferred movement pattern resembles that of a mealworm. This learning process interferes with a genetically determined development which consists in a general preference for continuous or smooth, discontinuous movement, a preference that is typical for nondeprived *Salamandra*.

In addition to the development of prey orientation and prey movement-pattern preferences, the rigidity or plasticity of the interaction between the olfactory and visual system was tested in several series of experiments (Roth and Luthardt 1980, Luthardt and Roth 1983, Luthardt-Laimer and Roth 1983).

As previously described, in all salamanders tested so far a visually induced inhibition of the snapping response to stationary odorous prey occurs. This inhibition seems to be due to the necessity of prey movement for visual induction of snapping. A conflict arises between the positive olfactory information about presence of prey, and the "negative" visual information about the absence of movement. This often leads to very prolonged fixation of the prey object and a rather substantial percentage of complete suppression of the final snapping response. Luthardt and Roth tested whether the necessity for prey movement is innate or whether it is, at least to some degree, experience-dependent.

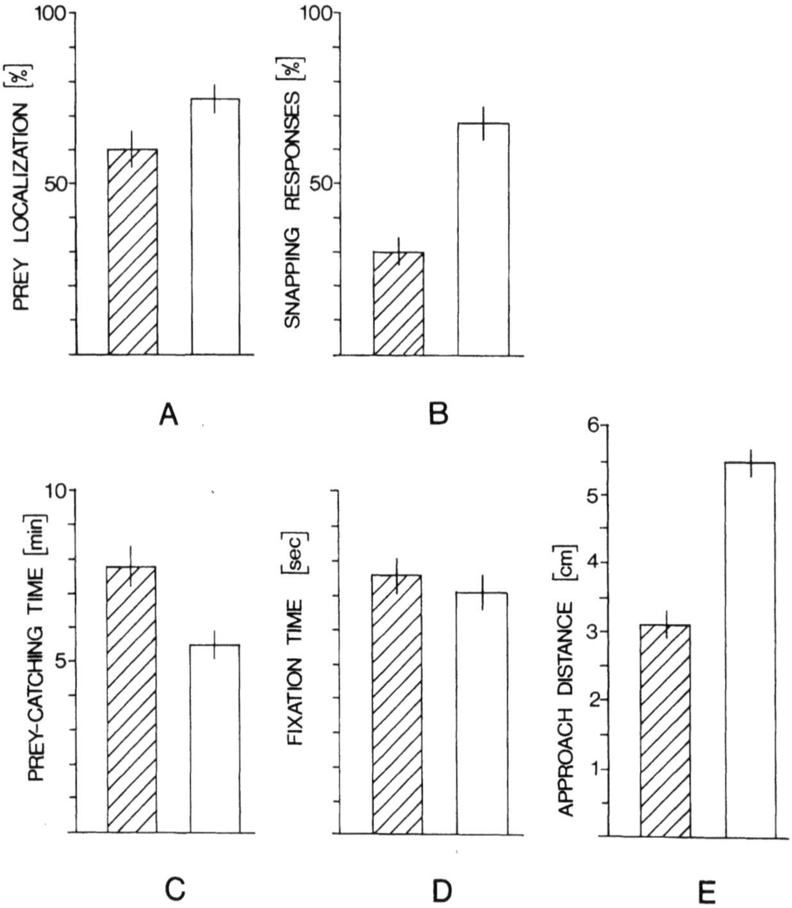

Fig. 2.31 A–E. Influence of early visual prey experience on feeding behavior. Responses of juvenile *S. salamandra* raised with live prey (*hatched columns*) and with stationary prey (*white columns*) to stationary prey under light. **A** Percentage of prey localization; **B** percentage of snapping; **C** prey-catching time; **D** fixation time; **E** approach distance. *Vertical bars* indicate standard error of the mean. (From Roth and Luthardt 1980)

The authors raised two groups of newly metamorphosed *Salamandra*. One group was fed stationary food (comparable to group 1 in the previously described experiments; STAT group), and the other live food (crickets as in the above group 2; MOT group). Both groups were tested 10 months after metamorphosis with regard to their responses to stationary food (dead crickets). The following data were measured: (1) percentage of prey localization; (2) percentage of snapping at the prey object; (3) prey catching time; (4) fixation time; (5) distance from where direct approach of the prey occurred.

The results are given in Fig. 2.31. The STAT subjects localized the stationary prey and snapped at it significantly more frequently than the MOT subjects (localization: 75.6% vs 59.4%, $p < 0.01$; snapping: 67.8% vs

30.6%, $p < 0.001$). Furthermore, the STAT animals localized the stationary prey in much shorter time and approached it directly from a significantly greater distance than the MOT animals (prey-catching time: 5.6 min vs 7.8 min, $p < 0.001$; approach distance: 5.5 cm vs 3.1 cm, $p < 0.001$). Only the fixation time did not differ significantly (71 s vs 76 s, $p > 0.5$).

These results clearly show that those salamanders that had been fed exclusively stationary prey since metamorphosis localized stationary prey much more easily, and almost all of them (nearly 90%) snapped at stationary prey after complete approach, whereas only 50% of the salamanders accustomed to moving prey did this. Our interpretation is that the STAT subjects had "learned" that motionless objects which smell like prey, but do not move, can nevertheless be prey.

In these experiments, it was still unclear whether the increase in response to stationary prey, observed in the STAT group, is caused by a reduction of the "visual inhibition" of the snapping response to stationary prey due to exclusive early experience with that type of food, or caused by a sensitization of the olfactory system. In order to decide this question, Luthardt and Roth raised two groups of salamanders again under STAT and MOT conditions. The two groups were tested with the following types of prey:

1. odorous contrast prey (dead crickets, = OC);
2. visible prey dummies of neutral odor (pieces of black plastic, the size of a dead cricket; = C);
3. odorous, invisible prey (mass of squashed crickets masked by a black background, = O).

The detailed results are given in Fig. 2.27 I–II. With regard to localization and snapping, both groups preferred OC to C and to O, when presented pairwise. However, the STAT animals localized and snapped more frequently at C than at O, whereas the MOT subjects showed the opposite behavior.

Comparing OC and C, the prey-catching time of the STAT group was significantly shorter and the percentage of prey localization and snapping significantly greater than that of the MOT group ($p < 0.001$). However, no differences occurred between the STAT and MOT group with regard to prey type O. With regard to fixation time, no differences were found between the two groups with respect to all three types of prey, whereas approach distance to OC and C was significantly larger in the STAT animals compared to the MOT group.

Since there were significant differences between the two groups with respect to visible stationary prey, with or without odor, but not with respect to invisible, odorous prey, it can be concluded that the continuous experience of the STAT group with stationary visual prey indeed affects the visual system, reducing the inhibition of the response to stationary prey, but not, or not significantly, the olfactory system, for example by enhancing the sensitivity of olfaction. The ability to follow an odor gradient was not increased in the STAT compared to the MOT group.

The question remains whether or not the modification of the feeding behavior observed in salamanders raised with stationary prey was a transient one and reversible by later experience, or a longer lasting one and rather resistant to later experience. This question was investigated in another series of experiments (Luthardt-Laimer and Roth 1983). Here, several groups of salamanders, raised under different conditions, were compared:

1. salamanders (n = 60) raised under MOT conditions, i.e., fed live crickets (MOT-1);
2. salamanders (n = 60) raised under STAT conditions, i.e., fed dead crickets (STAT-1);
3. salamanders (n = 180) raised first under MOT conditions, and converted to STAT conditions 1, 3, and 6 months after metamorphosis (CON-1, CON-3, CON-6);
4. salamanders (n = 32) raised exclusively under MOT conditions (MOT-2);
5. salamanders (n = 34) raised first under STAT conditions, and converted to MOT conditions 2 years after metamorphosis (STAT-2).

All salamanders were tested with regard to their responses to stationary prey according to the following schedule:

1. 70 subjects were tested 8–10 days after metamorphosis prior to their first terrestrial feeding action;
2. MOT-1 and STAT-1 subjects were tested 1 month after metamorphosis;
3. CON-1, -3, and -6 subjects were tested 9–10 months after metamorphosis;
4. MOT-2 and STAT-2 subjects were tested 4 and 9 months after conversion of STAT-2; STAT-2 subjects were additionally tested 1.5 months after conversion to live prey.

Therefore, if any sensitive period of imprinting existed, it could be narrowed down by this experimental schedule.

The results of the experiments are given in detail in Table 2.2A–D.

The salamanders tested 8–10 days after metamorphosis prior to their first terrestrial feeding showed very little response to stationary prey: there was only one prey localization, but no snapping in 70 trials. All of the subjects, however, responded to live prey.

Table 2.2A gives a comparison between MOT-1 and STAT-1 groups. STAT-1 subjects were significantly better in prey localization, snapping, and prey-catching time. There were no significant differences between STAT-1 and MOT-1 groups in fixation time and approach distance.

In Table 2.2B the results of the experiments with the salamanders converted from live to stationary food 1, 3, and 6 months after metamorphosis

Table 2.2 A–D. Influence of early sensory experience on feeding on stationary prey in *Salamandra salamandra.* (From Luthardt-Laimer and Roth 1983)

A Comparison between behavior of MOT 1 and STAT 1 salamanders toward stationary prey objects. The test was done 1 month after metamorphosis (for further explanation see text)

Response	MOT 1	STAT 1
Prey localization (%)	26	43*
Snapping responses (%)	6.6	35**
Prey-catching time (min)	10.6	6.3***
Fixation time (s)	3.5	4.4
Approach distance (mm)	20	23

* $p < 0.001$; ** $p < 0.01$; *** $p < 0.05$.

B Behavior toward stationary prey objects in salamanders "converted" from moving to stationary prey at ages of 1 (CON 1), 3 (CON 3), and 6 (CON 6) months after metamorphosis. The test was done 9 months after metamorphosis. The differences between the groups are not significant

Response	CON 1	CON 3	CON 6
Prey localization (%)	28	33	28
Snapping responses (%)	18	27	19
Prey-catching time (min)	9.6	7.3	8.4
Fixation time (s)	82	77	71
Approach distance (mm)	26	27	27

C Comparison between "converted" subjects (cf. B) and MOT/STAT 1980 (From Roth and Luthardt 1980)

Response	CON-1, 3, 6	MOT-'80	STAT-'80	CON-1, 3, 6 vs MOT '80	CON-1, 3, 6 vs STAT '80
Prey localization (%)	30	52	86	$p < 0.01$	$p < 0.001$
Snapping responses (%)	21	32	66	$p < 0.01$	$p < 0.001$
Prey-catching time (min)	8.4	7.8	5.6	n.s.	$p < 0.01$
Fixation time (s)	77	76	71	n.s.	n.s.
Approach distance (mm)	27	31	55	$p < 0.05$	$p < 0.001$

D Behavior toward stationary prey objects in salamanders fed exclusively moving prey (MOT 2), or exclusively stationary prey for 2 years and then "converted" to moving prey (STAT 2), tested at different times after "conversion"

Response	Prior to conversion		After conversion 1.5 mo.	4 mo.		9 mo.	
	MOT-2	STAT-2	STAT-2	MOT-2	STAT-2	MOT-2	STAT-2
Prey localization (%)	52	76*	75	52	93*	46	92*
Snapping responses (%)	33	66*	63	32	63*	40	80*
Prey-catching time (min)	7.8	5.6**	6.8	7.8	5.9**	6.4	6.2
Fixation time (s)	69	71	60	68	63	66	63
Approach distance (mm)	31	55**	50	29	47	28	43**

* $p < 0.01$; ** $p < 0.05$.

(CON-1, CON-3, CON-6) are shown. It is evident that the date of conversion had no significant influence on the response to stationary prey. When the results from all CON groups are taken together and compared with the data from the studies of 1980 (MOT-1980, STAT-1980, Roth and Luthardt 1980), as done in Table 2.2C, then it becomes clear that the CON groups showed significantly fewer prey localization and feeding responses to stationary prey than the STAT 1980 subjects, but also fewer ones than the MOT 1980 subjects which had had no conversion to stationary prey and had been tested 2 years after metamorphosis.

A comparison between the salamanders converted from stationary to live food 2 years after metamorphosis (STAT-2) and the MOT-2 salamanders is given in Table 2.2D. The results show that the conversion had no influence on the STAT-2 subjects: the significant differences between STAT-2 and MOT-2 existing before the conversion of STAT-2 remained except with regard to prey-catching time where the differences disappear. However, this was not due to a change in the behavior of the STAT-2 animals, but to an "improvement" of the MOT-2 subjects. There was no significant difference in fixation time between STAT-2 and MOT-2 before or after conversion of STAT-2.

The experiments show that the modification of the feeding behavior to stationary prey objects due to early experience fits two essential criteria for imprinting (Immelmann 1975, Meyer 1976, Apfelbach and Döhl 1976, Tembrock 1978): (1) the modification was effective only during a specific limited period shortly after metamorphosis (sensitive period), and (2) the modification was not reversed by subsequent changes of feeding conditions. The sensitive period took place within the first month after metamorphosis. If the experience with stationary prey was acquired later than that period, an increase in response frequency to stationary prey did not occur regardless of the date of conversion.

Taking into account the fact that feeding does not usually occur during the first 9–10 days after metamorphosis, then in those experiments the effective sensitive period amounts to 3 weeks, with a total of nine feeding reactions. It is remarkable that afterward the STAT-2 subjects did not lose their specific behavior despite exclusive experience of live prey during a period of 6 months after their "conversion".

All ontogenetic experiments discussed so far show that salamanders possess an inborn scheme of prey recognition and feeding response which, however, is not rigid, but can be modified to a large degree by individual experience. Sensory prey experience acquired during the first 3 to 4 weeks after metamorphosis seems to be very important in this context. Experience acquired later has little or no effect on visual prey preferences. The experience-dependent prey preferences, however, are not simply specified and restricted inborn schemata, according to the concept of an experience-modified, innate

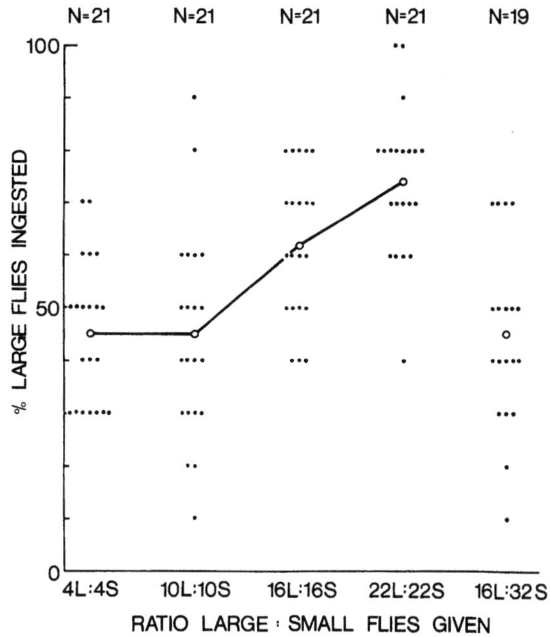

Fig. 2.32. Influence of visual experience on prey preferences in *Plethodon cinereus*. Percentage of large flies ingested as a function of the density of large (*L*) plus small (*S*) flies given. *Black dots* are responses of individual salamanders, *open circles* are mean values. (After Jaeger and Barnard 1981)

releasing mechanism. Early sensory experience can alter prey preferences rather drastically in opposite directions as was shown by the case of movement-pattern preferences in *S. salamandra* raised with mealworms and crickets, respectively.

In a series of elegant experiments Jaeger and Barnard (1981), Jaeger and Rubin (1982) and Jaeger et al. (1982) demonstrated that adult salamanders can change their feeding strategy according to individual prey experience. In these experiments adult *Plethodon cinereus* were offered fruit flies of two size classes, the large *Drosophila virilis* (2.9 mm) and the small *D. melanogaster* (0.9 mm), at different densities, 4:4, 10:10, 16:16, and 22:22 per experiment.

As shown in Fig. 2.32, at the two low prey densities the salamanders fed indiscriminately on both types of flies, but specialized on large flies at the two high densities. Specialization occurred by increasingly "ignoring" small flies that were encountered. Large flies were not similarly ignored. Interestingly, with increasing prey density not only prey preference, but also feeding strategy changed from "hunting" (or "pursuit") to "ambush" strategy, which is the less energy-consuming feeding method. Salamanders actually captured flies faster by ambush (1.5 min/fly) than by pursuit (2.4 min/fly). However, they did pursue more large than small flies at all but the lowest density of prey.

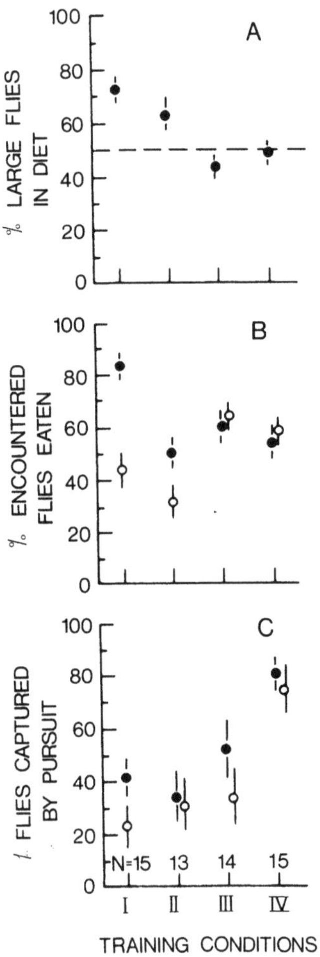

Fig. 2.33 A–C. Influence of visual experience on prey preferences in *Plethodon cinereus*. **A** Mean percent of the diet composed of large flies (*dashed line* denotes large and small flies equal in diet); **B** mean percents of the encountered large and small flies that were eaten; **C** mean percents of the large and small flies that were captured by pursuit (as opposed to ambush) after four conditions of training: *I* large and small flies; *II* large flies only; *III* small flies only; *IV* nondipteran prey. *Black circles*: large flies; *open circles*: small flies. (After Jaeger and Rubin 1982)

In another series of experiments, Jaeger et al. (1982) showed that *Plethodon* uses prey encounter rates, but not foraging rates, to assess prey densities. It uses visual information, probably by "counting" the number of prey items within the visual field.

Jaeger and Rubin (1982) could further demonstrate the role of learning in this process (Fig. 2.33). They formed four groups of adult *Plethodon cinereus* of which for 3 months the first group was provided an equal mixture of small and large fruit flies, the second only large fruit flies, the third only small flies, and the fourth were fed larvae and adults of the beetle *Tribolium castaneum*. In the test experiments the animals were given an equal number of small and large fruit flies. The salamanders specialized on the larger and more profitable flies when trained on both types of flies or on large flies only, but fed indiscriminately when trained on small flies only or on nonfly prey.

Furthermore, as shown in Fig. 2.33 C, during prey capture, the salamanders used a low-energy ambush strategy whenever they had had prior experience with any type of prey, and they switched to the more expensive pursuit tactic in the absence of such prior experience. However, here again large flies were captured significantly more often than small flies by pursuit than by ambush, if the animals had experience with the two types of fly. The authors interpreted this in the way that the animals had learned that larger and more profitable prey are more worthy of pursuit than the less profitable small flies. The authors could exclude the possibility of an inborn prey-preference scheme in favor of larger *Drosophila* on the basis that fruit flies of the family Drosophilidae are almost completely absent from the natural diet of *Plethodon cinereus*.

As already mentioned in this chapter, salamanders possess a well-developed olfactory system, too. In the olfactory system, individual experience seems to have a still stronger effect on prey recognition and feeding preferences than in the visual system. Experiments carried out in our laboratory in which newly metamorphosed *S. salamandra* were raised with different types of food and tested in darkness with regard to olfactory prey preferences show that juvenile salamanders learn new prey odor very rapidly, sometimes after one trial. This ability decreases after the first 12 months after metamorphosis, but during that period the animals relearn rather quickly. At an age of 3 years, however, a new prey odor is learned only very slowly or not at all (Becker 1985).

Both facts, higher speed of early learning and prolonged sensitive period, suggest that the olfactory system is more "plastic" than the visual one. This is in good accordance with the fact that for salamanders visual prey schemata can be rather simple. Most salamanders respond to moving objects that are not too small or too large. In contrast, in the feeding behavior guided by olfaction, e. g., in very dim light or darkness, a large variety of different prey odors has to be recognized.

Although it is clear that early postmetamorphic experience can be decisive for later feeding behavior, it is still disputed to what degree this is true for premetamorphic or larval experience. Only a few studies have dealt with the problem of retention of larval experience beyond metamorphosis. Flower (1927) observed that *Salamandra* larvae forget their prey experience during metamorphosis. He trained larvae to take food from his hand, and reported that these feeding habits were lost the last day of metamorphosis. The animals showed fear of being approached and tried to escape when picked up. It then took from 1 to 5 weeks for the adults to become tame enough to come to hand again.

In contrast, Hershkowitz and Samuel (1973) reported experiments in which *Triturus cristatus* larvae had to learn a simple visual discrimination task for food reward. After metamorphosis the trained larval stimulus preferences remained unchanged, i.e., the animals had not "forgotten" their larval experience. The authors also found that natural, untrained, visual-stimulus preferences of the newt larvae did not change during metamorphosis. Adult newts confronted with the same task as the larvae learned it much more slowly or not at all.

2.2 Optomotor Behavior

Vertebrates usually show compensatory movements with regard to motion of the visual environment relative to the retina, serving gaze stabilization during self-induced motion or passive drift within the environment (cf. Collewijn 1981). These compensatory movements consist of a smooth pursuit and a saccade directed in the opposite direction. Optomotor responses are performed by movement of the eyes relative to the head in most mammals, or by head movement in lower terrestrial vertebrates, often combined with a certain degree of eye movement. Salamanders show only optokinetic head movements. Optokinetic eye movements are either totally absent or undetectably small, in contrast to frogs where they play a considerable role (Dieringer et al. 1982).

The optokinetic nystagmus as well as its anatomical and electrophysiological basis and their neurocybernetic aspects have been studied in detail in a certain number of salamanders and anurans by Manteuffel and co-workers (Manteuffel 1979, 1984a, b, in press, Kopp and Manteuffel 1984, Manteuffel et al. 1983, 1986).

The optokinetic response of salamanders to a continuously moving, structured large-field stimulus consists of a slow compensatory pursuit phase of the head in the direction of stimulus movement, interrupted by occasional saccades in the opposite direction, which reset the head back into a more or less medial position. During slow pursuit movement the head accelerates for a few seconds, until maximum velocity is reached. After a saccade, there is again such a build-up phase.

In a recent comparative study, Manteuffel et al. (1986) measured the optokinetic nystagmus of three salamander species, *Salamandra salamandra*, *Tylototriton verrucosus*, and *Hydromantes italicus*, and of three anuran species, *Bufo bufo*, *Rana temporaria*, and *Bombina variegata*. In these experiments the subjects were placed into a optomotor drum (diameter 30 cm), the inner surface of which was covered with a black and white random-dot pattern with an average period of 6°. The optomotor behavior of the subjects was measured by means of a search coil set-up.

Figure 2.34a shows that all species tested showed a clear head nystagmus. With regard to the slow-phase head velocities the species *Hydromantes*, *Bombina*, and *Tylototriton* showed an increase in head velocity with increasing

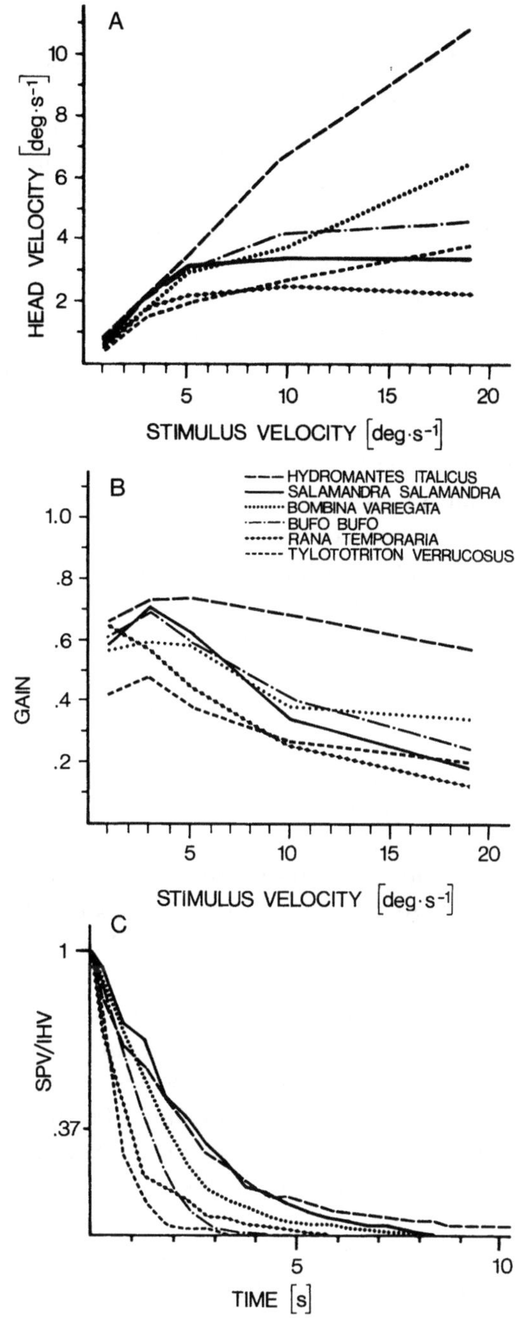

Fig. 2.34 A–C. Optokinetic behavior in five amphibian species. **A** Relationship between slow phase head velocity and stimulus velocity. **B** Relationship between compensation gain (head velocity/stimulus velocity) and stimulus velocity. **C** Time course of optokinetic afternystagmus (OKAN) at a stimulus velocity of 10°/s. *Ordinate* indicates ratio of slow phase head velocity (*SPV*) vs initial head velocity (*IHV*), at which OKAN started when the light was switched off. (From Manteuffel et al. 1986)

stimulus velocity up to a maximum velocity of 19°/s. Differences existed with regard to the slopes of the increment of head velocity. The ratio between head velocity increment and stimulus velocity was 0.48 in *Hydromantes*, 0.32 in *Bombina*, and only 0.12 in *Tylototriton*. The remaining three amphibian species showed a head velocity saturation with increasing stimulus velocity. The saturation was reached at 9.5°/s stimulus velocity with a head velocity of 4°/s in *Bufo*, at 5°/s stimulus velocity with a head velocity of 3°/s in *Salamandra*, and at 5°/s stimulus velocity with a head velocity of 2.2°/s in *Rana*.

In Fig. 2.34b, the relationship between the gain (ratio between head and stimulus velocity) and the stimulus velocity is given as a measure for compensatory optomotor behavior. Of all species tested, *Hydromantes italicus* showed the best compensation over the entire range of the tested velocities. Compensation of *Salamandra* was very similar to that of *Bufo*, showing a maximum at 3°/s stimulus velocity and a pronounced decrease toward higher stimulus velocities. *Bombina* showed a slightly weaker compensation; *Rana* showed good compensation at low stimulus velocities, but relatively weak compensation at higher velocities. The salamander *Tylototriton* showed the weakest compensation of all species tested.

Salamanders do not stop the slow nystagmic phase at termination of stimulus movement if the illumination is immediately switched off (Kopp and Manteuffel 1984). This poststimulus reaction usually consists of a persistent smooth head movement with decreasing velocity, but it may also be interrupted by saccades similar to the optokinetic afternystagmus (OKAN) known from higher vertebrates. In the six amphibian species, Manteuffel et al. measured the duration and the time constant (i.e., the time required for a reduction of the initial slow-phase head velocity, IHV, to a slow-phase head velocity SPV = 37% of IHV). The subjects were tested at a stimulus velocity of 10°/s, where a sufficiently high head velocity is reached by all species examined.

As shown in Fig. 2.34c, *Hydromantes* showed the longest OKAN, immediately followed by *Salamandra* (time constants of 2.30 s and 2.29 s, respectively). The three anuran species all showed shorter OKAN with a time constant of 1.69 s in *Bombina*, 1.06 in *Bufo*, and 0.98 in *Rana*. The salamander *Tylototriton* showed the weakest OKAN of all species tested with a time constant of only 0.59 s.

These results show that there is no simple interspecific correlation between head velocity and the time constant and duration of the OKAN. In *Hydromantes* the initial head velocity was more than twice as fast as in *Salamandra*, but OKAN was almost the same in both species. With regard to initial head velocity, *Bufo* was superior to *Bombina* and to *Salamandra*, but the time constants of these species show a reversed order. The existence of an OKAN is thought to be strong evidence for neuronal velocity storage in the optokinetic system of salamanders. However, effectivity and time constants of velocity storage elements in amphibians are significantly reduced compared to so-called higher vertebrates (Manteuffel 1984b; Manteuffel et al. 1986). In urodeles, as in mammals, the presence of an OKAN critically depends on vestibular inputs and the OKAN is abolished after bilateral labyrinthectomy.

The fact that *Hydromantes* showed the greatest ability to compensate relative background movement, is of great interest for object background discrimination during feeding. As mentioned above, tongue-projection salamanders are able to catch fast-moving prey by tongue projecting combined with a quick lateral head movement, often with lateral tongue extrusion. In order to determine the exact position of the prey object in such a situation, the visual system has to correlate the trajectory and velocity of the prey with the velocity of its own head, which may be determined by pretectal and mesencephalic units involved in optomotor response. This process will be further discussed in the last chapter.

2.3 Visual Mate Recognition in Newts

During courtship behavior of newts, visual as well as nonvisual stimuli are important. Experiments have shown that the first reaction of the male is visually directed, whereas subsequent behavior is mostly determined by the influence of pheromones (Zippelius 1948, Prechtl 1951, Gauss 1961, Himstedt 1972, Halliday 1974). The newts of the genus *Triturus* show conspicuous body coloration, and males and females are differently colored. Generally, males have better developed patterns that become more prominent during the breeding season.

Himstedt (1979) has studied the significance of color signals in mate recognition in the newt *Triturus alpestris*, a species that has the most distinct coloration in the genus. In *Triturus alpestris* both sexes have an orange-red belly. The dorsal side of the body and the legs of the females are bluish gray or in some cases brownish or olive-green. Males show additional yellow, white, and black markings contrasting with the blue body coloration.

In the experiments, pairs of models of females differing in body coloration were presented to a male during the breeding season. In order to increase the motivation of the subjects, water (3 ml) from a female's tank containing female sexual pheromones was pipetted in front of the male. The following responses of the males toward the female dummy were recorded: turning toward the dummy, approaching, and making snout contact with the cloacal region.

In the first series a black dummy was tested against monochromatic red, yellow and blue dummies and a dummy with a black dorsum and a gray belly. The black dummy proved to be superior to the red and the yellow dummy,

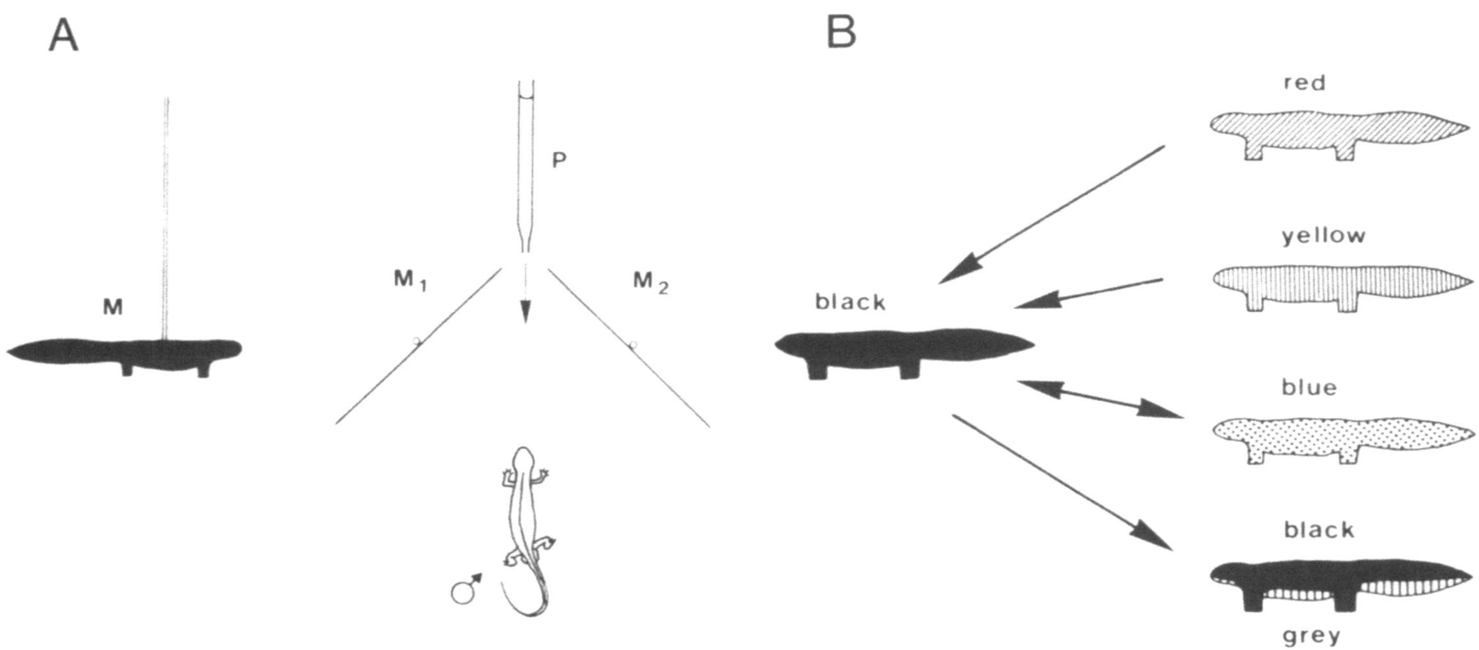

Fig. 2.35 A, B. Visual mate recognition in *Triturus alpestris*. **A** Experimental arrangement. *M* Female model attached to a thin glass rod. *Right*: Two models (M_1, M_2) are presented to a male. *P* Pipette from which water from the female's tank flows toward the male. **B** Choice test with a black model. In each series one model was black, the alternative models are shown on the *right*. *Arrows* point toward the significantly preferred model. *Arrows with two heads* indicate that the two models were chosen equally.

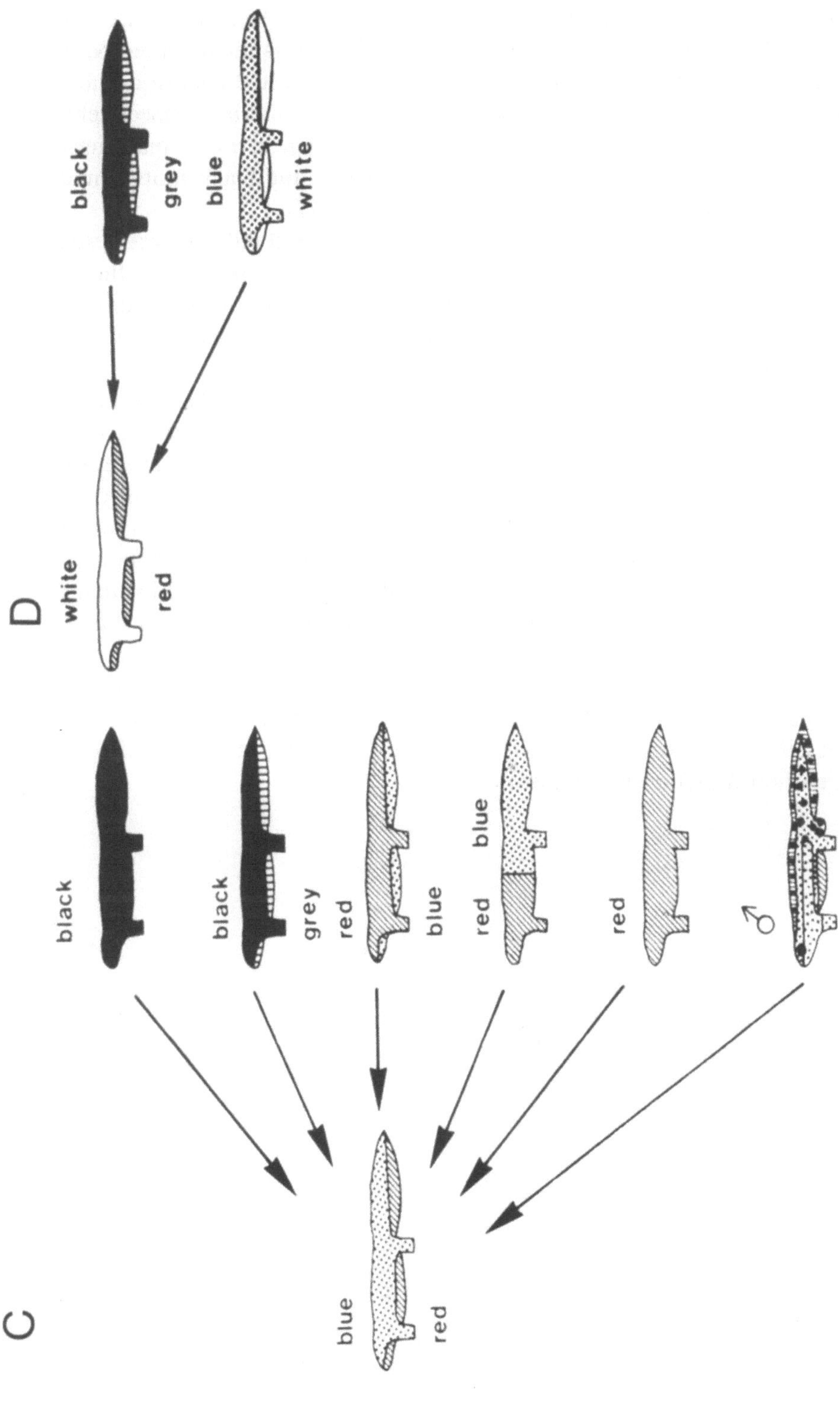

Fig. 2.35 C, D. C Preference of the blue-red model (*left*) was presented together with one of the alternative models shown on the *right*. **D** Importance of the red belly. In two test series, the white-red model was presented together with the black-gray one or the blue-white one, respectively. (From Himstedt 1979)

equal to the blue one, and less effective than the black and gray one (Fig. 2.35 B). In the next series a dummy with a blue dorsum and a red belly, corresponding to the natural female coloration, was tested against a black, a black-gray, a red-blue (dorsum-belly and anterior-posterior, respectively), a red dummy, and a dummy with a coloration typical for a male newt (blue-red with additional markings of yellow, white, and black spots). In all cases the "female" dummy was preferred (Fig. 2.35 C).

In the third series a dummy with a white dorsum and red belly was tested against a black-gray and a blue-white dummy. Here, the dummy with the red belly was preferred. Another experiment showed that the subjects did not distinguish between a "natural" blue-red dummy and a dummy with a red belly and a black dorsum (Fig. 2.35 D).

These experiments show that male *Triturus alpestris* use color vision for female recognition during the breeding season. The most effective dummies were those with a red belly. The color of the dorsum was rather irrelevant. This corresponds with the fact that the dorsal coloration is not uniform in females and may differ individually. The red belly of a female releases the initial response in the sequence of male courtship behavior. This response is inhibited if the red belly is combined with color markings typical for males.

Experiments with blinded *Triturus* (Zippelius 1948) have shown that visual stimuli are not necessary for mate recognition. However, visual stimuli may be of great importance in the natural habitat where as many as four species of newts may breed simultaneously in the same pond. Because these species differ in coloration, this visually guided behavior may be an effective species-discrimination mechanism.

3 Morphology and Function of the Salamander Eye

3.1 General Morphology

The eye of most salamanders has a large cornea and a large lens. There are differences between eyes of permanently or predominantly aquatic salamanders, on the one hand, and of predominantly or fully terrestrial salamanders on the other. In water, there is no refraction of light at the corneal surface, since cornea and external medium are of the same refractive index. Focusing of an image onto the retina is achieved by the lens only. Therefore, predominantly aquatic salamanders show more or less spherical lenses and rather flat corneae. In air, the situation is quite different. Because of the low refractive index of the outer medium, in terrestrial salamanders the cornea becomes an important part of the dioptric system and the lens tends to be more flattened.

Salamanders like *Ambystoma* or *Triturus* which return to the aquatic medium every year during the breeding season have had to find a compromise. Since amphibians have no soft lens which can change its shape as the lenses of diving birds or seals, they either could live with a terrestrial eye which is extremely far-sighted in water, or with an aquatic eye which is short-sighted on land.

Triturus has a nearly spherical lens which makes it better suited for aquatic vision. In contrast, *Salamandra*, which after metamorphosis is totally terrestrial, has a flattened lens. Similar aerial optics are found in the hynobiid salamander *Onychodactylus*, although these salamanders, like newts, enter the water for breeding. Possibly, aquatic vision is not as important in hynobiids as in *Triturus* since these animals stay in the water only for a short time, while *Triturus* lives in ponds for several weeks. In *Triturus* the mode of life is generally more aquatic than terrestrial. This can be concluded from the activity in both habitats. During water life, the locomotor activity in the *Triturus* species is more than ten times higher than during their terrestrial life (Himstedt 1971). Thus, the short-sightedness in air may be no great problem.

Hynobiidae, in contrast to Salamandridae or Ambystomatidae are quite agile on land, and visually guided behavior such as prey-catching appears to be more precise than in terrestrial newts or in *Ambystoma*. This is probably possible for hynobiid eyes because they have been adapted to aerial vision by

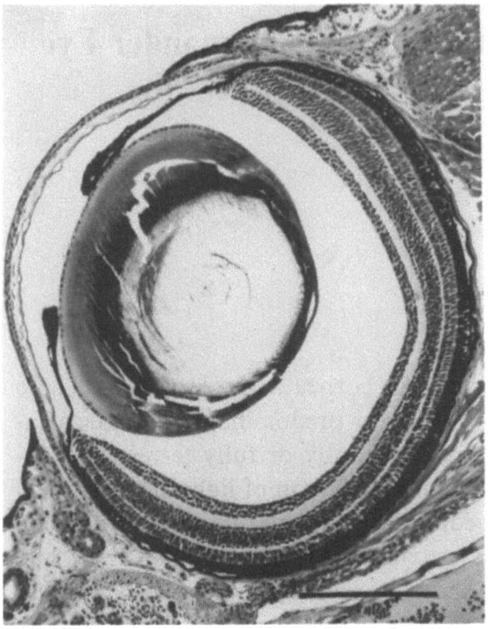

Fig. 3.1. Horizontal section through the eye of *Bolitoglossa subpalmata*. *Bar* = 500 μm. (From Linke et al. 1986)

flattening of the lens. Flattened lenses are also present in the eyes of terrestrial plethodontids such as *Hydromantes*, *Plethodon* or *Bolitoglossa* (Fig. 3.1). In the eyes of these species two further features are evident: the size of the lens in relation to the eye ball is very large and the posterior surface of the lens is much more arched than the anterior surface. The strong curvature of the posterior edge of the lens causes a very short focal length so that an image can be focused onto the retina at a short distance.

Haller's well-known allometric rule (cf. Rensch 1948) states that eye size relative to body size decreases with increasing body size. Möller (1951) compared eye morphology of five salamandrid species: *Triturus vulgaris*, *T. alpestris*, *T. cristatus*, *Salamandra atra*, and *S. salamandra* (= *S. maculosa* in Möller's study). Within the series of newts Haller's rule was confirmed: the smallest species, *T. vulgaris*, had the relatively largest, and the largest species, *T. cristatus*, the relatively smallest eyes. However, among the two species of the genus *Salamandra* the opposite was the case: *S. salamandra*, being considerably larger than *S. atra*, had relatively much larger eyes.

Linke et al. (1986) studied eye morphology in seven plethodontid species: *Thorius narisovalis*, *Batrachoseps attenuatus*, *Desmognathus ochrophaeus*, *Eurycea bislineata*, *Plethodon cinereus*, *Bolitoglossa subpalmata*, and *Hydromantes italicus* (Table 3.1 and Fig. 3.2). Here, according to Haller's rule, *Thorius* and *Batrachoseps*, being the smallest species, had the relatively largest eyes. However, as in the Salamandridae, there are immediate exceptions. *Hydromantes* and *Bolitoglossa*, as the largest and second largest animals, had

Table 3.1. Head volume (HV) and eye volume (EV) in absolute terms (mm^3) and eye volume in percent of head volume (EP). Mean values obtained from five animals per species

	HV	EV	EP
D. och.	59.97	2.09	4.24
E. bis.	57.45	2.86	5.22
P. cin.	48.42	2.36	4.90
B. sub.	119.33	8.28	5.60
T. nar.	11.70	1.06	10.90
Ba. att.	28.93	1.93	7.20
H. ital.	313.00	21.80	7.00

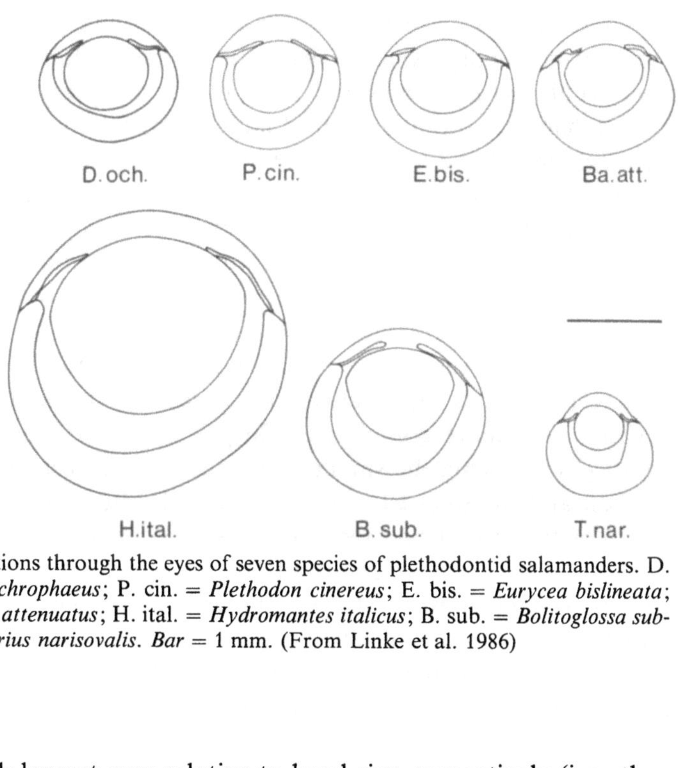

Fig. 3.2. Horizontal sections through the eyes of seven species of plethodontid salamanders. D. och. = *Desmognathus ochrophaeus*; P. cin. = *Plethodon cinereus*; E. bis. = *Eurycea bislineata*; Ba. att. = *Batrachoseps attenuatus*; H. ital. = *Hydromantes italicus*; B. sub. = *Bolitoglossa subpalmata*; T. nar. = *Thorius narisovalis*. *Bar* = 1 mm. (From Linke et al. 1986)

the third and fourth largest eyes relative to head size, respectively (i.e., they had much larger eyes than expected according to Haller's rule). This is mostly due to the large size of cells, including photoreceptors and neurons, in these animals, especially in *Hydromantes*, which strongly reduces the number of photoreceptors per unit area (cf. Table 3.5). On the other hand, *Desmognathus* had relatively smaller eyes than expected, and this species has rather small cells.

A similarly complex situation was found with regard to the size of the lens and the thickness of the retina. In the salamandrid species studied by Möller,

Table 3.2. Volume of retina and lens in percent of eye volume. Mean values obtained from five animals per species

	Retina	Lens
D. och.	38.95	26.08
E. bis.	34.48	19.19
P. cin.	37.56	20.81
B. sub.	39.86	23.98
T. nar.	51.34	11.04
Ba. att.	46.58	16.51
H. ital.	35.81	24.63

the larger species had relatively larger lenses, which is the opposite of that found in mammals (Rensch 1948). In the plethodontid species studied by Linke et al., *Hydromantes* and *Bolitoglossa*, the largest and second largest species, had the second and third largest lens relative to eye size. However, *Desmognathus*, the third largest plethodontid species with regard to head size, had the relatively largest lens. *Thorius* and *Batrachoseps*, the smallest and second smallest species, had the smallest and second smallest lenses (Table 3.2 and Fig. 3.2). Therefore, with the exception of *Desmognathus*, the same situation occurs in both salamandrid and plethodontid species investigated with respect to relative lens size.

As to the thickness of the retina in salamandrids, the smaller species had both relatively and absolutely thicker retinas. In the plethodontid species the situation was more complicate (Table 3.3, Figs. 3.2 and 3.3). *Thorius* and *Batrachoseps*, being the smallest and second smallest species, had the third thickest and the thickest retina, respectively. But *Hydromantes* and *Bolitoglossa* as the largest and second largest species, had the second and fourth thickest retina, and *Plethodon* and *Eurycea* which are medium- to small-sized species with medium-sized eyes, showed the thinnest and second thinnest retina. In both groups of salamanders, the increase in thickness of the retina was mainly due to an increase in thickness of the inner nuclear layer containing the interneurons (horizontal, bipolar, and amacrine cells), and of the layer of ganglion cells (Table 3.3). Whereas in large salamander eyes, the layer of ganglion cells usually consists of one to two rows of retina ganglion cells (rgc), in very small eyes, e.g. of *Desmognathus*, *Thorius* and *Batrachoseps*, the retina shows up to six rows of rgc (Fig. 3.3 A, C, D). The same complicated situation is found with regard to the volume of the retina relative to eye size: as could be expected, *Thorius* and *Batrachoseps* had the largest and the second largest retina. But *Bolitoglossa* had the third largest retina, and *Eurycea* showed the smallest retina (Table 3.2). It is clear that the enormous differences in size of the retinal cells play an important role for both thickness and volume of the retina (cf. Table 3.7 and below).

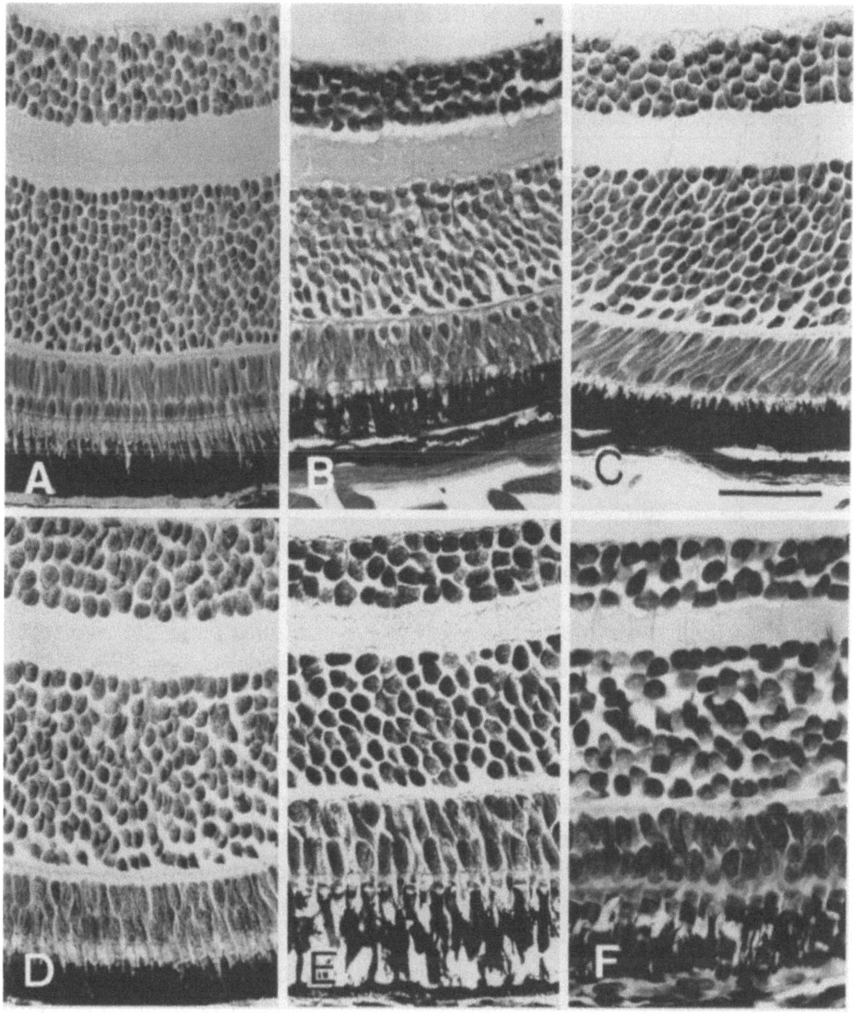

Fig. 3.3 A–F. Comparison of the structure of the retina in six plethodontid salamanders of different sizes. Note the increase in cell size from **A** to **F**, whereas the thickness of the whole retina changes very little. **A** *Desmognathus ochrophaeus*; **B** *Eurycea bislineata*; **C** *Thorius narisovalis*; **D** *Batrachoseps attenuatus*; **E** *Bolitoglossa subpalmata*; **F** *Hydromantes italicus*. *Bar* = 50 μm. (From Linke et al. 1986)

In addition, these data show that in the plethodontid species studied lens and retina strongly "compete" with each other for space. Whereas in *Desmognathus* the lens is rather large and the retina rather small, the opposite is the case in the miniaturized species *Thorius* and *Batrachoseps*. Here, the size of the lens is apparently sacrificed in favor of volume of the retina, in order to have space for a minimum number of photoreceptors, interneurons, and retina ganglion cells (see also below).

Table 3.3. Thickness of the whole retina (RET) and absolute (A) and relative (R) thickness of retina ganglion cell layer (RGCL), the inner plexiform layer (IPL), the inner nuclear layer (INL), the outer plexiform layer (OPL) and the outer nuclear layer (ONL). All absolute values are in µm, all relative values in percent of the thickness of the whole retina. The measurements are taken from three to five animals per species

		D.och.	*E.bis.*	*P.cin.*	*B.sub.*	*T.nar.*	*Ba.att.*	*H.ital.*
RET		224.5	210.3	181.3	264.2	277.3	289.9	282.8
RGCL	A	39.2	30.5	24.9	37.7	54.6	66.7	43.5
	R	17.5	14.5	13.7	14.3	19.7	23.0	15.4
IPL	A	32.1	23.2	23.2	30.5	33.4	29.0	26.3
	R	14.3	11.3	13.0	13.3	12.0	10.0	9.3
INL	A	74.0	82.7	60.9	82.7	99.9	118.9	111.7
	R	33.0	39.3	33.6	31.3	36.1	41.0	39.6
OPL	A	4.4	2.9	4.4	8.7	4.4	4.4	5.8
	R	1.9	1.4	1.4	3.3	1.6	1.5	2.1
ONL	A	22.9	35.7	27.7	44.1	50.0	36.4	42.7
	R	10.2	17.0	15.3	16.7	18.0	12.6	15.1

It is difficult to compare the relative eye size data from the two families. If we determine relative eye size by comparing absolute eye size with body length or snout-vent length, as Möller did, then the plethodontid species investigated by Linke et al. have much smaller eyes than the salamandrid species with the exception of *Hydromantes*. However, the plethodontid species investigated are usually much more slender than the salamandrids, with the extremes of *Batrachoseps* and *Thorius*; only *Hydromantes* has a more compact body shape. Therefore, in plethodontids salamanders relative eye size (as well as relative brain size) cannot reasonably be determined on the basis of body (or snout-vent) length. In those plethodontid species which were first miniaturized and then underwent extreme elongation of the body, as in *Batrachoseps*, and especially in *Lineatriton* and *Oedipina* (Wake, pers. comm.), so that they achieve the shape of long worms, a measurement based on body or snout-vent length would lead to absurd results. Linke et al., therefore, used head volume as the basis of determination of relative eye size.

In miniaturized plethodontids, eyes become enormously large with regard to head size and seem to compete with the brain for space (cf. Table 3.1). In *Thorius*, for example, the two eyes occupy 11.3% of the total head volume. Almost the same percentage is found for the brain, so that eyes and brain occupy about 22% of the total head volume. The lowest values are found in *Desmognathus*. Here the eyes occupy 4.1%, and the brain 2% of head size. An interesting case is *Hydromantes*, in which the eyes represent 7% and the brain 3.8% of the head volume. Here, the eyes are relatively very large with respect to the brain.

3.2 Morphology of the Retina

3.2.1 Light Microscope Overview of the Structure

The retina of salamanders, as that of all other vertebrates, shows two fiber layers and three cell layers (Fig. 3.4). The most distal layer of the neural retina contains the outer segments of the photoreceptors (PR) with rods (R), single cones (SC), and double cones (DC). The outer segments are separated from their nuclei by the outer limiting membrane (OLM) which is constituted by the terminal structures of the glial cells of the retina, the cells of Müller. Next to the OLM the nuclei of the rods and cones, constituting the outer nuclear layer (ONL), are found. In amphibians, this layer differs from the ONL of other vertebrate classes in that rod nuclei are generally more distal than cone nuclei (Menner 1929, Dowling and Werblin 1969, Dickson and Hollenberg 1971, Keefe 1971).

The next layer is the outer plexiform layer (OPL) which is usually very thin in amphibians. Here, the rods and cones are connected with bipolar and horizontal cells (Dowling and Werblin 1969, Lasansky 1973). The bipolar

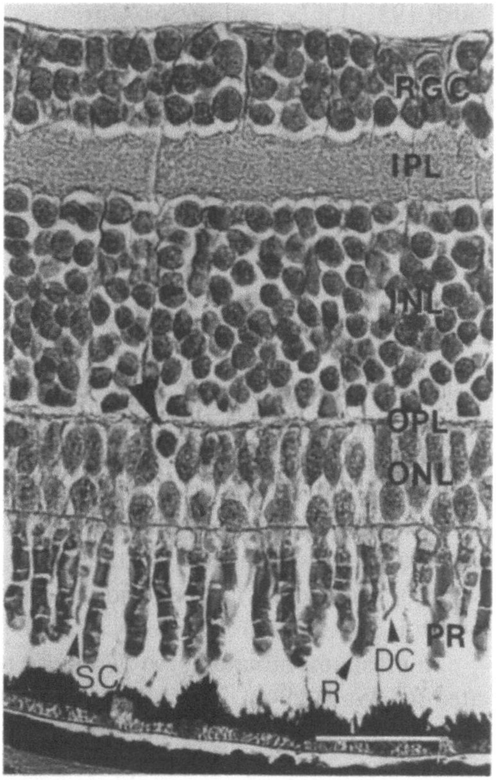

Fig. 3.4. Horizontal section through the central retina of *Plethodon cinereus*. Clearly identifiable are rods (*R*), single cones (*SC*), and double cones (*DC*). *Large arrowhead* points to a darkly stained cell lying among the photoreceptor nuclei which may be a displaced bipolar cell. *RGC* Retina ganglion cell layer; *IPL* inner plexiform layer; *INL* inner nuclear layer; *OPL* outer plexiform layer; *ONL* outer nuclear layer; *PR* layer of photoreceptor outer segments. *Bar* = 50 µm. (From Linke et al. 1986)

and horizontal cells belong to the inner nuclear layer (INL). This layer also contains amacrine cells, so called by Ramon y Cajal (1892) because they lack a true axon. Recently, in goldfish and other vertebrates, Dowling et al. (1976) discovered an interplexiform cell, which constitutes a feedback system between the OPL and the INL. This cell type has not yet been found in amphibians. In the INL the horizontal cells lie most distal, followed by bipolar cells. The most vitread cells are the amacrine cells. Vitread to the INL the second synaptic layer of the retina, the inner plexiform layer (IPL) is found, which is usually much thicker than the first one. Here, bipolar, amacrine, and ganglion cells are interconnected. The final nuclear layer is called the retina ganglion cell layer (RGCL). From this layer the ganglion cells send out axons which are bundled together at the inner surface of the retina and leave the eye through the optic disk constituting the optic nerve.

In salamanders, therefore, all retinal elements found in other vertebrate retinas are present: photoreceptors (rods, cones, and double cones), horizontal cells, bipolar cells, amacrine cells, and ganglion cells.

The retinas of *Necturus* (Brown et al. 1963, Dowling and Werblin 1969), *Triturus* (Keefe 1971, Dickson and Hollenberg 1971) and larval *Ambystoma* (Lasansky 1973, Vallerga 1981) are the best studied retinas among salamanders. In plethodontids only a few studies have been carried out (Burkhardt 1931, Linke et al. 1986). Curiously, special interest has focused on the degenerated eyes of the troglobitic species of plethodontid salamanders since the beginning of the century (see Brandon 1971, for a short review, Besharse and Brandon 1972, 1974a, b, 1976). In the following part, the morphological organization of the retina will be discussed in greater detail. The special features of the eye morphology of the troglobitic salamanders will be discussed further below. For physiological details of single elements in the salamander retina the reader is referred to the review of Grüsser-Cornehls and Himstedt (1976) and the work of Attwell and co-workers (1980, 1984). Response properties of retina ganglion cells are discussed in Chapter 5.

Receptors and Visual Pigments. As in other vertebrates, the photoreceptors in salamander retinas are of two morphological types: rods and cones (Figs. 3.4, 3.6, and 3.7). The structure of photoreceptors in *Necturus*, investigated by Brown et al. (1963), are used today as models of vertebrate rods and cones. The rod outer segment is of cylindrical shape and contains a pile of disks surrounded by the outer plasma membrane. The cone outer segment is conically tapered and the disks are part of the plasma membrane which is invaginated repeatedly (Fig. 3.5).

In most amphibians these two basic receptor types are further differentiated (cf. Walls 1942). There are two kinds of rods, red and green ones, and besides single cones there are double cones. The red and green rods were named due to their colored appearance in a freshly excised retina. In frogs

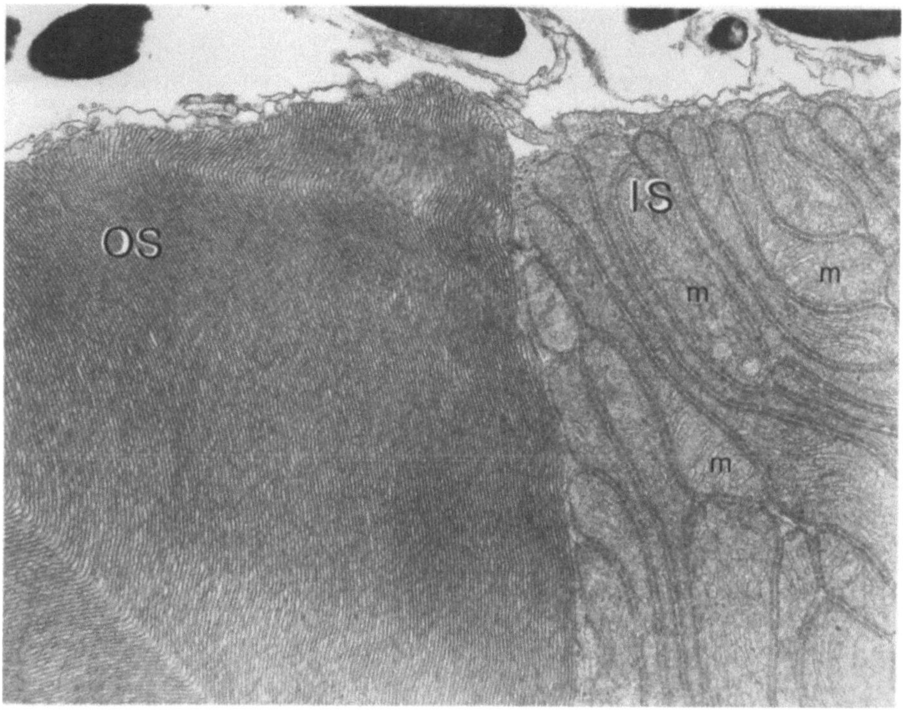

Fig. 3.5. Electron micrograph of the junction of an inner and outer segment of a photoreceptor in *Bolitoglossa subpalmata*. Mitochondria (*m*) in the inner segment (*IS*) and disks in the outer segment (*OS*) (55 disks per 1 μm length) are closely packed. Magnification 25,000 x. (Linke unpublished)

both types of rods show clear structural differences, the green rods having a shorter outer segment and a longer thinner myoid than the red rods. The double cones consist of two receptors, in which the membranes of the inner segments are fused and the outer segments separated. This set of receptors is typical for most amphibians with well-developed eyes, anurans as well as urodeles. The less differentiated retina in *Necturus* may be due to larval characteristics of this neotenic species.

The same receptor types as in frogs were found in Salamandridae (Ranvier 1875), Ambystomatidae (Mariani 1986) and Plethodontidae. The statement of Burkhardt (1931) that plethodontid salamanders lack cones is not correct. The study of Linke et al. (1986) revealed that all species investigated so far possess rods as well as cones and double cones (cf. Fig. 3.4).

In fixed salamander retinas the two types of rods are not as clearly distinguishable by their structure as in frogs. This is possible, however, in freshly excised retinas. By this method, Himstedt and Fischerleitner (1975) showed that *S. salamandra* does not lack green rods as was formerly stated

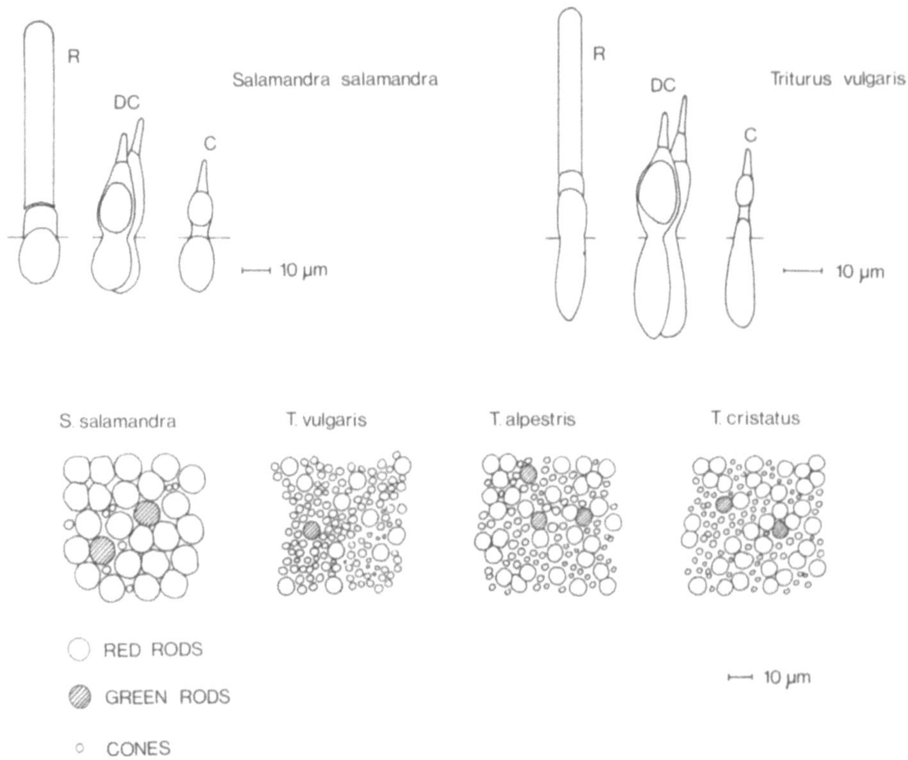

Fig. 3.6. Receptor types and density in *S. salamandra*, *Triturus vulgaris*, *T. alpestris*, and *T. cristatus*. *Above*: Schematic outlines of photoreceptor types found in *S. salamandra* and *Triturus vulgaris*. *R* Rods; *C* cones; *DC* double cones. *Below*: Schematic drawings of the photoreceptor mosaic of fresh retina whole mounts. Green rods are dispersed among the red rods. Note the high percentage of cones in the *Triturus* species. (After Himstedt and Fischerleitner 1975)

(Ranvier 1875, Walls 1942, Muntz 1963). Figure 3.6 shows the types of photoreceptors in *S. salamandra* and *Triturus vulgaris* taken from sections and from fresh retinas. The green rods in *Salamandra* do not look as grass-green as in frogs, their coloration is a paler gray-green. In *Triturus* the rod outer segment is more electron dense than the cone outer segment due to the more closely packed disks (Dickson and Hollenberg 1971).

Visual Pigments. Amphibian photoreceptors differ from those of other vertebrates with regard to the visual pigments they contain.

The green rods do not contain the typical rod pigment. Amphibian green rods have a spectral sensitivity similar to blue-sensitive cones in other vertebrates. On the other hand, one of the cones in amphibian retinas contains the rod pigment; this is the accessory member of the double cones.

There are many studies of visual pigments in frogs. In a review, Donner and Reuter (1976) described three different pigments in adult *Rana*. The first

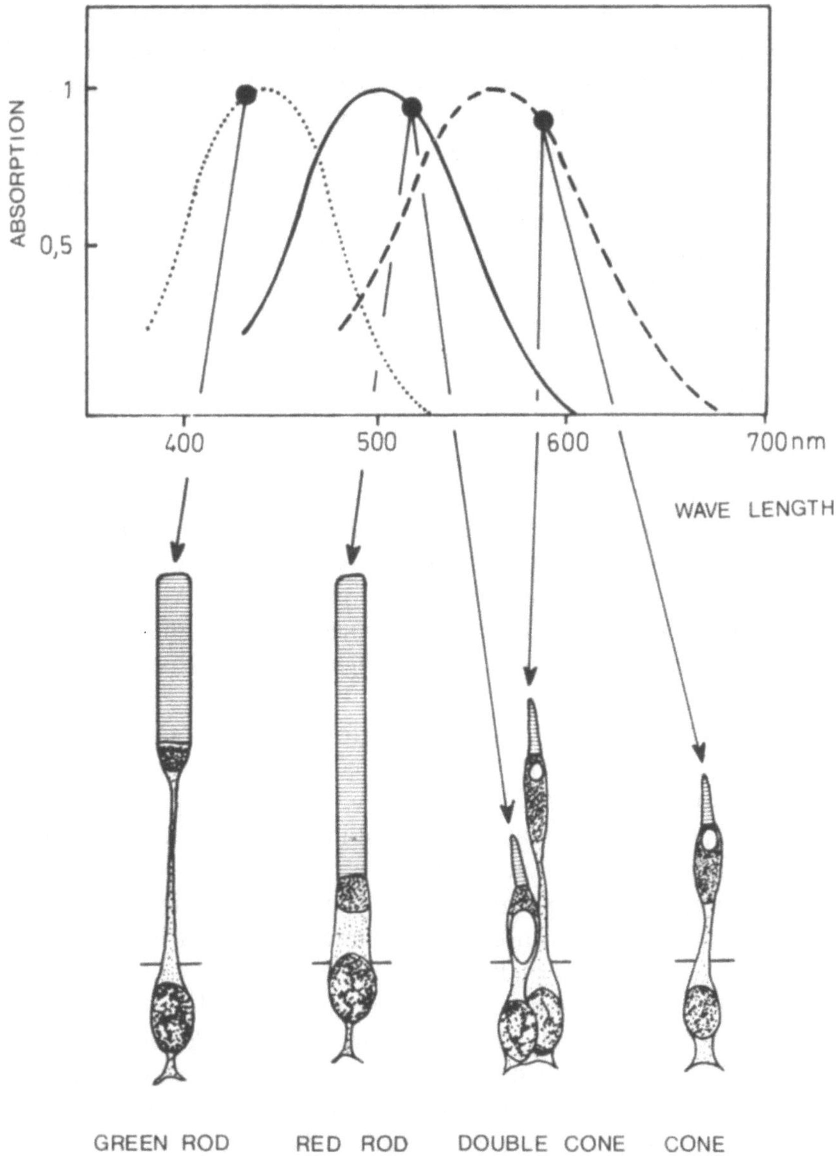

Fig. 3.7. Visual pigments (After Donner and Reuter 1976) and receptor types (After Walls 1942) of the frog. The same pigments and receptors are present in the retinas of terrestrial *S. salamandra*, *Triturus*, and *Ambystoma*. (After Himstedt 1982)

is most sensitive in the blue spectral region with maximal absorption at 433 nm. The second is rhodopsin which absorbs light maximally in the green region at 502 nm. The third is most sensitive at longer wavelengths absorbing maximally at 580 nm.

The distribution of these three pigments in the different morphological types of receptors was clarified by microspectrophotometry (Liebman and Entine 1968). By this technique Liebman (1972) investigated *Ambystoma tigrinum*. In this salamander the receptor types and the pigments were found to be the same as those in *Rana* (cf. Fig. 3.7). Thus, *Ambystoma* has green rods with P (= visual pigment) 433, red rods with P 502, single cones with P 580 and double cones with P 580 in the principal member and P 502 in the accessory member.

An additional complication in amphibian photopigments is caused by a change in the chemical structure of the pigments in a variety of species. The prosthetic group of the photopigments, the retinal, is derived either from vitamin A_1 (retinol) or from vitamin A_2 (3-dehydroretinol). Generally, vitamin A_2 pigments (porphyropsins) are typical for freshwater fish, vitamin A_1 pigments (rhodopsins) occur among terrestrial vertebrates. Wald (1947, 1952) was the first to investigate the retinas of aquatic and terrestrial urodeles. In *Necturus maculosus* and adult aquatic *Diemictylus* (= *Notophthalmus*) *viridescens*, he found vitamin A_2. In metamorphosed terrestrial juveniles of *D. viridescens*, however, he found vitamin A_1. Further investigations in urodeles have generally confirmed the rule that aquatic forms have A_2-based porphyropsins as rod pigments, whereas terrestrial forms have A_1-based rhodopsins.

In *Necturus*, Crescitelli (1958) found P 522_2 (a pigment with maximal absorption at 522 nm with a retinal based on vitamin A_2), while in the terrestrial *Taricha torosa*, P 502_1 (A_1-based rhodopsin) was present. Brown et al. (1963) stated that the pigment in *Necturus rods* is a porphyropsin with maximal absorption at 525 nm.

In species of the genus *Triturus* a regular alteration of the visual pigments can be recorded both during metamorphosis and in an annual rhythm in adult animals (Himstedt 1973a, b). The spectral sensitivities of the eyes were determined by threshold measurements with the help of electroretinography. In Fig. 3.8 the results from the newt *Triturus vulgaris* are shown.

The data for larval eyes fit the absorption curve of porphyropsin, P 522_2 (Reuter et al. 1971). During metamorphosis the maximum sensitivity is shifted towards shorter wavelengths. The data for the juvenile stage can be matched to the absorption curve of frog rhodopsin P 502_1 (Dartnall 1953) (Fig. 3.8, left). A corresponding conversion of retinal_2 to retinal_1 in the course of metamorphosis has been described for several anurans (for a summary see Crescitelli 1972). Furthermore, when adult *Triturus* return to water during the mating season in the spring, they resume larval spectral sensitivity. Open circles in Fig. 3.8, left, match the porphyropsin curve. If one studies animals of the same species in summer or fall, when they have left the water and returned to terrestrial life, their spectral sensitivity corresponds to rhodopsin P 502_1 (Fig. 3.8, right).

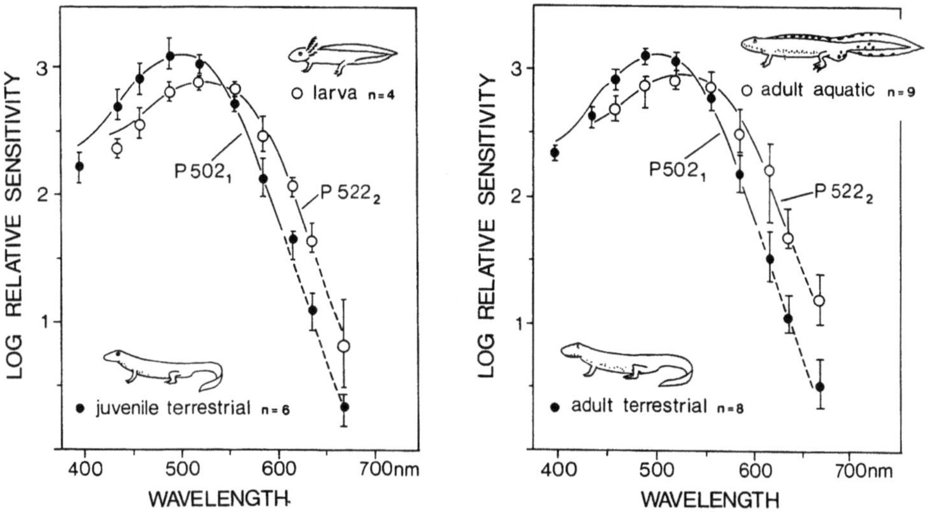

Fig. 3.8. Spectral sensitivity and seasonal changes in the retina of four stages (larval, juvenile, adult aquatic, adult terrestrial) of *Triturus vulgaris*. Larvae and aquatic adults are found during spring and summer, terrestrial adults occur in fall. The diagram shows that, in general, terrestrial animals are slightly more sensitive to shorter wavelengths than the aquatic ones. The peaks of the sensitivity curves correspond to two different retinal pigments. For further details, see text. (From Himstedt 1973a)

In *S. salamandra*, there is no shift of spectral sensitivity (Himstedt 1973a). Both before and after metamorphosis, the rods of this species evidently contain rhodopsin P 502_1. Nor is there any sign of a seasonal change in the visual pigments in the adult animals which never return to aquatic life.

In *Triturus* the alteration of spectral sensitivity may have significance as an ecological adaptation. Typical aquatic habitats of amphibians are often characterized by yellowish water (Reuter 1969, Bridges 1972, Himstedt 1973a, b). This yellow coloration is caused by humus colloids dissolved in the water. Animals living in these habitats can make better use of the available light when their receptors are more sensitive to long wavelength light, by virtue of $retinal_2$ pigments. The shift may also help in recognition of color signals for mating (Himstedt 1973b).

Neuronal elements. Little is known about the morphology of single neuronal elements in the salamander retina because of the difficulty of staining with the Golgi method. The few studies are mainly related to the IPL and the OPL and were carried out in *Necturus* (Dowling and Werblin 1969), *Triturus* (Hendrickson 1966) and *Ambystoma* (Lasansky 1973, Custer 1873, Wong-Riley 1974). Studies of Linke and Roth on the retina of plethodontid salamanders with HRP and Golgi techniques (Linke and Roth in prep.) are generally in accordance with the observations made by Dowling and Werblin

(1969) and Wong-Riley (1974). The following description is based mainly on the work of these authors and own studies.

1. Horizontal cells. The first neuronal elements, next to the OPL are the horizontal cells (HC). These cells usually have rounded somata with a flattend apical side. Numerous processes extend from the apical side and run into the OPL, sometimes for about several hundred microns (Dowling and Werblin 1969). In a recently published review on the OPL of the tetrapod retina, Gallego (1983) stated that there exist only two morphologically distinguishable horizontal cells. These are the short-axon HC and the axonless HC. Both types are found in anurans, e. g. in the frog (Stephan and Weiler 1981, Ogden et al. 1984, 1985). Whether these cells exist also in the salamander retina remains to be clarified.

The bodies of the HCs, like those in the teleost retina (Kaneko 1979, 1983), are electrically coupled as revealed by physiological studies and intracellular staining with a fluorescent dye (Skrzypek 1984), which stains not only the injected cell but also neighboring cells via gap junctions. In the OPL, acetylcholine (Hutchins et al. 1984) is discussed as neuromodulator with respect to HC arbors being involved in the control of receptive field size of HCs (Marshal and Werblin 1978).

2. Bipolar cells. The second cell type in the INL is the bipolar cell. Whereas Dowling and Werblin described only one type in *Necturus*, Wong-Riley describes two types of bipolar cells in larval *Ambystoma*. These cells differ in soma size, shape, and in dendritic arborization. Type-1, which seems to stain more frequently, has a rounded or slightly oval-shaped soma and lies more sclerad within the INL. The dendrites branch tangentially in the OPL. The axon divides within the sclerad, mid-, or vitread sublayer of the IPL in a unistratified fashion. Type-2 has a larger body and a more oval-shaped soma. The dendritic stratification is similar to type-1 bipolar cells, but the axonal arborization in the INL is somewhat different in that the axon usually traverses the whole depth of the IPL and gives off branches at various levels which extend for the same distance as in type-1 cells. Figure 3.9 shows four examples of bipolar cells from *S. salamandra* (A–D) and one from *Hydromantes italicus* (E). Whereas the axon of cell A branches extensively at the vitread margin of the IPL, the axons of cells B–D branch in more scleral parts of the IPL. The single example of a plethodontid bipolar cell (E) shows a similar branching pattern like cell A from *Salamandra*.

Both types of bipolar cell possess a so-called Landolt club. In *Triturus* (Hendrickson 1966) and *Salamandra* (Linke unpublished) this specific apical fiber is a process which is normally thicker than the other apical dendrites and runs in the longitudinal plane of the bipolar cell. It extends through the OPL and the cell bodies of the photoreceptors and ends at the outer limiting membrane. The terminals of this fiber normally have a bulbous swelling. For

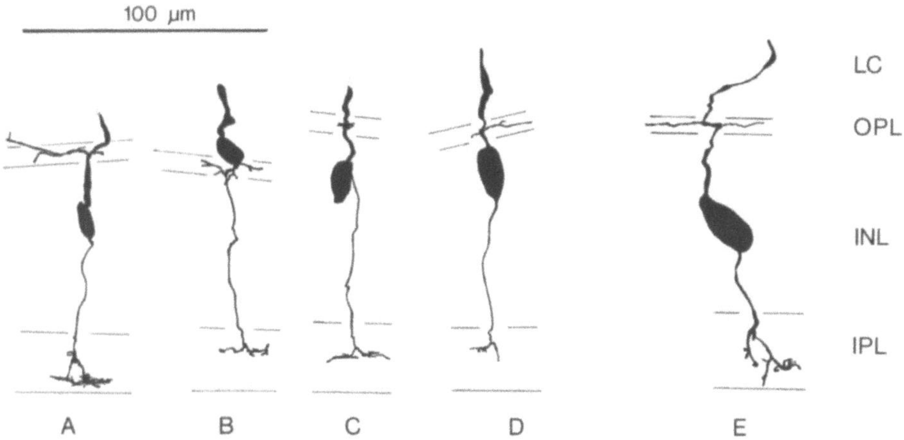

Fig. 3.9 A–E. Camera lucida drawings of Golgi-impregnated bipolar cells of *Salamandra salamandra* (**A–D**) and *Hydromantes italicus* (**E**). All bipolar cells in salamanders so far examined have a Landolt club (*LC*) which, in plethodontids, is rather thin with one or two swellings along its course (E). Bipolar cells seem to differ in their arborization both in the outer plexiform (*OPL*) and inner plexiform (*IPL*) layer and in soma size (**A–D**). In some of the bipolar cells the soma is displaced in the outer nuclear layer (**B**). *Bar* = 100 μm. (Linke and Roth in prep.)

further details, especially electron microscope observations of Landolt clubs in salamander retina, the reader is referred to Hendrickson (1966). In plethodontids, in contrast, the Landolt club is not much thicker than the other processes. Sometimes, it has a bulbous swelling in the middle and at the end (Fig. 3.9 E). Occasionally, the somata of bipolar cells could be found among the perikarya of the photoreceptors (cf. Fig. 3.9 B).

The two morphological types are assumed to divide the input of the photoreceptors into independent ON and OFF channels, similar to the situation in the teleost retina (Kaneko 1979, 1983). For these cells, amino acids like glutamate or its analogs are assumed to be excitatory transmitters (Slaughter and Miller 1983).

3. Amacrine cells. The third cell type within the INL is the amacrine cell, which normally lies at the vitread border of the INL. In *Necturus*, Dowling and Werblin (1969) described two types of amacrine cells, one with a single thick process, which branches laterally in the IPL, and a second one, which has several small processes that branch in the IPL. Two types of amacrine cells also exist in *Ambystoma* (Wong-Riley 1974). Both have rounded somata but differ in their dendritic branching and the position of their somata. The soma of type-1 usually lies more vitread and has several independent processes that extend into the IPL. The body of type-2 has a more sclerad position and a single large process which divides into several branches within the IPL. According to Wong-Riley, these two classes could be further subdivided into two groups according to their branching in the IPL which is either

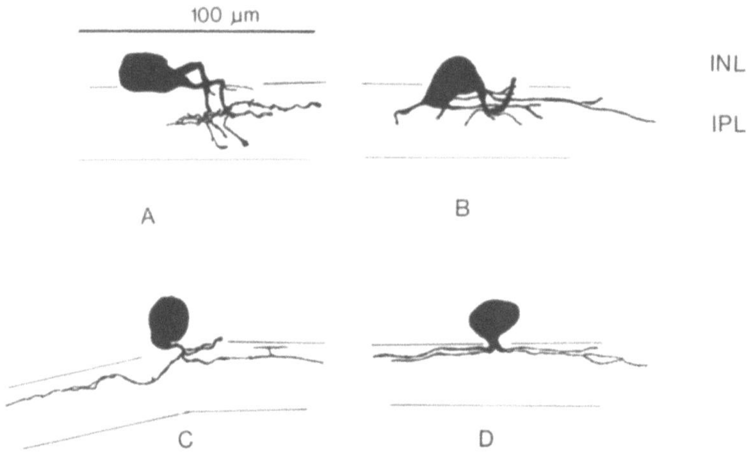

Fig. 3.10 A–D. Camera lucida drawings of Golgi-stained amacrine cells of *Hydromantes italicus*. *INL* Inner nuclear layer; *IPL* inner plexiform layer. *Bar* = 100 µm. (Linke and Roth in prep.)

radial or horizontal in a stratified fashion. This corresponds to the description of diffuse and stratified amacrine cells by Ramon y Cajal (1872) and by Boycott and Dowling (1969) in other vertebrates.

In plethodontid salamanders, amacrine cells show a similar morphology (Linke and Roth in prep.). At least, two types could be distinguished (Fig. 3.10). Type-1 cells (Fig. 3.10 A, B) have a rounded to oval-shaped soma with a flattening at the border of the IPL. They always lie at the outermost (sclerad) border. Many small dendrites extend from the soma and spread, sometimes widely in the IPL. In type-2 (Fig. 3.10 C, D), the perikaryon normally lies somewhat deeper in the INL and has a single, often very thick, primary dendrite, from which many small processes branch in the outer laminae of the IPL.

4. Ganglion cells. Detailed information on the morphology of single retina ganglion cells in the salamander retina is scarce. Some observations are given by Dowling and Werblin (1969) on the ganglion cells of *Necturus*, an aquatic and neotenic species with poorly differentiated eyes. On the basis of electron microscope analysis of the IPL, the authors distinguished two types of retina ganglion cells that differ with regard to their inputs. Type-1 was driven directly by bipolar cells, whereas type-2 was mainly driven by amacrine cells. Most cells showed a single thick process that branched in the IPL. A similarly short description can be found in the work of Wong-Riley (1974), but she described a different morphological type with several primary dendrites. Again, no detailed investigation of dendritic arborization was made.

Linke and Roth (in prep.) carried out a study on the morphology of retina ganglion cells in several plethodontid species (primarily *Hydromantes ita-*

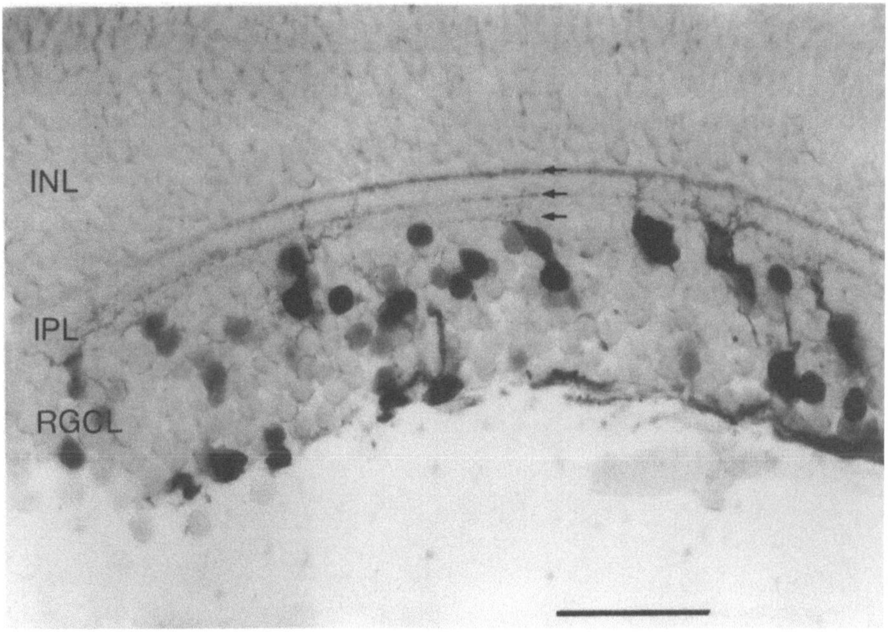

Fig. 3.11. Photograph of part of the central retina in *Hydromantes italicus* after injection of HRP into the chiasma. In the retina ganglion cell layer (*RGCL*) many filled cells are visible. In the inner plexiform layer (*IPL*) three darkly stained layers (*arrows*) of different intensities separated by unstained layers can be seen. *Bar* = 100 μm. (Linke and Roth in prep.)

licus, *Batrachoseps attenuatus*, and *Bolitoglossa subpalmata*) by means of retrograde filling of rgc with horseradish-peroxidase (HRP) and the Golgi method. HRP was injected into the chiasma opticum and different target sites of retinofugal projections such as the tectum, the thalamus, and the basal optic neuropil.

Somata of retina ganglion cells often have a piriform, more rarely a rounded shape. About half of the rgc possess a thick, primary dendrite, from which finer processes extend. The remaining cells are "ganglionic" in the sense that in addition to one or two thick main dendritic processes, they possess one to several thin processes extending directly from the perikaryon. The somata of these ganglionic rgc often have a more rounded shape.

Stratification of the IPL. In nearly all HRP-labeled retinas a fivefold sublayering was observed which consists of two brighter and three darker zones. The darker zones lie vitread, sclerad, and in the middle of the IPL and are separated by the brighter ones (Fig. 3.11). This is in good accordance with the structure of the IPL described by Ramon y Cajal (1892) in the frog and the findings of Ball and Dickson (1983) in the newt retina. In the following description, zone (or sublayer) 1 is most sclerad and 5 most vitread.

3.2.2 Classification of Retina Ganglion Cells

In plethodontid salamanders, retina ganglion cells could be classified according to their dendritic field size, their pattern of dendritic arborization, and to their lamination into at least four classes. Figure 3.12 shows examples of these classes.

Class-1 neurons (Fig. 3.12A) always have small dendritic fields. From the perikaryon which always lies in the vitread division of the rgc layer, one primary dendrite extends into the inner plexiform layer where it usually divides into thin and short branchlets. In general, these branchlets show small

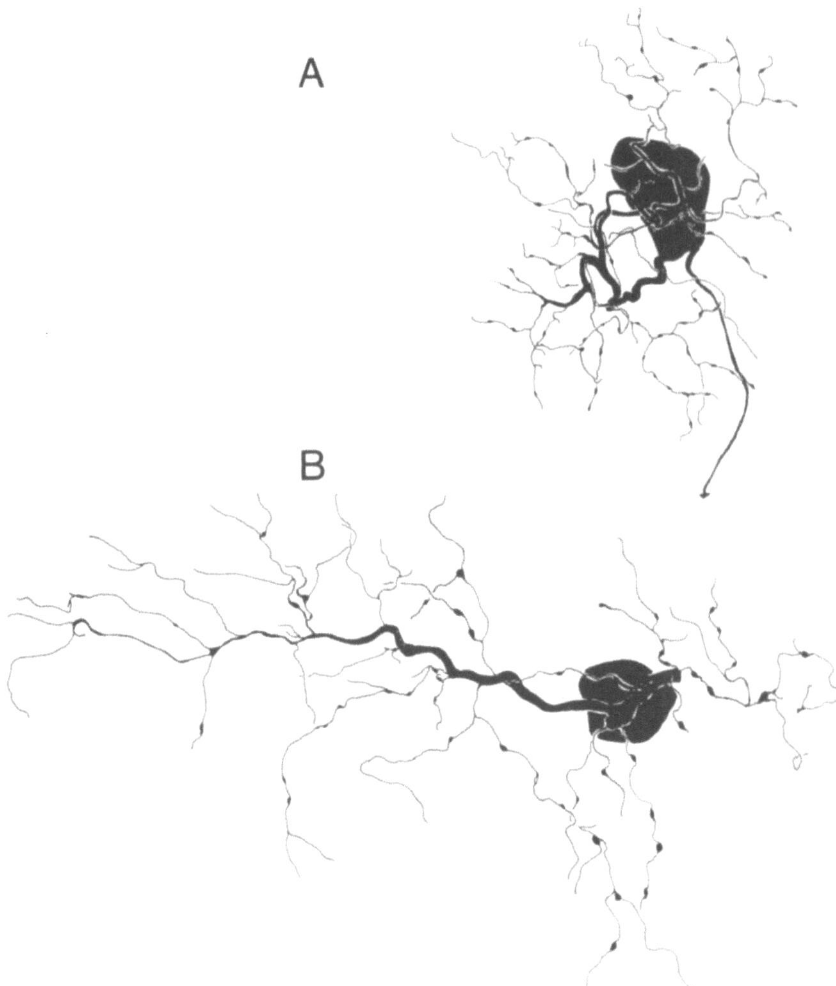

Fig. 3.12 A, B. Camera lucida drawings of HRP-filled retina ganglion cells of *Bolitoglossa subpalmata*. **A** Class-1 cell; **B** class-2 cell

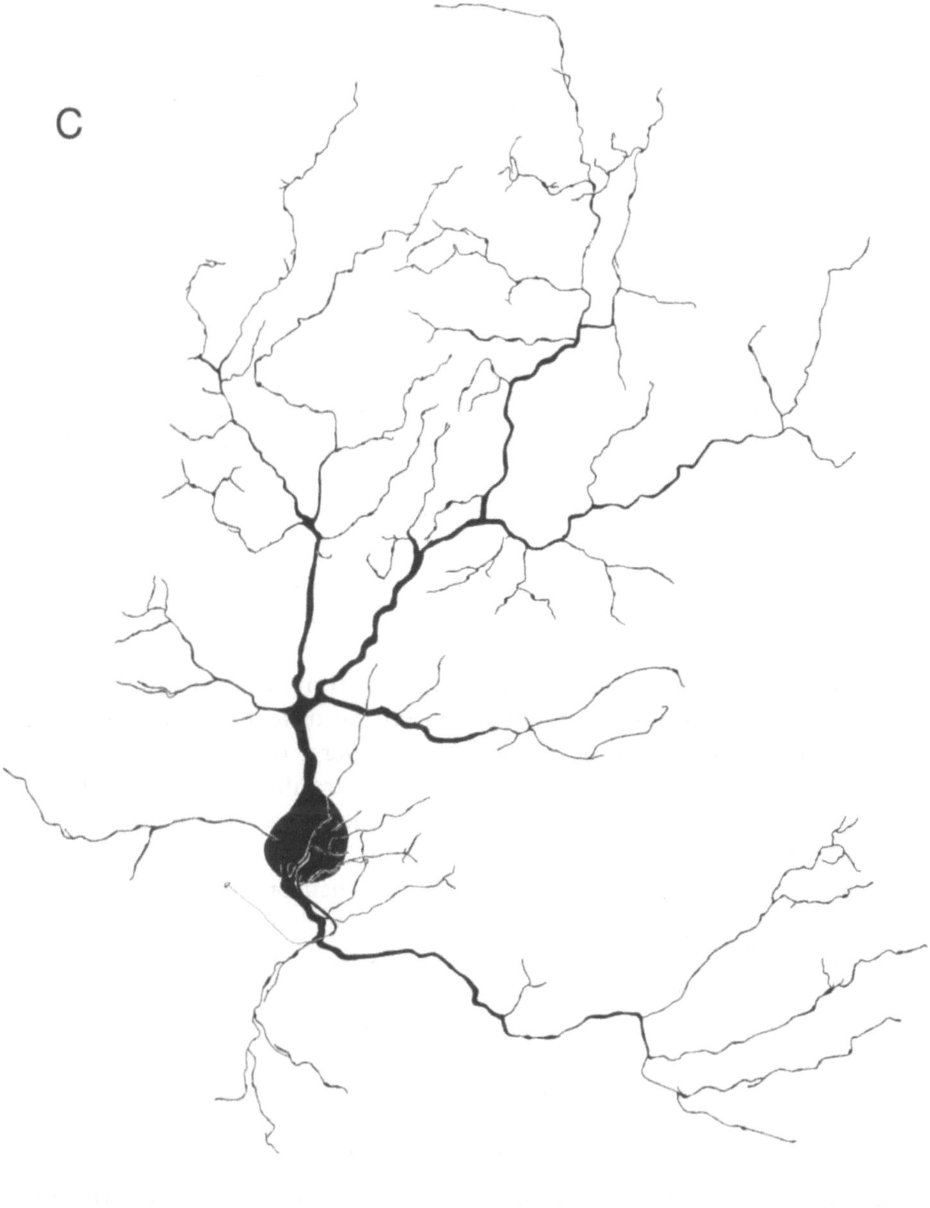

Fig. 3.12 C. Class-4 cell. *Bar* = 100 μm. (Linke and Roth in prep.)

varicosities along their course. The dendritic tree branches throughout the whole depth of the IPL which, in transverse sections, gives them a bushy appearance. In some cases the primary dendrite extends in an oblique manner into the IPL, with a length up to 100 μm, so that distal dendritic branching lies at a considerable distance from the soma.

Class-2 cells (Fig. 3.12B) have larger dendritic field sizes compared to class-1 cells. Their dendrites show a lower, if any, amount of varicosities. In the central part of the retina, the dendritic area is often round to slightly oval and becomes more oval in shape in more peripheral parts. From the soma, which often occupies a peripheral position within the dendritic field, one or two rather thick primary processes arise from which thin secondary branches derive which are much longer than in class-1 cells. In some of these cells, a clear lamination into two laminae can be observed. Others show a more diffuse lamination pattern.

Class-3 neurons were only rarely found. They have medium to large dendritic fields, but show a less dense arborization pattern compared to class-2 or class-4 neurons, and their primary dendrites are not as thick as in those cells. Furthermore, they tend to arborize in only one lamina of the IPL. However, due to the insufficiency of some HRP stainings, it cannot be excluded that these neurons do not represent a separate class but are only incompletely filled cells.

Class-4 cells (Fig. 3.12C) have larger somata and always much larger dendritic fields than cells of the other classes. They are scattered over the whole retina, and their somata always lie in close proximity to the IPL. Two, seldom one, very thick primary processes extend from the soma. Sometimes, small spinelike processes can be observed on the dendrites. The dendritic processes are always found to extend into sublayer 1 of the IPL.

In transverse sections of the retina, a distinct stratification pattern of rgc dendrites is visible. According to their stratification pattern, at least three different types of rgc can be distinguished (Fig. 3.13).

The first type (Fig. 3.13A) includes all those neurons which arborize in one sublayer. Clearly identifiable are those rgc which arborize only in sublayer 1 of the IPL. Somata of these cells mostly lie in close proximity to the IPL. One, seldom two, thick primary dendrite leaves the soma and runs straight or in an oblique manner to sublayer 1, where it further arborizes without leaving the sublayer. Sometimes, one to several very small additional processes are observed originating from the soma. These processes run straight forward to sublayer 1 where they arborize or run in a horizontal direction near the border of the INL without arborizing. Subtypes of these single-layered neurons have their dendrites only in the middle of the IPL or at the vitread margin.

A second type (Fig. 3.13B) of rgc found in transverse sections is represented by those neurons which arborize within two sublayers of the IPL.

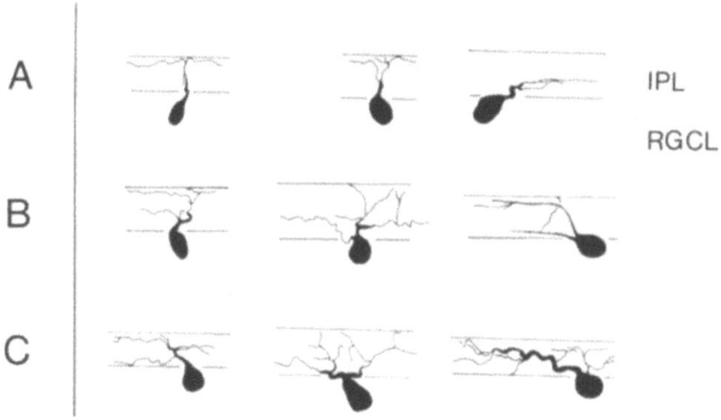

Fig. 3.13 A–C. Lamination patterns of retina ganglion cell dendrites in the inner plexiform layer in plethodontid salamanders. **A** Monostratified cells; **B** bistratified cells; **C** diffuse cells. *IPL* Inner plexiform layer; *RGCL* retina ganglion cell layer. For details, see text. (Linke and Roth in prep.)

Here, at least three subtypes are distinguishable arborizing in sublayers 1 and 3, 3 and 5, and 1 and 5, respectively. The most common subtype arborizes in the more distal sublayers of the IPL, i.e., sublayers 1 and 3. These neurons send one to three primary dendrites to sublayers 3 and 1 where the dendrites further divide and ramify. If only one primary process occurs, it often runs obliquely into sublayer 3, and very fine processes extend perpendicularly to sublayer 1. Neurons arborizing in sublayers 3 and 5 have a similar arborization pattern. Neurons sending their dendrites to the marginalmost sublayers of the IPL (sublayers 1 and 5) are rare.

A third type (Fig. 3.13C) of stratification pattern of dendrites is shown by those rgc which arborize in more than two layers. In contrast to Ramon y Cajal (1892) who put multistratified and diffusely arborizing neurons into separate classes, Linke and Roth considered both groups as being a variation of a single type because many intermediate types can be found. Two examples are presented here. The first has a clear multistratified dendritic structure. Frequently, these neurons arborize in sublayers 1, 3, and 5, although other combinations, e.g., 1, 4, and 5 are possible. Sometimes it was difficult to assign the dendrites to a specific sublayer.

Neurons with a diffuse arborization pattern in the IPL have pear-shaped somata with one primary dendrite, or more rounded somata with two to three primary dendrites. Whereas in the first type of neurons the dendritic area is always small, it is generally large in the latter.

In summary, the study of whole-mount preparations of the retina strongly suggests, too, that different morphological classes of retina ganglion cells exist in salamanders. In contrast to all other vertebrate retinas investigated so far (including the frog retina), cell size does not seem to be an appropriate criterion for the classification of rgc. For example, no rgc with extraordinary large somata were found which corresponded to α-cells in mammals (Boycott and Wässle 1974) and giant cells in chondrichthyes (Stell and Witkovsky 1973) and teleosts (Dunn-Meynell and Sharma 1986) and which in the frog are quite obvious (Kalinina 1974, and Linke pers. comm.).

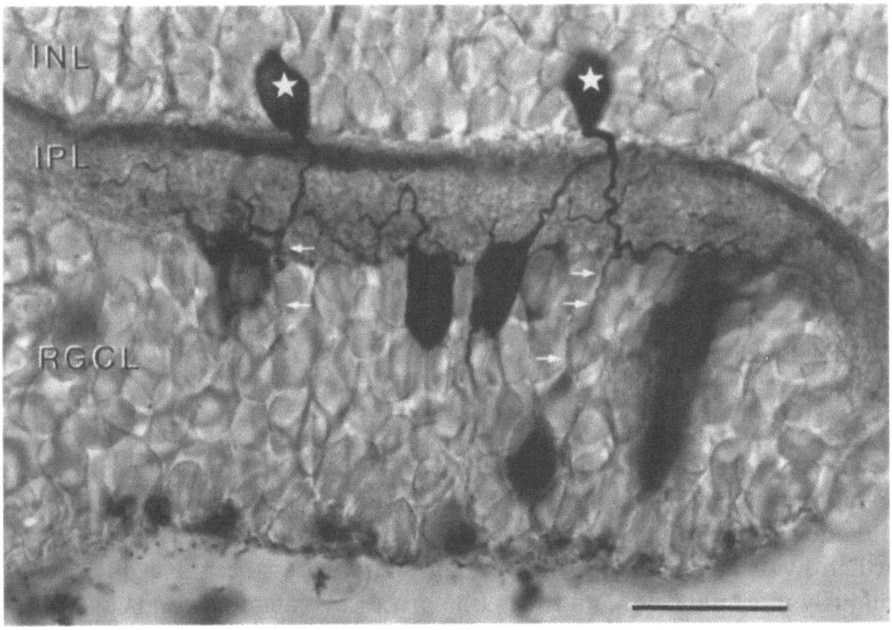

Fig. 3.14. Photograph of part of the peripheral retina in *Hydromantes italicus* after injection of HRP into the chiasma. *Stars* indicate two displaced retina ganglion cells. *Arrows* point to axons of the two cells which originate from the primary dendrite. *Bar* = 50 µm. (Linke and Roth in prep.)

However, clear differences in arborization pattern exist between retina ganglion cells which justify the above classification.

Displaced Ganglion Cells in the Salamander Retina. Many authors have shown that not all bipolar cells and amacrine cells lie within the INL and not all ganglion cells lie within the ganglion cell layer. These are the so-called displaced cells. Figure 3.14 shows examples in the retina of *Hydromantes.* To a certain percentage, which differs among the vertebrate classes and among genera within a class, ganglion cells lie in the INL, amacrines lie in the ganglion cell layer, and perikarya of bipolar cells lie among the somata of the photoreceptors. In salamanders, only one attempt has been made to count the number of displaced retina ganglion cells and displaced amacrine cells (Ball and Dickson 1983). From differences in the number of somata counted in the rgc layer and the number of optic nerve fibers, the authors concluded that there are 39% displaced amacrines and 2.5% displaced ganglion cells in the retina of *Notophthalmus viridescens.* The displaced amacrine cells were found throughout the entire ganglion cell layer. The displaced rgc were found primarily at the vitread margin of the INL and mainly in peripheral parts of the retina.

3.2.3 Neuronal Connections in the Retina

The neuronal connections in the salamander retina have been studied in detail only in *Necturus* (Dowling and Werblin 1969, 1970) and larval *Ambystoma* (Lasansky 1973, for the OPL, and Wong-Riley 1974, for the IPL) (Fig. 3.15).

Photoreceptor Contacts. Photoreceptors were originally thought to be totally isolated from each other. The outer segments were found to be enve-

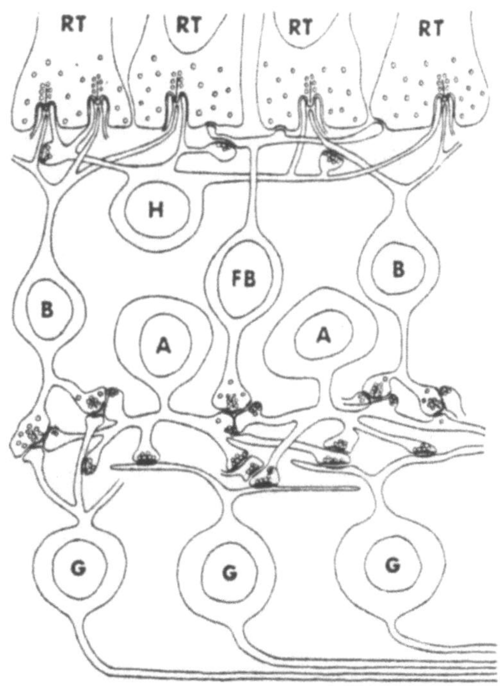

Fig. 3.15. Summary diagram of the arrangement of synaptic contacts found in vertebrate retinas. In the outer plexiform layer, processes from bipolar (*B*) and horizontal (*H*) cells penetrate into invaginations in the receptor terminals (*RT*) and terminate near the synaptic ribbons of the receptor. The processes of flat bipolar cells (*FB*) make superficial contacts on the bases of some receptor terminals. Horizontal cells make conventional synaptic contacts onto bipolar dendrites and other horizontal cell processes (not shown). Since horizontal cells usually extend further laterally in the outer plexiform layer than do bipolar dendrites, distant receptors can presumably influence bipolar cells via the horizontal cells. In the inner plexiform layer, two basic synaptic pathways are suggested. Bipolar terminals may contact one ganglion cell dendrite and one amacrine process at ribbon synapses (*left side* of diagram) or two amacrine cell (*A*) processes (*right side* of diagram). When the latter arrangement predominates in a retina, numerous conventional synapses between amacrine processes (serial synapses) are observed, and the ganglion cells (*G*) are contacted mainly by amacrine processes (*right side* of diagram). Amacrine processes in all retinas make synapses of the conventional type back onto bipolar terminals (reciprocal synapses). (From Dowling 1970)

loped by the processes of the pigment epithelial cells and the perikarya by the distal processes of the Müller cells. However, several exceptions are known in different vertebrate groups (Stell 1972). In the axolotl (larval *Ambystoma mexicanum*), Custer (1973) described at least four types of contacts, two at the level of the inner segments of the photoreceptors and two at the OPL level. At the level of the inner segments, gap junctions between adjacent rods and between rods and cones were found. In the OPL, synaptic terminals are often found in a postsynaptic position relative to rod ribbon terminals. Between photoreceptor telodendra a few tight junctions occurred. Extensive interreceptor contacts are also mentioned by Keefe (1971) in *Notophthalmus viridescens*.

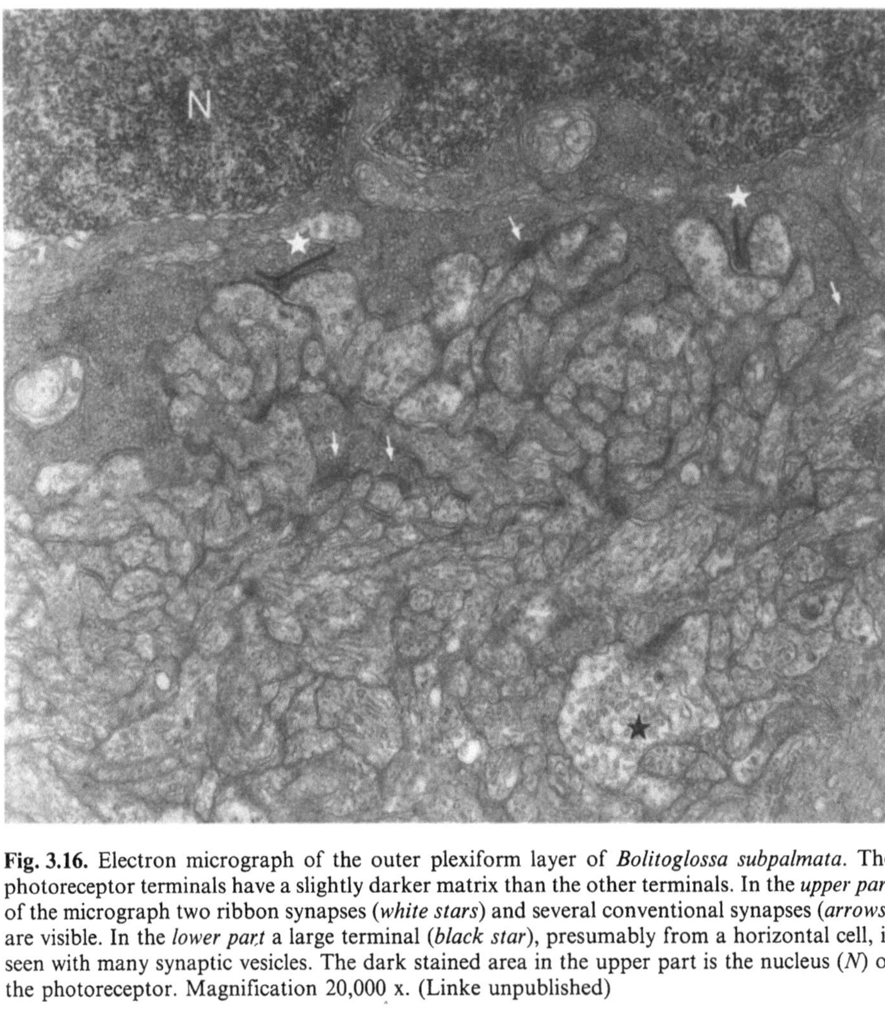

Fig. 3.16. Electron micrograph of the outer plexiform layer of *Bolitoglossa subpalmata*. The photoreceptor terminals have a slightly darker matrix than the other terminals. In the *upper part* of the micrograph two ribbon synapses (*white stars*) and several conventional synapses (*arrows*) are visible. In the *lower part* a large terminal (*black star*), presumably from a horizontal cell, is seen with many synaptic vesicles. The dark stained area in the upper part is the nucleus (*N*) of the photoreceptor. Magnification 20,000 x. (Linke unpublished)

Lasansky (1973) described several types of contacts between rod and cone pedicles in the OPL. The rod and cone pedicles were sometimes found as lateral processes in ribbon synapses, but no ribbon synapse was found between rod pedicles. If a process of a rod pedicle invaginated a cone pedicle no ribbon synapse was found.

More recently, Attwell et. al (1984) gave physiological evidence for electrical coupling between rods and rods, and rods and cones in *Ambystoma tigrinum*. Electrical coupling between cones was not found. There might be chemical synapses between photoreceptors as described by Lasansky (1973) in the larval tiger salamander and by Mariani and Lasansky (1984) in the turtle.

The outer plexiform layer contains the receptor terminals which in *Necturus* are arranged in two or three levels. In the distal part, mostly cone synapses are situated, while the proximal part contains mostly rod synapses. The dendrites of both bipolar and horizontal cells invaginate the cone terminals which contain typical ribbon synapses at the site of invagination (Figs. 3.15 and 3.16). Ribbon synapses are also formed by the rod terminals, but their contacts with bipolar cells and horizontal cells are flat. The horizontal cells form synapses with other horizontal cell dendrites, but mainly with the dendrites or the cell somata of bipolar cells.

Contacts at the Level of the IPL. The inner plexiform layer is characterized by a greater thickness and by higher density of synaptic contacts. The terminals of the bipolar cells form ribbon-type synapses with the dendrites of the ganglion cells and amacrines. These synapses are arranged in dyads with a variable combination of postsynaptic elements. The dyads are of amacrine-amacrine type as well as of amacrine-ganglion cell type. Dowling and Werblin (1969) described reciprocal synapses that transmit signals from the amacrines back to the bipolar cells, and serial synapses that are contacts between amacrine cell dendrites which in turn transmit the signals to the dendrites of ganglion cells.

An arrangement similar to that described in *Necturus* by Dowling and Werblin (1969) was found in the IPL of the larval *Ambystoma mexicanum* by Wong-Riley (1974). Here, the terminals of bipolar cells made synapses of the ribbon type and regular asymmetric synapses with the processes of ganglion, amacrine, and bipolar cells. No bipolar cell was found to contact the perikarya of the cells. In contrast, amacrine cells made contact on both processes and somata. Gap junctions were also found, mainly between two processes of bipolar cells, more seldom between two amacrine processes and very rarely between bipolar and amacrine cells. It seems that electric coupling occurs at this level of the retina as is the case in photoreceptors (Attwell et al. 1984) and horizontal cells (Skrzypek 1984).

3.2.4 Size, Number, and Density of Photoreceptors and Retina Ganglion Cells

Photoreceptors. Himstedt (1973a) determined size, number and ontogeny of photoreceptors in the salamandrid species *S. salamandra* and *Triturus alpestris* using freshly dissected retinae. The results are given in Fig. 3.17. In *S. salamandra* (Fig. 3.17A), rod outer segments have a length of 55 μm and a diameter of 11 μm. The total number of photoreceptors in the eye of *Salamandra* is 630,000. From hatching to the adult stage, rods double both in diameter and length. During this time the number of photoreceptors increases 27 times, but the anatomical receptor density decreases by about 36%. Due to the increase of the diameter of the eye, the physiological receptor density, i.e., the number of receptors per 10° retina area, however, increases from 140 to 2080 rods. In *T. alpestris* (Fig. 3.17B), rod outer segments are smaller than in *Salamandra* with a length of 30 μm and a diameter

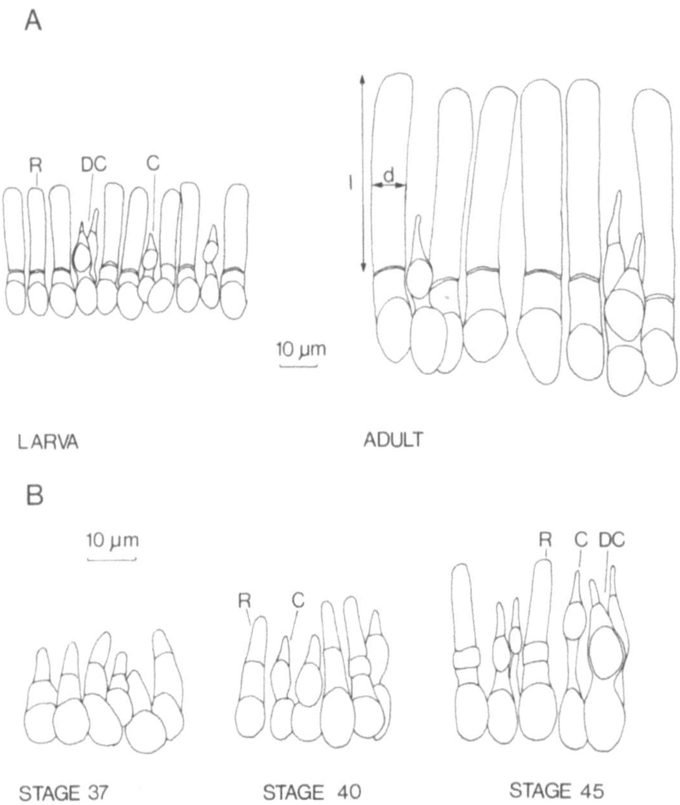

Fig. 3.17. A Photoreceptor elements of larval and adult *S. salamandra*. *l* and *d* refer to length and diameter of photoreceptor outer segments. **B** Photoreceptor elements of three stages of larval *Triturus vulgaris*, showing much greater increase in the size of rods as compared to cones. *R* Rods; *C* cones; *DC* double cones. (After Himstedt 1973a)

Table 3.4. Density of photoreceptors (PHOT), retina ganglion cells (RGC) and relative portions of cones and double cones (C) in percent of the total number of photoreceptors per unit area (147 μm) in the nasal (n), central (c) and temporal (t) area of the retina. Measurements taken from five animals per species

		D.och.	*E.bis.*	*P.cin.*	*B.sub.*	*T.nar.*	*Ba.att.*	*H.ital.*
PHOT	n	45.3	33.1	31.6	25.9	32.8	32.5	25.6
	c	51.3	39.2	35.1	27.9	32.4	37.6	27.7
	t	46.8	37.7	33.9	30.9	41.9	39.9	27.0
RGC	n	75.5	55.8	41.3	36.5	72.5	55.6	27.0
	c	82.5	61.7	44.7	32.2	60.8	59.3	29.6
	t	72.7	60.1	44.7	42.5	59.8	56.9	30.7
C	n	59.5	59.3	54.0	53.4	56.6	55.8	51.3
	c	58.1	59.6	55.2	54.9	56.7	59.6	55.7
	t	56.6	60.0	53.3	55.0	63.0	58.4	51.7

of 7.5 μm. The total number of photoreceptors is 480,000. During development, the rods increase almost threefold in size, and the number of photoreceptors increases 160 times. The anatomical density decreases by 20%, whereas the physiological density of rods increases from 9 per 10° to 360 per 10°.

The density and the number of photoreceptors and the sizes of rod outer segments are given in Tables 3.4 (PHOT), 3.5 (PHOT), and 3.7 (ROS), respectively, for the seven plethodontid species studied by Linke et al. (1986). *Desmognathus* and *Eurycea* have the smallest and *Hydromantes* the largest diameter of rod outer segments; interestingly, while *Desmognathus* has the shortest rod outer segments, *Eurycea* has the second longest ones. Similarly, *Thorius*, having the smallest eye, possesses rather long rod outer segments. This increases the probability of absorption of light quanta. Since most of the plethodontids investigated are much smaller than the two salamandrid species mentioned above, it is not surprising that their photoreceptors are much smaller. An exception is *Hydromantes*, which has rod outer segments almost as large as that of *Salamandra*, although the eyes of *Salamandra* are about ten times larger in volume. The total number of photoreceptors ranges from 25,705 in *Thorius* to 57,900 in *Hydromantes* which is less than one-tenth of the number of photoreceptors found in salamandrids (Table 3.5).

Rod-Cone Relationship. As in other vertebrates, nocturnal salamanders generally have a rod-dominated retina, while diurnal species have increased cone portions. This is evident in the European Salamandridae where a continuous series from diurnal eyes to nocturnal eyes can be observed (Möller 1951). In the retina of *Triturus vulgaris* 16% of the photoreceptors are rods, in *T. alpestris* 26%, in *T. cristatus* 44%, and in *S. salamandra* 61% of the receptors are rods.

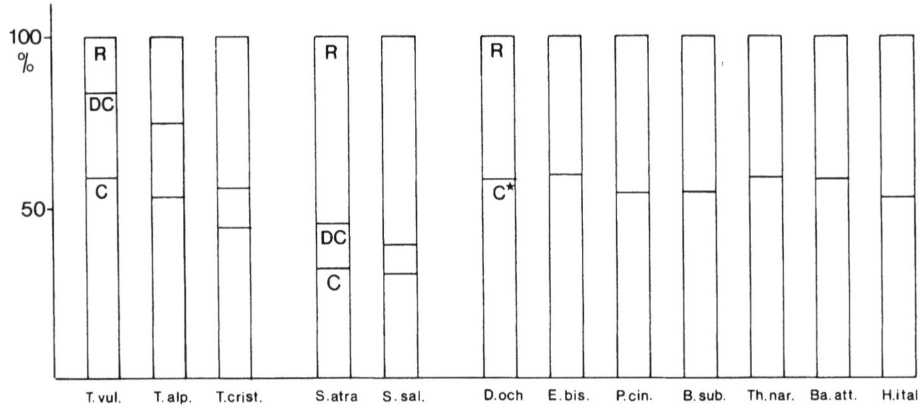

Fig. 3.18. Percentage of rods (*R*), single cones (*C*), and double cones (*DC*) in three species of *Triturus* (*T. vulgaris*, *T. alpestris*, *T. cristatus*), two species of *Salamandra* (*S. atra*, *S. salamandra*), and seven plethodontid species (see Fig. 3.2 for abbreviations). In plethodontid species, single and double cones (*C**, mean of the values of temporal, central, and nasal retina C from Table 3.5) were not counted separately. Data on salamandrids from Möller (1951), data on plethodontids from Linke et al. (1986)

As can be seen in Fig. 3.18, the portion of double cones increases with the increasing number of cones. In *S. salamandra* 9% of the outer segments are double cones, in *Triturus vulgaris* 25%. However, it is not clear how many of the rods are green rods. Since there is evidence that amphibian green rods are involved in hue discrimination (e.g. Muntz 1963, Reuter 1969, Himstedt and Fischerleitner 1975) it might be concluded that green rods are functionally more conelike than rodlike, so that in consideration of diurnal and nocturnal adaptations, green rods should be classified as photopic elements. Recent electrophysiological recordings from single green rods in toad retina (Matthews 1983) showed, however, that green rods were similar to red rods in all physiological characteristics except spectral sensitivity. Thus, toad green rods may not be considered as photopic receptors. They are highly sensitive cells working at low light intensities. There are no similar studies concerning salamander green rods so that it is not possible to decide whether these visual cells are more conelike or more rodlike.

The above statement about rod-cone relations is not affected by this uncertainty. From the studies of Himstedt and Fischerleitner (1975) it can be concluded that in salamandrids with high numbers of rods most of them are red rods (in *S. salamandra* only about one-thenth of the rods are green), so that it can be stated that the portion of scotopic receptors increases within the rank shown in Fig. 18 even if a small part of them should be classified as photopic elements.

The rod-cone relation may also be correlated with eye size (cf. Fig. 3.18). In Salamandridae the small species *T. vulgaris* with an eye diameter of

Table 3.5. Total number of retina ganglion cells (RGC) and photoreceptors (PHOT) and the ratio RGC/PHOT. The values are taken from four to six eyes and are corrected for double counting

	RGC	PHOT	RGC/PHOT
D. och.	37.207	44.562	0.84
E. bis.	51.742	55.838	0.93
P. cin.	32.452	41.898	0.77
B. sub.	38.462	44.111	0.87
T. nar.	28.427	25.705	1.11
Ba. att.	26.555	31.820	0.84
H. ital.	46.436	57.900	0.80

2.2 mm has a cone-dominated retina, while the large *S. salamandra*, with an eye diameter of 5.5 mm, has more rods than cones. Möller (1951) assumed that in the smaller species the relative increase in the number of cones may compensate for the loss of visual acuity due to the relatively small retinal image. Since the outer segments of cones are thinner than those of rods, the number of cones in the same area can be higher and thus the resolution power of the receptor mosaic can be improved. The same seems to be true in plethodontids. In the retinas of the plethodontids studied by Linke et al. (1986), the number of cones is rather high, but the highest values of about 60% cones occur in the smallest eyes of *Thorius* and *Desmognathus*, while the larger eyes of *Bolitoglossa* and *Hydromantes* have 51% to 55% cones (Table 3.5).

Linke et al. compared different parts of the retinas of the mentioned plethodontid species with regard to the number of photoreceptors and retina ganglion cells. There are no consistent differences in the density of both photoreceptor elements and rgc throughout the retina. Also, there is no morphologically detectable fovea or area centralis in the plethodontid retina.

Retina Ganglion Cells. Linke et al. (1986) also determined the number of retina ganglion cells in the plethodontid species. The results are shown in Table 3.5. The lowest number of retina ganglion cells is found in *Batrachoseps* with 26,555, and the highest number, surprisingly, in *Eurycea* (51,742), which has more rgc than *Hydromantes* (46,436), although this species has a much larger eye. This is due to the fact that *Eurycea* has much smaller and more densely packed retina ganglion cells. The ratio between retina ganglion cells and photoreceptors ranges from 0.77 in *Plethodon* to 1.11 in *Thorius*. Thus, *Thorius* has more retina ganglion cells than photoreceptors, which may be unique among vertebrates. Preliminary counts of optic nerve fibers of *Bolitoglossa subpalmata* using EM sections (Linke and Roth unpubl.) reveal an about 25% lower amount of fibers than the counts of rgc in transverse sections of the retina. This is consistent with the suggestion that there exists a considerable amount of other cells in the rgc layer that are not ganglion cells. The majority of the axons in the optic nerve have a small diameter

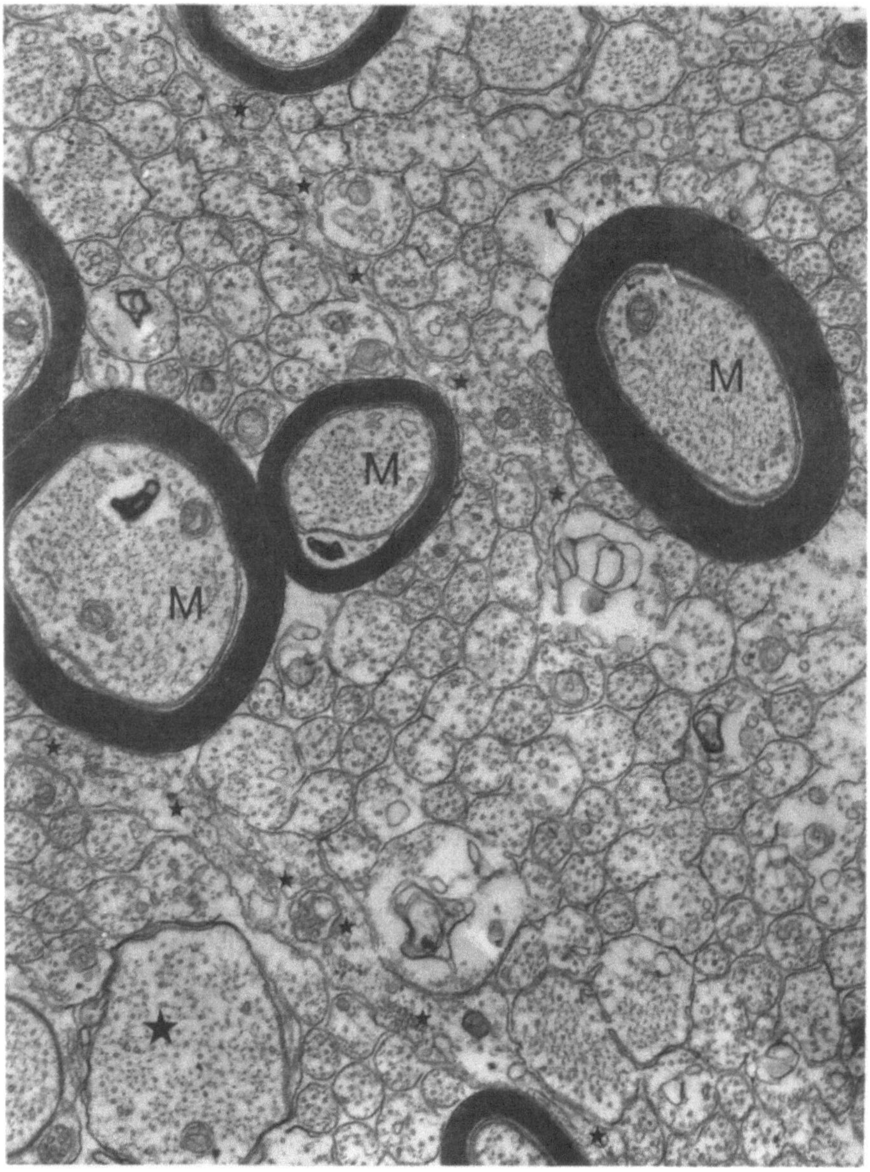

Fig. 3.19. Part of the nervus opticus of *Bolitoglossa subpalmata*. Three large-caliber optic axons in the middle of the micrograph are heavily myelinated (*M*). The smaller unmyelinated axons are packed close together. Two glial processes (*small stars*) running from *upper left* to *lower right* separate the optic nerve into small fascicles. The large axon in the *lower left* part (*big star*) is probably cut in close proximity of a node of Ranvier and, therefore, appears unmyelinated. Magnification 25,000 x. (Linke unpublished)

(< 0.3 μm) and are unmyelinated. Only about 3 % of all fibers have a diameter larger than 0.8 μm. These fibers are myelinated (Fig. 3.19).

Corresponding reliable data from salamandrids are mostly lacking. The data from Möller are uncorrected for double counting. For *S. salamandra*, he counted a total number of 308,300 ganglion cells. Fritzsch (pers. comm.) using EM sections of the optic nerve, counted a total of about 53,000 fibers which is equal to the number of retina ganglion cells. Since the number of receptors counted by Möller (533,000), roughly corresponds with that of Himstedt (630,000), we have to assume a ratio rgc/photoreceptors of 0.1, which means that an average of ten photoreceptors converge on one retina ganglion cell.

Thus, plethodontids, being mostly much smaller than salamandrids and having much smaller eyes, apparently compensate for the low number of photoreceptors with an increase in the relative number of retina ganglion cells. This increased number of retina ganglion cells may be the basis for neural processes that secondarily improve visual acuity, e. g., through lateral inhibition.

3.3 Functional Aspects

3.3.1 Dioptrics of the Salamander Eye

There are only a few refraction measurements for urodele eyes. The commonly applied technique of retinoscopy is rather limited for small eyes because the amount of light projected into the pupil is very low and the reflected light is too weak for reliable measurements. Nevertheless, by this method Beer (1899) found that *Triturus cristatus* and *T. alpestris* are nearly normal-sighted in water and become strongly short-sighted in air. *S. salamandra*, however, was normal-sighted in air. Manteuffel et al. (1977) applied a new method for the determination of the refraction power in urodele eyes by means of laser light. If a laser beam is projected into an eye, the reflected light produces an interference pattern. Such laser speckles can be used to determine refraction and focal distances of optical systems.

In Table 3.6 the far point distances and the refractive power (in diopters D) of the eyes of *T. cristatus* and *S. salamandra* are listed. The far point denotes the greatest distance at which an object produces a sharp image on the retina.

According to these results, *T. cristatus* is slightly short-sighted in water with a far point at about 20 cm. This distance decreases strongly to about 3 cm with the change to aerial vision. In *S. salamandra*, measurements were done in both larval and adult animals. Both appeared to possess eyes adapted to the respective media. Before and after metamorphosis the far point distance is at about 30 cm. This is achieved by altering ocular structures during

Table 3.6. Mean far point distances (in cm) and refractive power (in diopters) of the eyes of *Triturus cristatus* and *S. salamandra* (from Manteuffel et al. 1977)

	Far point distance	Refractive power
T. cristatus air n = 4	2.8	− 35.21
T. cristatus water n = 4	18.6	− 5.38
S. salamandra larval n = 1	32.6	− 3.07
S. salamandra adult n = 2	31.8	− 3.14

metamorphosis. Sivak and Warburg (1980) showed this by studying freeze-sectioned eyes of *S. salamandra* at different developmental stages. In larval eyes, the lens is spherical and in close proximity to the relatively flat cornea. During metamorphosis, the lens becomes flattened, the corneal curvature becomes more pronounced, and the lens moves toward the retina. These changes continue throughout the juvenile to the adult stage.

In the laser-speckle method, as in other refractometric techniques, it is uncertain at which retinal layer the light is reflected. If the reflecting plane is the membrana limitans interna, as it is presumed by most authors (cf. Glickstein and Millodot 1970), this is not the plane on which a focused image is produced. The outer segments of the receptors are situated about 100 to 200 μm more distal. In small eyes with short focal lengths this difference may result in a larger error of measurement compared to large eyes.

A further uncertainty of the laser-speckle method concerning the refractive power is derived from changes in the angle of incidence. The data of Table 3.6 were determined in the optical axis of the eyes. Great variations occur if the light beams striking the eye are not parallel to this axis. A salamander may be more short-sighted in the frontal direction and more far-sighted in the lateral direction. This is probably caused by the oval shape of the lens and eyeball and by the fact that the retina is not precisely concentric to the lens. Such an irregular shape is evident in *Triturus* and also in many plethodontid salamanders (Fig. 3.1) where different parts of the retina are at different distances from the lens. Ströer (1940) hypothesized that this structure might partly compensate the change in refraction during the transition from aquatic to aerial vision. In the different media different parts of the retina receive a focused image. We do not know, however, whether a newt really makes use of this possibility. In those plethodontids which are permanently terrestrial, the nonspherical shape of the lens and retina might be used in order to substitute accommodation processes. Such an eye is focused in the frontal direction at a shorter object distance than in the lateral direction. Thus, one could interpret this construction as a fixed-focus optic adjusted to far objects in the lateral visual field and to frontal prey objects at snapping distance. As far as these data are based on sections of preserved animals this must remain a speculation.

3.3.2 Accommodation

In amphibians, focus regulation is accomplished not by a change in the shape of the lens as in amniotes, but by a change of its position, as in fishes. In anurans, a dorsal and ventral protractor lentis muscle move the lens outward during contraction, not inward as do the lens muscles of fishes. In urodeles, there is only one single ventral muscle connected to the lens (Tretjakoff 1906). Because of this construction the salamander lens cannot be shifted very precisely.

In excised amphibian eyes, Beer (1899), by means of electrical stimulation, forced the protractor lentis to contract. He observed an outward movement of the lens and found that this movement can be asymmetrical in *Triturus* and *Salamandra*. In order to obtain more insight into the accommodation mechanism of the salamander eye, Werner and Himstedt (1984) determined the refractive power in *S. salamandra* before and after accomodation and studied the change in lens position.

In anaesthetized animals, refraction measurements were done by means of the laser-speckle method (Manteuffel et al. 1977). Contraction of the lens muscle was caused by applying succinyl-choline onto the cornea or by means of electrical stimulation. Succinyl-choline causes a spasmodic contraction of muscles and a subsequent total relaxation. After instilling this drug an immediate shortening of the outer focus distance was recorded. The eyes became more short-sighted, probably caused by contraction of the protractor lentis muscle.

Changes in refraction were recorded after a 10-min period of drug influence. In the lateral direction, corresponding to the optical axis of the eye, the effect of accommodation was about 3 diopters (D) with a mean error of 0.6 D. In a more nasal or frontal direction (60° to the sagittal plane) the effect was less with about 1.5 D (mean error 0.3 D). During the following 50 to 60 min the eyes remained in the state of near accomodation, but later returned slowly to the original focus distance.

When a stimulating electrode was placed onto the cornea ventral to the pupil, the motion of the lens could be observed under a microscope. Only the ventral part of the lens moves toward the cornea while the dorsal part stays in its position. In three salamanders the accommodation range during electrical stimulation was determined. The data were 2.8, 4, and 11 D, respectively.

The shift of the salamander lens could also be demonstrated in eye sections. In Fig. 3.20 two drawings of frozen sections from eyes of *Salamandra*

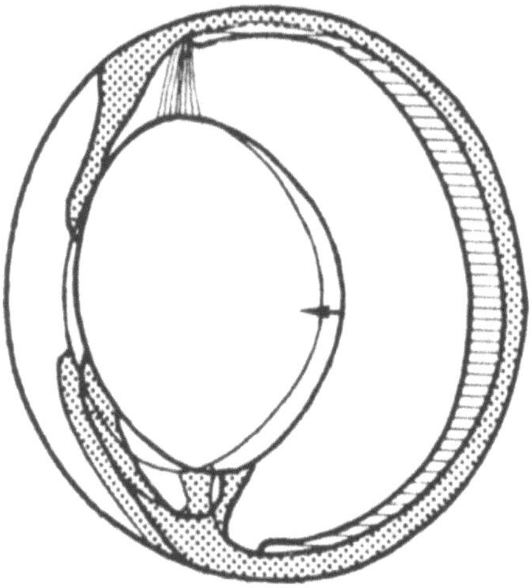

Fig. 3.20. Accommodation mechanism in the eye of *S. salamandra*. The diagram shows forward movement of the lens (*arrow*) during contraction of the protractor lentis muscle. (From Werner 1983)

are superimposed. The first is from a normal eye, the second is from an eye treated with succinyl-choline. The drug caused a deformation of the ventral ciliary fold where the protractor lentis is situated. Dorsally, the lens is held by the zonula fibers (which in freeze sections are not visible); thus, the contraction of the lens muscle results in a tilting movement of the lens.

This tilting probably causes some optical insufficiencies since the shift is larger in the ventral part compared with the dorsal part. It seems impossible that in all states of accommodation the whole retina receives a well-focused image. A salamander might solve this problem by focusing only a single object (e.g., a prey) on a central region of the retina.

It has to be asked whether these data obtained from experiments with anaesthetized salamanders could fully explain the mechanism of accommodation. Anaesthesia may have influenced the tension of the protractor muscle, and the dosage of the drug as well as the electrical stimulus might have been of unphysiological intensity. The experiments show that the eye can change its refraction. The question whether or not this ability is used in normal behavior was studied in freely moving salamanders (Werner and Himstedt 1984).

Refractive changes were measured during prey capture behavior by the sciascopy method which works as follows. Parallel light is projected into the eye, the pupil is illuminated by the light reflected from the retina. When the light beam is moved horizontally, the reflection disappears laterally followed by a shadow. If the shadow follows the disappearing light in the same direction, then the outer focus lies behind the observer; if it moves in the opposite direction, then the focus lies between observer and animal. If the observer is standing right in the focus, no defined movement can be observed (flicker-point). The problem of this method is that it needs a considerably

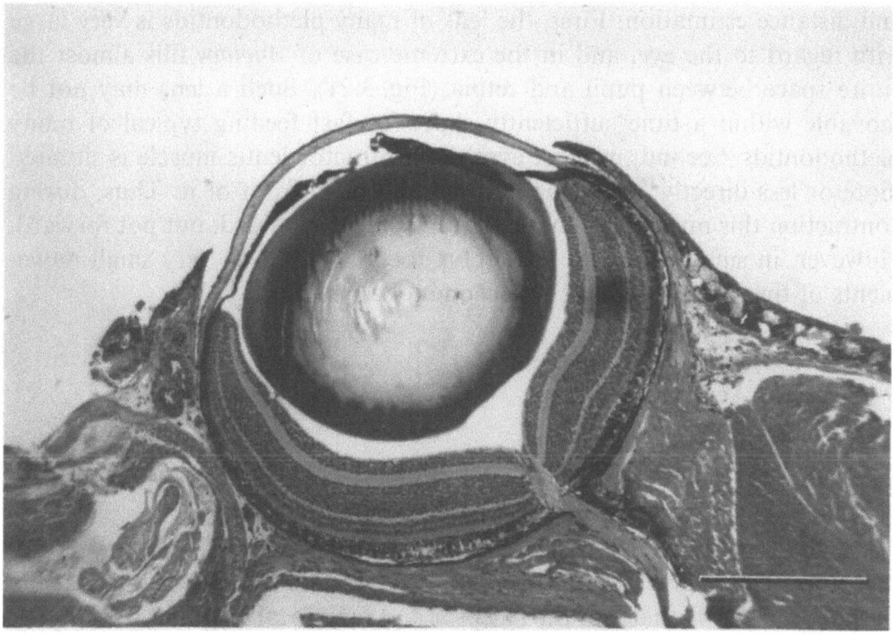

Fig. 3.21. Horizontal section through the eye of *Aneides lugubris* at the level of the exit of the optic nerve. Note the extraordinarily large lens. *Bar* = 1 mm. (Linke and Roth in prep)

high light intensity which may disturb normal behavior. For those reasons the sciascopy was done by the authors with infrared light.

It was possible to measure eye refraction before, during, and after feeding behavior in *Salamandra*. In these experiments, the quietly sitting salamanders were normal-sighted or slightly far-sighted. During fixation of a prey object a change toward short-sightedness could be observed. This means that the eye had accomodated. In slowly reacting salamanders it was evident that accommodation occurred before the first movement of the head was performed. Subsequently, bending down of the head to the typical fixation position took place. In this posture, too, short-sightedness could be recorded. After snapping and swallowing the prey the refraction of the eye returned to the original value. In these recordings the range of accommodation was 2 to 4 D focused on the retina.

These experiments show that *Salamandra* does in fact accommodate during fixation of prey objects. This focusing mechanism may help to identify objects and additionally may be used to measure object distance.

Unfortunately, attempts to measure accommodation in plethodontid salamanders by means of the laser speckle method have proven unsuccessful so far, due to the smallness of the eyes of these species (Himstedt and Roth, unpublished). However, the morphology of the eyes of plethodontids suggests that accommodation may play only a minor role in object identification and distance estimation. First, the lens of many plethodontids is very large with regard to the eye, and in the extreme case of *Aneides* fills almost the entire space between pupil and retina (Fig. 3.21). Such a lens may not be movable within a time sufficiently short for fast feeding typical of many plethodontids. Second, in most eyes the protractor lentis muscle is situated more or less directly below the lens, instead of in front of it. Thus, during contraction this muscle can move the lens only downward, but not forward. However, in small eyes with very short focal length even very small movements of the lens could have an accommodative effect.

3.3.3 Visual Acuity

With respect to the visual properties of an eye we have to distinguish between:

(1) Light sensitivity, i.e., the absorption probability of light in photoreceptors. This is given by the product of $A^2 x d^2/f^2$, where A is the aperture of the eye, d the diameter of the receptor and f the focal length of the dioptric system.
(2) Optical resolution $1/\Delta r$ which is determined only by the aperture A of the eye, since Δr, the radius of Airy's disk, $= 1.22\ \lambda/A$ (rad) in diffraction-limited eyes.
(3) Anatomical resolution which is determined by the angular distance between receptors.

As has been shown by Kirschfeld (1976, 1984) in contrast to a common belief, light sensitivity and angular resolution are not necessarily lower in small eyes compared to large eyes. This means that a small eye is equivalent to a large one if the pupil diameter (A) remains constant and if the angular separation of the photoreceptor outer segments has proportionally decreased such that the ratio between the angular diameter of Airy's disk and the angular separation of the receptors remains constant. In other words: small eyes should have large apertures and small photoreceptor diameters for the maintenance of good resolution.

Linke et al. (1986) have calculated the visual acuity of seven plethodontid species, *Desmognathus ochrophaeus*, *Eurycea bislineata*, *Plethodon cinereus*, *Bolitoglossa subpalmata*, *Thorius narisovalis*, *Batrachoseps attenuatus*, and *Hydromantes italicus*, based on morphological data. As shown in Figs. 3.1 and 3.2, the eyes of plethodontid salamanders have very wide maximal apertures, due to large corneae and large absolute lens diameters. Since they have short focal length as well, the ratio A/f is also large. Therefore, in these small animals, small absolute size of the eyes is not per se of disadvantage. The critical factor, then, is only the angular separation of the photoreceptors.

Table 3.7. Comparison of length (L) and diameter (D) of rod outer segment (ROS), diameter of cone outer segment (COS) and diameter of retina ganglion cells (RGC) in plethodontid salamanders. All values in μm. Measurements taken from five animals per species

		D.och.	*E.bis.*	*P.cin.*	*B.sub.*	*T.nar.*	*Ba.att.*	*H.ital.*
ROS	L	23.52	29.99	26.46	26.46	27.30	24.15	36.25
	D	4.41	3.68	5.15	5.88	4.41	4.35	9.25
COS	D	1.47	1.47	1.47	2.94	2.21	2.21	2.94
RGC	D	6.62	8.09	8.09	12.50	12.12	11.60	13.50

Table 3.8. Determination of anatomical resolution based on rod vision (ARR) and cone vision (ARC) and of optical resolution based on rod vision (ORR) of the species investigated. Values in degrees of visual angle and in minutes and seconds of arc

	ARR	ARC	ORR
D. och.	0.505 (30′18″)	0.169 (10′08″)	0.070 (4′12″)
E. bis.	0.322 (19′19″)	0.140 (8′24″)	0.054 (3′14″)
P. cin.	0.490 (29′24″)	0.140 (8′24″)	0.070 (4′12″)
B. sub.	0.374 (22′26″)	0.187 (11′13″)	0.047 (2′49″)
T. nar.	0.505 (30′18″)	0.253 (15′11″)	0.070 (4′12″)
Ba. att.	0.402 (24′07″)	0.204 (12′14″)	0.070 (4′12″)
H. ital.	0.424 (25′26″)	0.138 (8′17″)	0.035 (2′04″)

Table 3.9. Maximum distance of detectability of a small prey object of 0.5 mm in length based on the data from Table 3.9 on anatomical and optical resolution of the eyes. Values in mm. For abbreviatons see Table 3.8

	ARR	ARC	ORR
D. och.	56.69	170.07	409.17
E. bis.	81.52	204.08	1472.32
P. cin.	58.25	204.08	409.28
B. sub.	76.53	153.03	609.75
T. nar.	56.69	113.12	409.17
Ba. att.	71.26	140.27	408.97
H. ital.	67.57	212.59	781.25

Therefore we should expect that smaller eyes have smaller receptor diameters. Table 3.7 shows that smaller eyes indeed have smaller rod outer segment diameters, although this diameter is not strictly correlated with eye size: *Eurycea* with a medium-sized eye has the smallest diameter.

On the basis of the data from Table 3.7 we can calculate the anatomical resolution power of the eyes which is determined by the angular distance between photoreceptors (Table 3.8). For rods we find values between 0.322° (*Eurycea*) and 0.505° (*Thorius* and *Desmognathus*). This means that under conditions of rod vision (twilight or night) an object of a length of 0.5 mm which corresponds to the size of small insects eaten by Bolitoglossini (e.g., collembolans) can be detected at a maximum distance between 57 mm (*Desmognathus*) and 82 mm (*Eurycea*) (Table 3.9 ARR). If we assume light conditions at which cones are sufficiently active, the maximum distance will further increase (Table 3.9 ARC).

Under sufficient light conditions the detectability of an object is ultimately limited by the optical resolution of the eye which, as shown above, is dependent only on the eye aperture. Table 3.9 (ORR) gives the respective values calculated on maximal pupil diameter. For example, for *Hydromantes* a value of 0.035° is found. An object of this angular size would cover about half of Airy's disk and would certainly be detectable. A small insect of a length of 0.5 mm would be detectable from a distance of 78 cm (Table 3.9 ORR).

In contrast to the situation in the human eye, where optical and anatomical resolution are more or less equal, in plethodontid salamanders there is, thus, a difference between optical and anatomical resolution of about an order of magnitude. It is known, however, that anatomical resolution can be improved by temporal scanning induced by either eye or object movement and appropriate neural processing.

Involuntary eye movements have been shown at least in *Salamandra salamandra* (Manteuffel et al. 1977). Furthermore, it is known that salamanders, as well as amphibians in general, respond preferentially to moving objects. We may, therefore, assume that visual resolution of the plethodontid species studied is above anatomical resolution.

Himstedt (1967) determined visual acuity in Salamandridae by means of behavioral studies. In these experiments behavioral responses such as fixation, partial approach, and complete approach to moving stimuli, consisting of vertical lines of different width, were used. The subjects had a distance of 3 cm from the stimulus. Additionally, ambient illumination was changed within a range of 0.005 to 100 lx in order to determine the dependence of visual acuity on this parameter.

The best results are found in juvenile *Salamandra* with a minimum visual angle of 6′, followed by larvae of this species. Newts are somewhat inferior: here the metamorphosed (aquatic) forms are slightly better than the larvae (10′ : 11′ and 8′ : 9′). This optimum visual acuity is found down to an illumination level of 10–30 lx. In all species tested below that illumination level a strong decline occurred, followed by a less steep decrease in visual acuity between 0.1 and 0.001 lx. Below that illumination level, again a strong decrease takes place which presumably corresponds to the transition from photopic vision (activity of cones) to scotopic vision (activity of rods).

Experiments with newts at a distance between 14 and 9 cm in which optomotor instead of feeding responses were recorded at an illumination level 30 lx revealed somewhat higher values for the minimum separabile (aquatic stage: 8′, terrestrial stage: 12′). These values are comparable to those known from other lower vertebrates such as *Rana temporaria* (6′53″, Birukow 1937), the teleost *Phoxinus laevis* (10′50″, Brunner 1934) and in lizards (11′28″, Ehrenhardt 1937).

Systematic behavioral experiments based on visual acuity of plethodontid salamanders are still lacking. Experiments by Wiggers and Roth (in prep.) showed that adult *Bolitoglossa subpalmata* are able to recognize prey objects of a length of 0.5 mm and a width of 0.2 mm from a distance of 5 cm. During recording from the tectum opticum of *Batrachoseps attenuatus*, neurons were found that responded well to a 2 × 2 mm stimulus presented at a distance of 50 cm which corresponds to an angular size of 0.23° or 13′5″.

If we take the values for anatomical resolution power based on cone vision from plethodontid salamanders for comparison with the values for salamandrids from Himstedt's experiments, then the range of visual acuity in plethodontids from 8′24″ in *Eurycea* to 15′11″ in *Thorius* are well comparable to the values found in salamandrids.

3.4 Eye Degeneration in Troglobitic Salamanders

Remarkable and well-studied examples of the transition to neotenic morphology are the troglobitic salamanders, especially with respect to eye morphology. All cave salamanders, except *Proteus anguinus*, belong to the family Plethodontidae. They are all neotenic species (except for *Typhlotriton spelaeus* which reaches sexual maturity only after metamorphosis; Brandon 1971), and all show a high degree of eye degeneration. Plethodontid cave salamanders are closely related to the genus *Eurycea* of which several species are known to rely heavily on visual cues in prey-catching behavior, although a few species of *Eurycea* are also cave dwellers with degenerated eyes.

The degeneration of the eyes affects the size and shape of the eye as a whole, and additionally in some species simplification of neural and optic systems occurs. The eyes of *Typhlotriton spelaeus* have been thoroughly investigated among cave salamanders, and they are of special interest, because this species possesses functional eyes during its larval period (Stone 1964, Besharse and Brandon 1974a, b). Behavioral experiments show that larval *T. spelaeus* respond to optomotor stimuli (Stone 1964, Besharse and Brandon 1972, 1974a, b). In larval *Typhlotriton spelaeus*, visually guided prey catching is reported (Besharse and Brandon 1974a).

Morphological investigations on larval eyes of these animals reveal that although the eyes are small in relation to their epigean relatives, they show characteristics of a functional eye found in larvae of epigean species. Degeneration or cessation of growth begins with metamorphosis, so that adults often have smaller eyes than larvae of the same size. Degeneration takes place at various points, but begins mostly with degeneration of photoreceptor outer segments and the pigmented choroid layer. The second step or the consequence of the first is the disappearance of the outer plexiform layer and often a thinning of the layer of retina ganglion cells which results in a thinning of the retina as a whole (Besharse and Brandon 1974a, b). As an obvious process of degeneration after metamorphosis, the eyelids begin to close and in many instances overlap the whole eye. It was shown by Besharse and Brandon (1976) that in animals which were forced to metamorphose, degeneration occurs in darkness as well as in light but seems to be facilitated by the former. In some cases these eyes remained somewhat functional because optomotor responses could be elicited (Stone 1964, Besharse and Brandon 1976). Stone (1964) reported also, that some adults behave as if blind, although the eyes have a histologically normal appearance. In fact, eye degeneration in the left and right eye of a single animal can be very asymmetrical (Stone 1964, Besharse and Brandon 1974a, b) so that optomotor behavior could be elicited by the right eye but not by the left eye, which was functionally blind (Stone 1964).

In other species of troglobitic salamanders, such as the proteid *Proteus anguinus*, eye degeneration is much stronger compared to *Typhlotriton*. In this species retinal and also optical structures are affected by degenerative processes (cf. Durand 1971).

4 Anatomy of the Visual System

4.1 General Anatomy of the Salamander Brain

The brain of salamanders, like that of all vertebrates, is divided into five major parts: (1) forebrain, or telencephalon; (2) diencephalon; (3) midbrain, or mesencephalon; (4) cerebellum; (5) hindbrain, or medulla oblongata (Fig. 4.1).

In the following, these five parts of the urodele brain will be presented briefly with respect to their gross morphology. In the second part of this chapter, details will be discussed only with respect to visual functions or to visually guided motor functions. Accordingly, this presentation does not

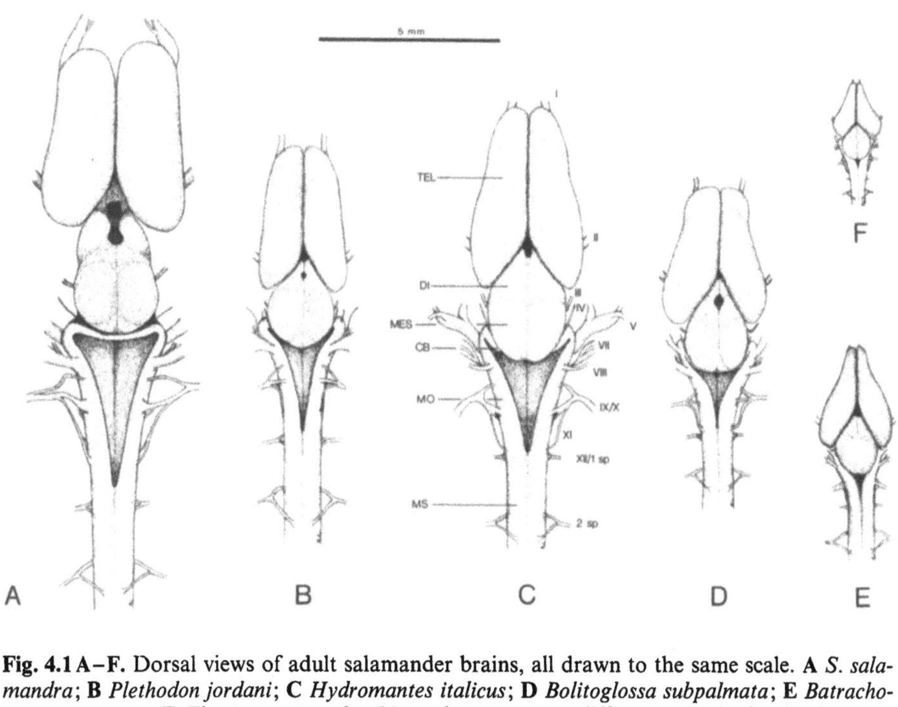

Fig. 4.1 A–F. Dorsal views of adult salamander brains, all drawn to the same scale. **A** *S. salamandra*; **B** *Plethodon jordani*; **C** *Hydromantes italicus*; **D** *Bolitoglossa subpalmata*; **E** *Batrachoseps attenuatus*; **F** *Thorius narisovalis*; Note the enormous differences in brain size between *Salamandra* and *Thorius*. For abbreviations, see list

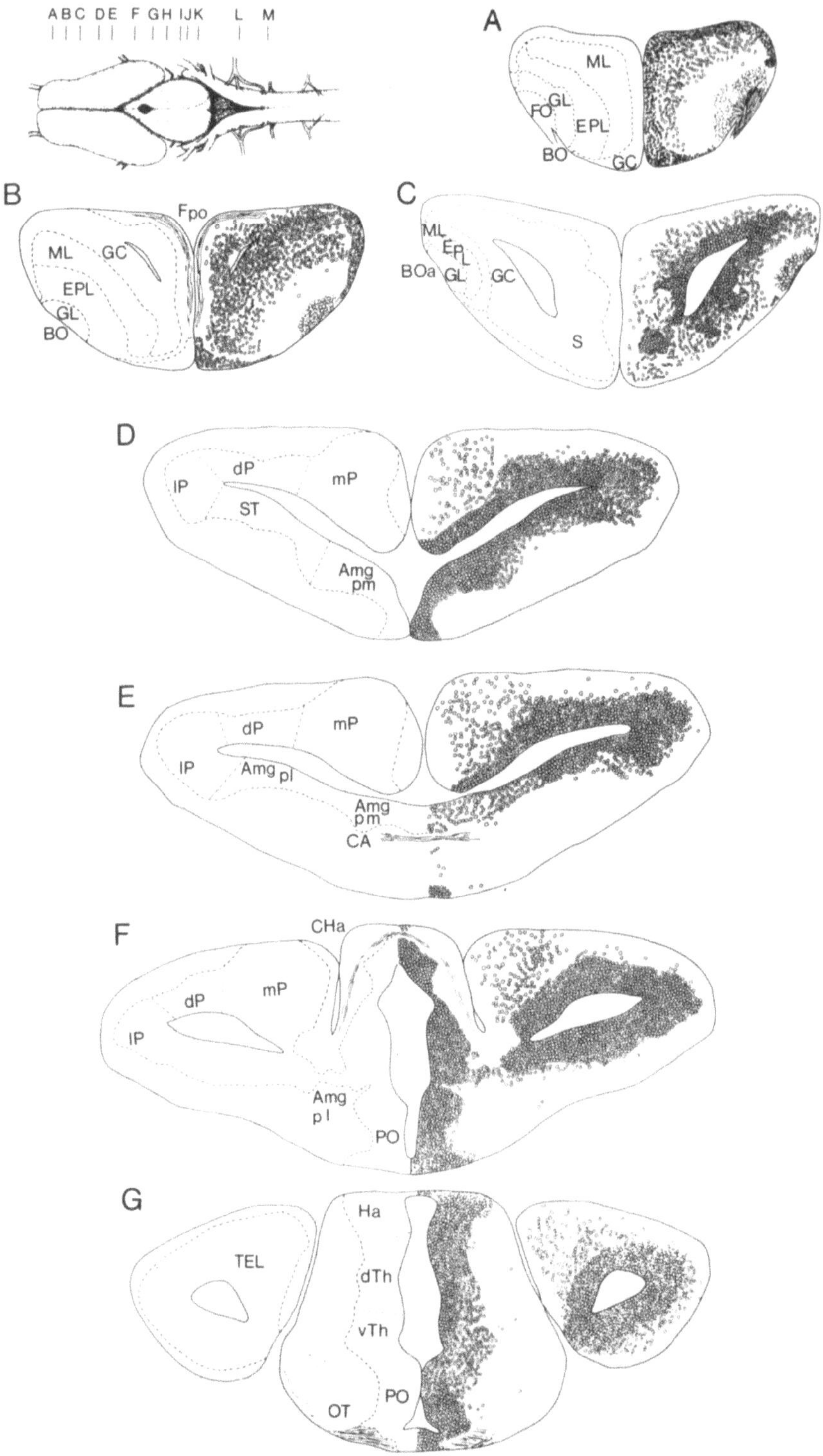
A B C D E F G H I J K L M
A
ML
FO
GL
EPL
BO
GC
B
Fpo
ML
GC
EPL
GL
BO
C
ML
EPL
BOa
GL
GC
S
D
dP
mP
lP
ST
Amg pm
E
dP
mP
lP
Amg pl
Amg pm
CA
F
CHa
dP
mP
lP
Amg pl
PO
G
Ha
TEL
dTh
vTh
PO
OT

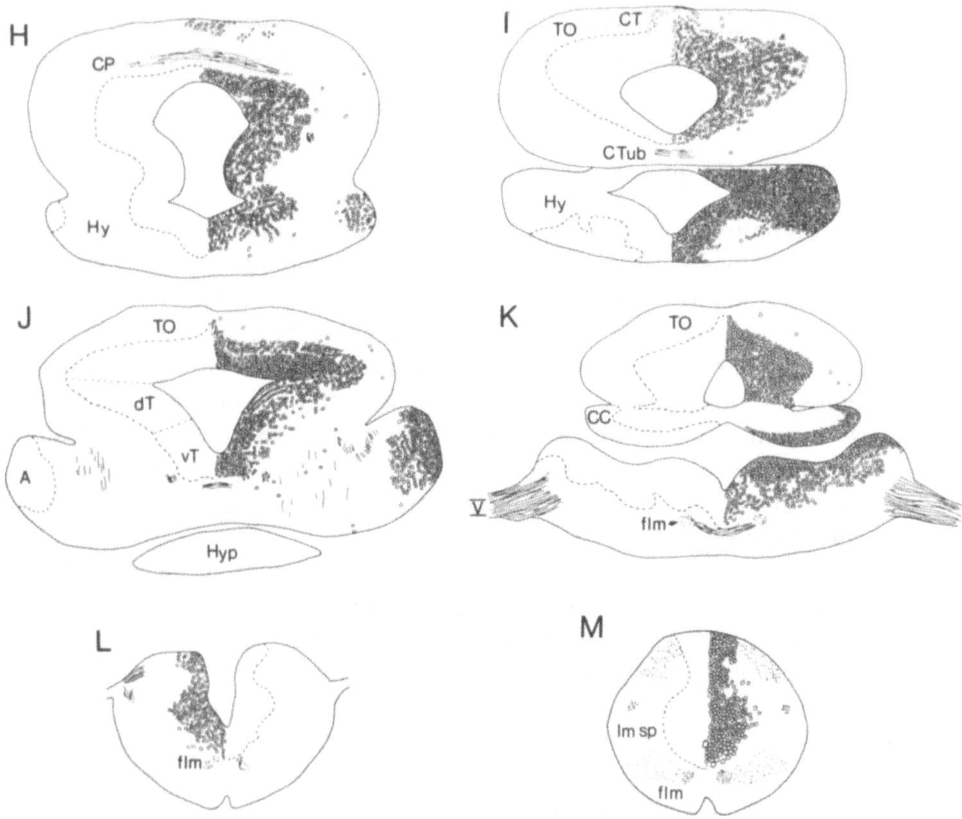

Fig. 4.2 A–M. Drawings of transverse sections through the brain of *Bolitoglossa subpalmata*. **A–B** Telencephalon at the level of the olfactory bulb; **C** telencephalon at the level of the accessory olfactory bulb; **D** mid-telencephalon; **E** section through the caudal telencephalon at the level of the commissura anterior; **F** caudal telencephalon/rostral diencephalon at the level of the commissura habenulae; **G** section through the mid-diencephalon; **H** praetectum/rostral mesencephalon at the level of the commissura posterior; **I** mid-mesencephalon/hypothalamus at the level of the commissura tuberculi posterioris; **J** mesencephalon and beginning medulla oblongata (auricles); **K** caudal mesencephalon/cerebellum/rostral medulla oblongata at the level of 5th nerve root; **L** mid-medulla oblongata at the entrance of 9th–11th nerve roots; **M** rostral cervical spinal cord

claim to be of any completeness with regard to the general anatomy of the urodele brain. For this, the reader is referred to the monograph of Herrick (1948).

4.1.1 Forebrain (Figs. 4.1 and 4.2A–F)

The forebrain of salamanders, like that of amphibians in general, consists of paired, evaginated cerebral hemispheres and an unevaginated portion, the telencephalon impar. Unlike the anuran brain in which the olfactory bulbs are clearly separated from the rostral telencephalon by a characteristic lace, boundaries cannot be recognized externally in the urodele brain. The olfactory bulbs are an integrated part of the two hemispheres. In anurans the two hemispheres themselves are clearly separated from each other representing two tubular formations. In urodeles the separation of the hemispheres is much less distinct or, as in many plethodontids, is indicated externally only by a shallow groove.

The cerebral hemispheres join the telencephalon impar at the level of the foramen interventriculare. Caudally the telencephalon is separated from the diencephalon by the large nucleus praeopticus which, however, is considered by some authors as being part of the diencephalon.

The forebrain is divided into the following structures: (1) olfactory bulbs; (2) pallial structures: (a) dorsal pallium, (b) lateral pallium, (c) medial pallium, (d) amygdala pars lateralis; (3) subpallial structures: (a) nuclei septi, (b) amygdala pars medialis, (c) striatum.

4.1.1.1 Olfactory Bulbs (Fig. 4.2A–C)

The olfactory bulbs form the anterior part of the hemispheres. They consist of five distinct layers: olfactory nerve fibers and glomerular layer as the most superficial layer; external plexiform layer composed of secondary olfactory fibers; external cellular layer containing small granule and larger mitral cells; internal granule layer; layer of ependymal cells as the innermost layer. The glomeruli of the first layer are spherical aggregates formed by olfactory nerve terminals interwoven with dendrites of mitral cells situated in the third layer. One mitral cell sends dendrites to several glomeruli. The axons of the mitral cells constitute the secondary olfactory pathways. The granule cells of the fourth layer possess no axons but send processes toward the glomerular layer where they arborize among the mitral cells. There are some fascicles running between the mitral and the granule cell layers. These fibers do not form a distinct layer. An inner plexiform layer can be distinguished only in a few species.

Three olfactory tracts, a medial, a lateral and a ventral one, leave the bulbs (Fig. 4.3A, B). In *Ambystoma tigrinum* (Royce and Northcutt 1969),

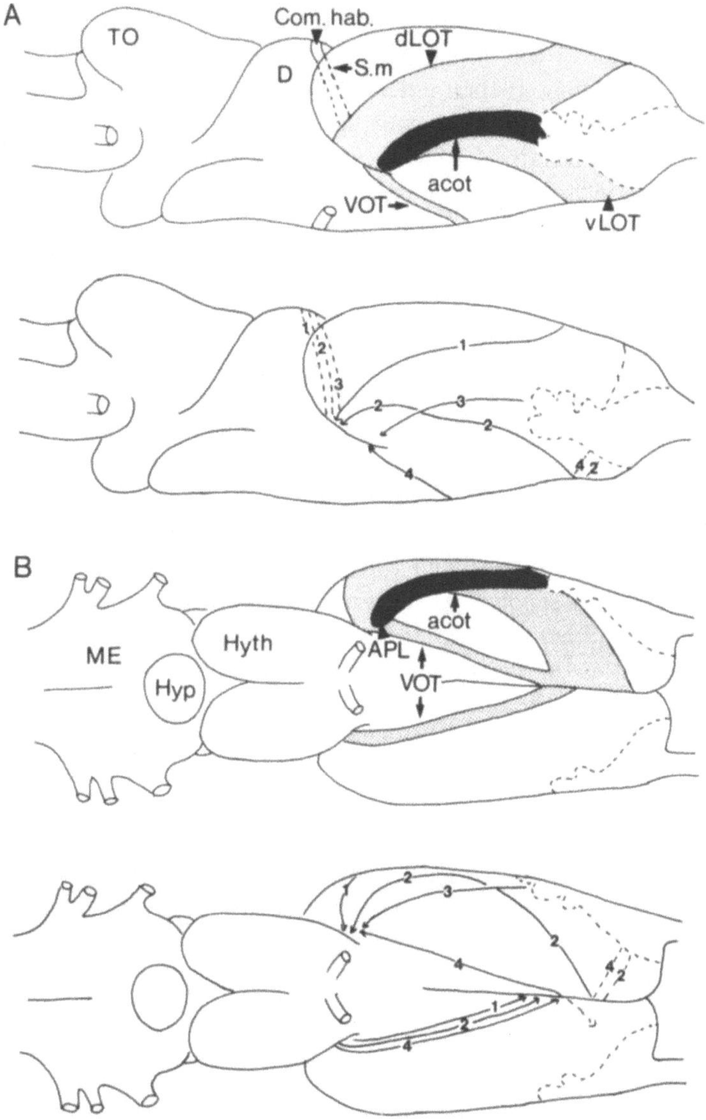

Fig. 4.3 A–B. Olfactory and accessory olfactory pathways in the forebrain of *S. salamandra*. **A** Lateral view; **B** ventral view of the brain. Olfactory pathways: *gray area* and *lines 1, 2* and *4*; accessory olfactory pathways: *black area* and line *3*. For abbreviations, see list. (From Schmidt and Roth in prep.)

Salamandra salamandra, and a variety of plethodontid salamanders (Schmidt and Roth, in prep.), the medial olfactory tract projects to the rostral ipsilateral medial pallium and the septal nuclei. The lateral tract divides into a dorsolateral and a ventrolateral portion that course around the accessory olfactory bulb. Behind the accessory olfactory bulb these two portions of the lateral olfactory tract merge. The lateral tract then gives off fibers to the ventral half of the dorsal pallium, the lateral pallium, and probably to the dorsal striatum. It then enters the stria medullaris, crosses the commissura habenulae and enters the contralateral stria medullaris. Some of these fibers join the contralateral ventral olfactory tract while others seem to terminate in the stria medullaris. Furthermore, there are fibers which project to the neuropil of the ipsi- and contralateral habenula (Schmidt and Roth in prep.). According to Northcutt and Kicliter (1980), the course of the medial and the dorsolateral olfactory tract closely resembles the situation found in *Rana*, whereas a tract comparable to the ventrolateral one in salamanders does not exist in ranids. The ventral tract separates from the medial tract and passes the ventral striatum and the pars medialis amygdalae. Behind the striatum the ventral tract and the lateral tract merge, and together they enter the stria medullaris, cross in the commissura habenulae, and join the corresponding contralateral ventral olfactory tract.

Herrick (1948) described an "anterior olfactory nucleus" as "a zone of relatively undifferentiated cells interpolated between the bulbar formation and the more specialized areas posteriorly of it", where part of the secondary olfactory fibers arise. After Northcutt and Kicliter (1980), this area has to be regarded as the caudal internal granule layer of the olfactory bulb.

Caudal to the olfactory bulbs, the accessory olfactory bulb is situated at the ventrolateral wall of the hemispheres. It receives fibers from the vomeronasal organ, which is considered to play an important role in courtship, partner recognition, and territoriality. In contrast to the statement of Herrick (1948), the accessory olfactory bulb is well developed in many urodeles (Schmidt et al. submitted). It shows the same laminar structure as the olfactory bulb. Its fibers project exclusively to the amygdala, pars lateralis (Northcutt and Kicliter 1980, Schmidt and Manteuffel 1984, Schmidt and Roth in prep.).

4.1.1.2 Pallial Structures (Fig. 4.2 D–F)

a) Medial Pallium. The medial pallium (= archipallium, Ariens Kappers et al. 1936, primordium hippocampi, Herrick 1933, 1948) represents the dorsomedial wall of the hemisphere. It is separated from the ventrally located septum by the zona limitans medialis. In plethodontid salamanders, there is no sharp boundary to the dorsal pallium except that in the medial pallium the cells show a strong increase in number and space and a pronounced cell migration. This cell migration could be interpreted as an incipient lamination of cells in the forebrain.

Inside the forebrain the medial pallium projects to the contralateral medial pallium, the ipsilateral and contralateral septum, the eminentia postolfactoria, and the bed nucleus of the pallial commissure. The medial pallium projects outside the telencephalon to the nucleus praeopticus, the ventral thalamus, and the hypothalamus via the medial forebrain bundle (Schmidt and Roth in prep.).

b) Dorsal Pallium. The dorsal pallium of Northcutt and Kicliter (= primordium pallii dorsalis, Herrick 1948) includes the pallial parts dorsal and caudal to the most rostral extent of the medial pallium. In salamanders it includes the more lateral field, too. The dorsal pallium is an important integration area: it receives both olfactory fibers from the olfactory bulb and afferents from the thalamus through the medial forebrain bundle. In addition to the striatum and the amygdala, it represents the third visual area within the telencephalon. It is separated from the lateral pallium by the sulcus rhinalis. The dorsal pallium projects to the medial pallium only and does not project outside the telencephalon.

c) Lateral Pallium. This is the "primordium piriforme" or "nucleus olfactorius dorsolateralis" of Herrick (1948). Northcutt and Kicliter (1980) distinguished a pars dorsalis and pars lateralis. The lateral pallium projects to the dorsal and medial pallium, the striatum, and the septum as well as to the contralateral lateral pallium. Its projections, however, do not leave the telencephalon except when crossing to the contralateral side of the telencephalon in the habenula (Schmidt and Manteuffel 1984).

d) Pars Lateralis Amygdalae. Herrick (1948) considered the amygdala as a single cell mass, as a caudal continuation of the striatum, and thus representing a subpallial cell group. Other authors like Källén (1951) and Northcutt and Kicliter (1980) recognized two distinct cellular nuclei: one, the pars lateralis, of pallial origin, and one, pars medialis, of subpallial origin. In salamanders the pars lateralis can be seen as a distinct lateral cellular prominence dorsal to the striatum. The amygdala pars lateralis receives afferents from the accessory olfactory bulb, from the hypothalamus via the medial forebrain bundle, and perhaps from the main sensory nucleus of the trigeminal nerve. It projects to the ventral thalamus, the hypothalamus and more caudally at least to the medulla oblongata.

4.1.1.3 Subpallium

a) Septal Nuclei (Fig. 4.2C). The septal complex occupies the ventromedial wall of the cerebral hemispheres below the medial pallium between the olfactory bulb (the anterior olfactory nucleus of Herrick) and the lamina terminalis. The septal complex is believed to receive direct olfactory input from the nervus terminalis (McKibben 1911), secondary fibers from the olfactory bulb through the medial olfactory tract, and fibers from the pallium and the hypothalamus. Within the forebrain it projects only to the contralateral septum. Its chief efferent pathway is the fasciculus medialis telencephali, or "medial forebrain bundle". Northcutt and Kicliter (1980) distinguished three areas: the rostral septal area, a dense cellular plate occupying the same topographic position as the postolfactory eminence of anurans; the lateral and the medial septal nuclei which are, however, only poorly developed in salamanders compared to anurans.

b) Pars Medialis Amygdalae. This subpallial part of the amygdala in salamanders, unlike that of anurans, probably receives olfactory input from the olfactory bulb through the ventral olfactory tract. It projects to the medial pallium, the dorsal and ventral thalamus, the hypothalamus and the spinal cord (Northcutt and Kicliter 1980, Finkenstädt et al.).

c) Striatum (Fig. 4.2D). The striatum occupies the ventrolateral wall of the telencephalic hemisphere. In urodeles it can be divided into a lateral striatum proper and a medially situated nucleus accumbens adjacent to the septal nuclei. The striatum occupies a position immediately caudal to the accessory olfactory bulb. The lateral striatum is divided by a striatal sulcus into a pars dorsalis and a pars ventralis. The striatum, like the amygdala, receives both olfactory and visual input and seems to be a major coordination center in the forebrain. Within the forebrain it projects only to the contralateral striatum. The chief efferent pathway is the fasciculus lateralis telencephali, or "lateral forebrain bundle", which contains strio-thalamic and strio-tegmental fibers. Herrick (1948) compares this pathway with the mammalian extrapyramidal system.

4.1.2 Diencephalon (Figs. 4.2G–I and 4.4)

Unlike the diencephalon of anurans that of salamanders consists mostly of a more or less continuous band of an outer fiber layer and an inner (periventricular) cell layer with no clear cell-free parts dividing the diencephalon into distinct nuclei. Instead, a subdivision of the salamander diencephalon has to be based on cues such as the position of sulci, cell density and cell diameter and is, thus, not totally free of arbitrary decisions. The following description partly differs from that of Herrick (1948) and Kuhlenbeck (1975, 1977) and is based on recent experimental, mostly HRP material (Wicht 1984, and our own unpublished data).

The diencephalon of salamanders, like that of amphibians in general, is divided into four parts: (1) epithalamus, (2) dorsal thalamus, (3) ventral thalamus, (4) hypothalamus.

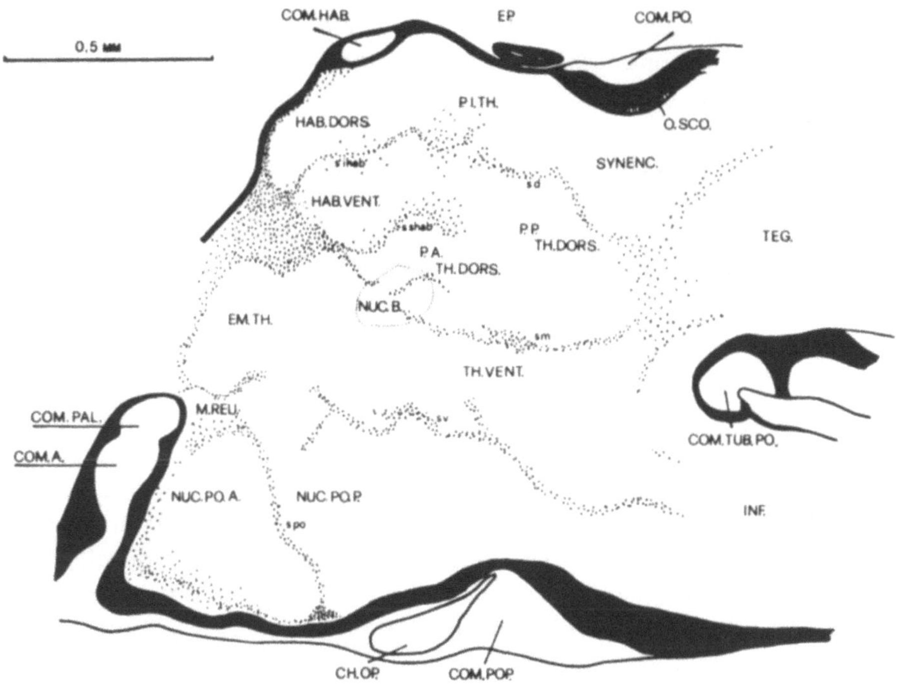

Fig. 4.4. Reconstruction of the structure of the diencephalon in *Triturus alpestris*. The diagram shows the distribution of the sulci (*stippled*) as natural boundaries of diencephalic nuclei. For abbreviations, see list. (From Wicht unpublished)

4.1.2.1 Epithalamus

The epithalamus represents the dorsal part of the diencephalon. Its anterior part is occupied by the habenular nuclei (nucleus dorsalis habenulae, n. ventralis habenulae) including the commissura habenulae. In the posterior direction, between the commissura habenulae and the commissura posterior, the pars intercalaris diencephali (or thalami) is situated. Rostrally, it is bound by the n. dorsalis habenulae and caudally by the n. praetectalis which descends around the commissura posterior (see below). A variety of authors consider the pars intercalaris thalami to belong to the dorsal thalamus.

4.1.2.2 Dorsal Thalamus

The dorsal thalamus is separated from the epithalamus by the sulcus dorsalis. Most authors divide the dorsal thalamus into two parts:

a) thalamus dorsalis pars anterior, which Herrick (1942) calls "nucleus Bellonci", because it is situated near the neuropil Bellonci, as an important termination field of visual fibers. However, the nucleus Bellonci, as determined here, forms the anteroventral part of the dorsal thalamus.

b) thalamus dorsalis pars posterior or caudalis. This caudal part of the dorsal thalamus is considered by several authors (e.g. Finkenstädt et al. 1983) as the nucleus praetectalis or as "rostral praetectum". Naujoks-Manteuffel and Manteuffel (1986), however, according to recent experimental data, considered the nucleus praetectalis proper as being the zone around the commissura posterior and caudal of the pars intercalaris thalami. The thalamus dorsalis pars caudalis receives afferents from the tectum and projects ipsilaterally to the striatum (Wicht and Himstedt 1986).

4.1.2.3 Ventral Thalamus

The ventral thalamus is separated from the dorsal thalamus by the sulcus medialis, and from the hypothalamus by the sulcus ventralis. It is subdivided by Herrick (1942) into an anterior and a posterior sector, separated by a wide and shallow depression. The anterior part, the "eminentia thalami" is the bed nucleus of the "hippocampal" or pallial commissure and of the olfacto-habenular tract. The posterior section of the ventral thalamus is continuous with the ventral tegmentum ("peduncle" sensu Herrick 1948).

4.1.2.4 Hypothalamus and Nucleus Praeopticus

The hypothalamus is divided by the sulcus hypothalamicus into a dorsal and ventral part. The dorsal part is connected with the ventral thalamus and the rostral ventral tegmentum, the ventral part with the nucleus praeopticus. This nucleus is unusually large in amphibians. It is bound anteriorly by the fibers of the fasciculus medialis telencephali partially crossing within the commissura anterior, and posteriorly it merges with the hypothalamus through the chiasma opticum. It is morphologically assigned to the forebrain, but its connections and its functions connect it more closely to the hypothalamus. It is divided into an anterior and a posterior part by the sulcus praeopticus. This nucleus is one of the endstations of the nervus terminalis and of all components of the medial forebrain bundle.

4.1.2.5 Chiasma Ridge

The chiasma ridge represents a transverse deviation of the ventricular floor of the diencephalon between the nucleus praeopticus and the ventral hypothalamus. Anteriorly, it contains the optic fibers decussating here in four compact bundles, together with the noncrossing fibers. The caudal part consists in the commissura postoptica, where tecto-tectal, tecto-isthmic and tecto-thalamic fibers cross.

4.1.3 Praetectum (Figs. 4.2 H and 4.5)

On the basis of recent neuroanatomical and neurophysiological evidence, Naujoks-Manteuffel and Manteuffel (1986) considered the praetectum as a region of its own. It partly overlaps with the anterior mesencephalon, and accordingly, Kuhlenbeck (1975) has used the term "synencephalon" for this region.

The praetectum extends around the fibers of the commissura posterior rostrally adjacent to the pars intercalaris thalami and descends ventrocaudally. Dorsally, it consists of the nucleus praetectalis, which is divided into the n. praetectalis profundus, situated within the periventricular gray matter, and the n. praetectalis superficialis, embedded within the white matter and, thus, forming one of the few migrated nuclei in the salamander brain. Ventrally and more caudally, the n. Darkschewitsch is found, separated from the n. praetectalis by the sulcus synencephalicus (sensu Kuhlenbeck) and more caudally by the sulcus lateralis mesencephali. Ventrally, the praetectum extends to the level at which the pars ventralis thalami and the mesencephalic tegmentum merge.

Comparison with the Anuran Diencephalon. In contrast to salamanders the anuran diencephalon shows a substantial morphological differentiation of the periventricular gray matter into separate nuclei and an incipient lamination within these nuclei. The most recent and most complete anatomical description of the anuran diencephalon was given by Neary and Northcutt (1983) in the bullfrog *Rana catesbeiana*. Here, the epithalamus shows the large habenular complex, divided into dorsal and ventral habenular nuclei (Scalia and Fite 1974).

In the dorsal thalamus of anurans, three periventricular nuclei and one migrated nucleus can be distinguished. In the periventricular zone the anterior, central, and posterior nucleus are found of which the central and the posterior nucleus show compact lamination. The nucleus migrated from the gray into the white substance is the lateral nucleus which is further divided into an anterior, posterodorsal, and posteroventral portion.

The ventral thalamus also shows both periventricular and migrated nuclei. Periventricularly, the ventromedial nucleus is found, which is replaced rostrally by the thalamic eminence, and caudally by the tuberculum.

The migrated nuclei are the ventrolateral nucleus, the nucleus Bellonci, situated immediately below the corresponding neuropil, and the superficial nucleus situated ventromedial of the corpus (= neuropil) geniculatum thalamicum.

The praetectum as transitional zone between caudal thalamus and optic tectum consists of three nuclei: the nucleus lentiformis mesencephali which is situated laterally to the lateral nucleus of the dorsal thalamus in a vertical orientation; the nucleus of the commissura posterior capping this commissure; and the pretectal gray dorsal of the commissura posterior as a medial continuation of tectal laminae 2 and 4.

The hypothalamus is divided into a preoptic and an infundibular part. The preoptic part is further divided into an anterior and posterior part; the latter contains the magnocellular preoptic nucleus and the suprachiasmatic nucleus. The infundibular hypothalamus possesses two periventricular nuclei, the dorsal and the ventral hypothalamic nucleus.

Compared to the diencephalon of *Rana catesbeiana*, that of the leopard frog, *Rana pipiens* seems to be less differentiated. According to Neary (1975) and Wilczynski and Northcutt (1977), the dorsal thalamus is again divided into an anterior, central, and posterior nucleus and a migrated lateral nucleus, in which the pretectal areas, the posteromedial and posterolateral pretectal nuclei of Trachtenberg and Ingle (1974) are included in the posterior and lateral nucleus.

4.1.4 Midbrain (Mesencephalon; Figs. 4.2I–K and 4.5)

The midbrain of urodeles includes:

1. The tectum mesencephali

After Herrick (1948) the tectum mesencephali consists of a large anterior part, the tectum opticum, and a small posterior, nonoptic part, the "nucleus posterior tecti". Herrick sees in both parts of the tectum mesencephali primordia of the mammalian superior and inferior colliculi, respectively. In larval and neotenic urodeles the optic fibers do not cover the whole tectum mesencephali, whereas in all adult salamanders studied so far, the tectum is nearly or completely covered by optic fiber terminals, though differences exist between the tecta of different salamander groups. The adequacy of Herrick's terminology may thus be questioned (Rettig et al. 1981).

While the anuran tectum shows a very conspicuous multilayered structure, in the salamander tectum the sharp separation of the inner layer of cell

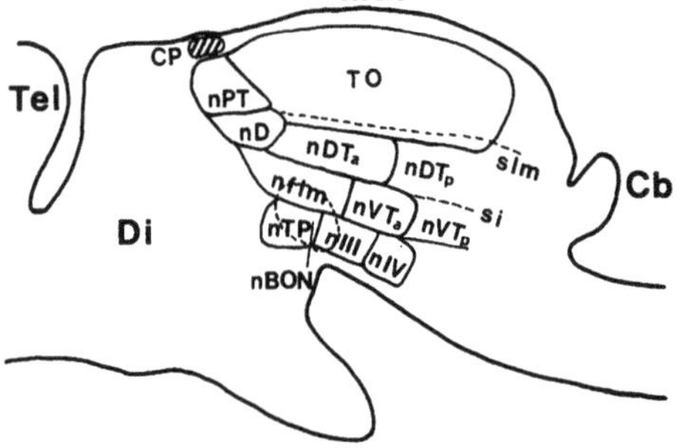

Fig. 4.5. The distribution of pretectal and mesencephalic nuclei in the brain of *S. salamandra*. For abbreviations, see list. (From C. Manteuffel 1985)

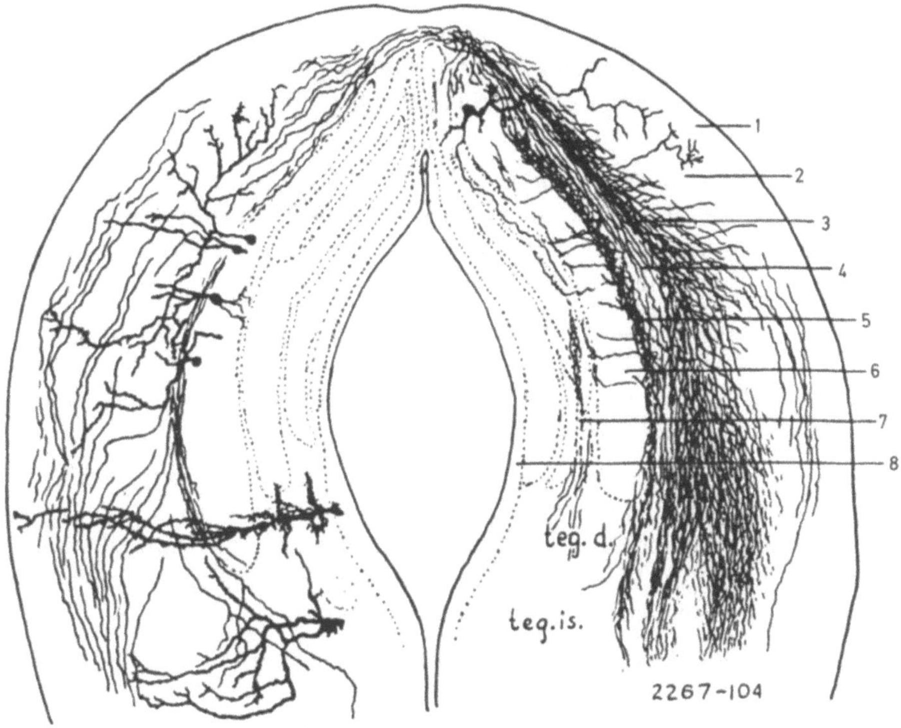

Fig. 4.6. Lamination of the tectum opticum of *Ambystoma tigrinum* according to Herrick. *Numbers* refer to the laminae described in the text. *teg.d.* Tegmentum dorsale; *teg.is.* tegmentum isthmi. (From Herrick 1942)

somata and the outer layer of fibers and dendrites is the only well-defined lamination. Only a few cells can be found dispersed in the outer layer. These cells, however, do not seem to be distributed in a completely random fashion; they are often arranged along a line immediately above the layer of commissural fibers (layer 4 of Herrick). Whether this formation indicates an incipient stratification of the urodele optic tectum, is still unclear.

Herrick (1942), on the basis of Golgi material, divided the optic tectum of *Ambystoma tigrinum* into eight layers which, however, do not correspond to the layers described in the anuran tectum (Fig. 4.6):

Layer 1: Superficial Layer. At the anterior end of the tectum this layer is completely constituted by visual afferent fibers. Most of them turn inwards and arborize in layer 2, but some of them arborize immediately below the tectal surface.

Layer 2: Intermediate Neuropil Layer. This layer contains a wide zone of dense neuropil and represents the most important synaptic field of the tectum. It receives terminals from the optic tracts, the lemniscus (somatosensory afferents), isthmo-tectal and thalamo-tectal tracts. The anterior part is closely connected with the pretectal neuropil, the posterior part with the nucleus posterior tecti through the brachium of the colliculus inferior. Primarily, this layer is the place of intrinsic dendritic interaction among tectal cells.

Layer 3: Layer of Intermediate Fibers. This layer is well separated from layer 2; it is a rather narrow zone of densely packed, chiefly unmyelinated fibers which are mostly efferent. They leave the tectum as tecto-thalamic, tecto-tegmental, and tecto-bulbar tracts.

Layer 4: Commissural Fiber Layer. This layer contains myelinated fibers from the commissura tecti mesencephali, the thick fibers of the nucleus mesencephalicus nervi trigemini, and several kinds of unmyelinated fibers. The commissural fibers are mostly dorsal.

Layer 5: Deep Layer of the Alba. This layer represents a narrow zone of unmyelinated fibers and neuropil immediately above the layer of cell bodies. Its fibers seem to connect parts of the tectum and extend into the epithalamus and dorsal thalamus. Ventrally and posteriorly, the fibers continue into the tractus tecto-peduncularis et tegmentalis profundus.

Layer 6: Outer Gray Layer. The cells of this layer are closely packed and arranged rather uniformly throughout the tectum. The cells are somewhat larger than those of the deeper layers, especially in the ventral half of the layer. The ventral border of this layer is marked by a sudden change to a looser arrangement of the cells of the dorsal tegmentum.

Layer 7: Intermediate Layer of the Gray. The cells of this layer are more loosely and irregularly arranged in lentiform masses which are separated by sheets of dense neuropil.

Layer 8: Deep Layer of Ependymal Cells. This layer contains one to two layers of ependymal cells which send their processes radially to the surface of the tectum.

On the basis of recent experimental Golgi and HRP material further distinctions of the optic tectum are made which are presented below.

The tectum contains conspicuously large cells belonging to the mesencephalic nucleus of the trigeminus. These cells are irregularly distributed in all parts of the tectum and in all layers of the gray substance. Another conspicuous structure of the tectum is the dorsal thickening which receives terminals from the medial optic tract and from the pretectal nucleus.

2. The tegmentum dorsale

The dorsal tegmentum is distinguished from the tectum by a change in cellular texture: the cells of the dorsal tegmentum with the n. dorsalis tegmenti pars anterior and pars posterior are more loosely and irregularly arranged. Ventrally, the dorsal tegmentum is separated from the ventral tegmentum by the "limiting sulcus of the nucleus tuberculi posterioris" of Herrick and posteriorly from the isthmic tegmentum by the sulcus isthmi.

3. The tegmentum ventrale

The ventral tegmentum or "peduncle region" of Herrick is a conglomerate of several nuclei. Since salamanders do not possess a structure equivalent to the peduncle of higher vertebrates, the notion "peduncle" should be avoided here (cf. C. Manteuffel 1985). Below Herrick's "limiting sulcus", the nucleus fasciculi longitudinalis medialis is found anteriorly with a primordial nucleus ruber ventral and caudal to it, and the nucleus ventralis tegmenti posteriorly (Naujoks-Manteuffel et al. 1986b). Ventral to these nuclei, in a rostrocaudal sequence, the nucleus tuberculi posterioris, the nucleus of the oculomotor nerve (third cranial nerve), the nucleus of the trochlear nerve (fourth cranial nerve) and the nucleus interpeduncularis are found. The latter two nuclei are considered by Herrick as belonging to the isthmic tegmentum. Lateral to the n. fasciculi longitudinalis and the n. tuberculi posterioris, the n. opticus tegmenti is found, which, together with the n. praetectalis superficialis, is another migrated nucleus in the salamander brain.

The ventral tegmentum contains the chief nuclei of origin of strong descending tracts to the motor fields of the medulla oblongata and the spinal cord (cf. Naujoks-Manteuffel et al. 1986a).

4. The tegmentum isthmi

This region is separated from the dorsal tegmentum by the sulcus isthmi. Ventrally, it contains the nucleus of the trochlear nerve which, according to the nomenclature used here, belongs to the ventral tegmentum, and a mass of cells, the tegmentum isthmi. At the dorsocaudal end of the dorsal tegmentum, immediately rostral to the corpus cerebelli, the nucleus isthmi is found, and immediately dorsal to it the nucleus visceralis secundarius (Rettig 1984 and in prep.). The presence of a differentiated nucleus isthmi in salamanders as known from other vertebrates was long denied. However, recent HRP studies clearly show its existence (see below).

Between the whole length of the isthmic tegmentum, the large nucleus interpeduncularis is situated. It receives mostly olfactory inputs via the fasciculus retroflexus which descends obliquely from the habenulae of the epithalamus. The terminals of the fasciculus retroflexus show a very conspicuous form of a flattened spiral.

4.1.5 Cerebellum (Fig. 4.2K)

In salamanders the cerebellum is small and simply organized. It consists of three parts: (1) the median corpus cerebelli; (2) the auricula cerebelli, which after Larsell (1967) are enlargements of the sensory zones of the medulla oblongata; (3) a nucleus cerebelli, situated ventral to the corpus cerebelli and which, after Herrick, is the primordium of the deep cerebellar nuclei of mammals.

In ambystomatids as well as many other salamander groups the cerebellum is located caudal to the mesencephalon. In most plethodontids, due to the forward movement of the brain stem, it comes to lie completely under the caudal portion of the mesencephalon.

4.1.6 Medulla Oblongata and Cervical Spinal Cord (Figs. 4.2L–M, 4.7 and 4.8A, B)

The medulla oblongata, or "bulb", includes the brain between the isthmus and the obex, the latter being the posterior end of the fourth ventricle, except the cerebellum. There is no pons in the amphibian brain. The rostrolateral wall and the rostral part of the bottom constitute the auricula cerebelli. The position of the auricles at the anterior pole of the rhombencephalon varies greatly among urodeles. Whereas in Ambystomatidae and Salamandridae the auricles are situated at the level of or even posterior to the caudal edge of the midbrain, in most plethodontid salamanders they lie far forward beside the midbrain (cf. Fig. 4.1). In some cases, the rostral pole of the auricles lies at the level of the posterior commissure between diencephalon and tectum, and almost comes into contact with the caudal pole of the cerebral hemispheres which extend far backward (e.g., in the species *Batrachoseps* and *Thorius*). So, the diencephalon and mesencephalon are almost surrounded by the auricles and the cerebral hemispheres.

The medulla oblongata is the area of termination and origin of the cranial nerves V to XII (cf. Figs. 4.7, 4.8, and 4.9). Its dorsal part, the alar plate of His, receives all sensory fibers from the head (except the optic and olfactory fibers), fibers from the lateral-line system, when present, and general visceral sensory and gustatory fibers. In the ventral part the motor nuclei of nerves V to X, and partly that of XII, are situated. This motor system is the coordination center of head and neck motor function including mouth and tongue movements involved in feeding.

The fifth nerve, trigeminus, leaves the medulla oblongata at or near its rostral pole, in ambystomatids and salamandrids at the level of the cerebellum, in plethodontids at the level of the rostral or mid mesencephalon. The seventh (facialis) and eighth (stato-acusticus) nerves leave the medulla oblongata together in a ventral and a medial position, respectively, somewhat behind the trigeminal root, in plethodontids at the level of the mid mesencephalon.

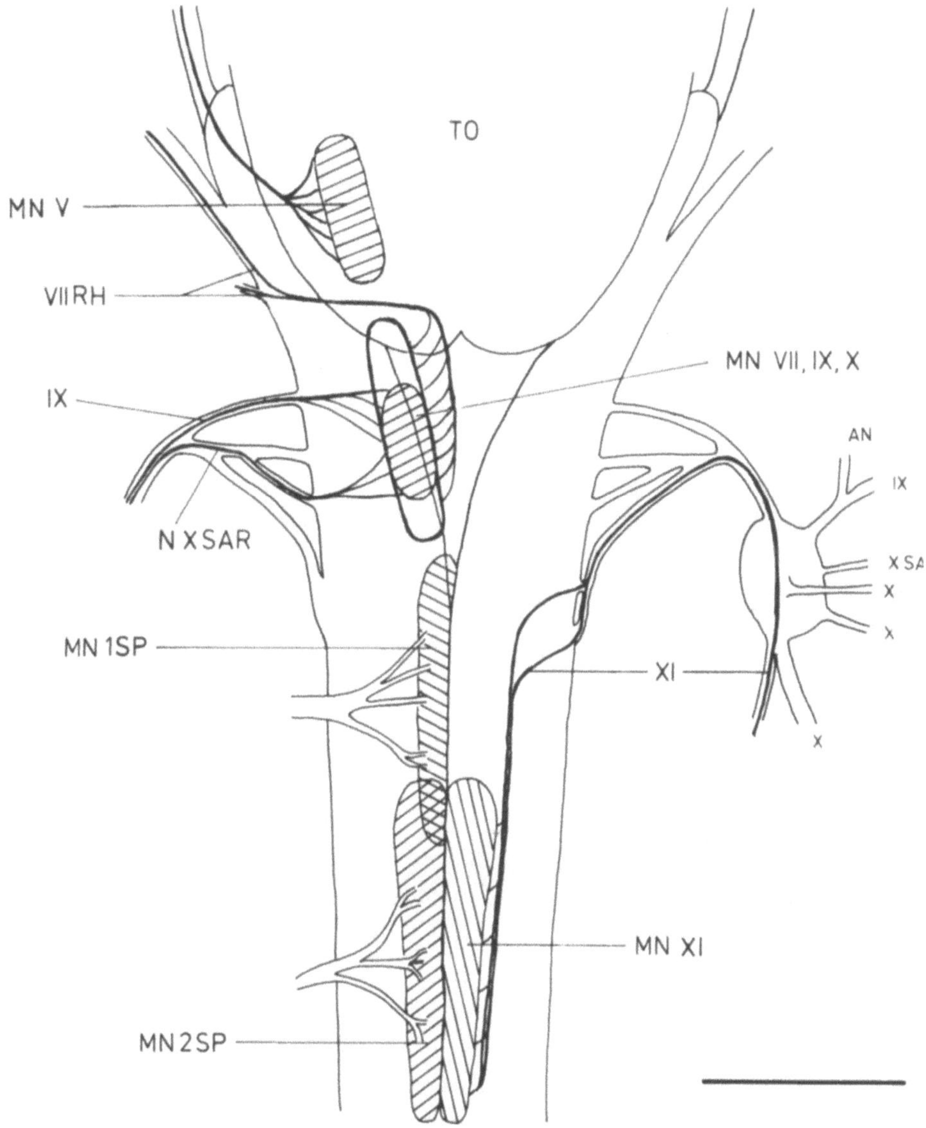

Fig. 4.7. Distribution of motor nuclei related to feeding in the brainstem and cervical spinal cord of *Batrachoseps attenuatus*. For further explanation see text. For abbreviations, see list. *Bar* = 500 μm

The next three cranial nerves, the ninth (glossopharyngeus), the tenth (vagus), and the eleventh (accessorius vagi or spinalis), in adult salamanders possess a common ganglion from which four main branches enter the brain. Additionally, there is a rather large anastomosis between the seventh and the ninth nerve distal to this ganglion through which sensory fibers pass from the ramus hyomandibularis of the seventh nerve entering the ganglion (Roth and

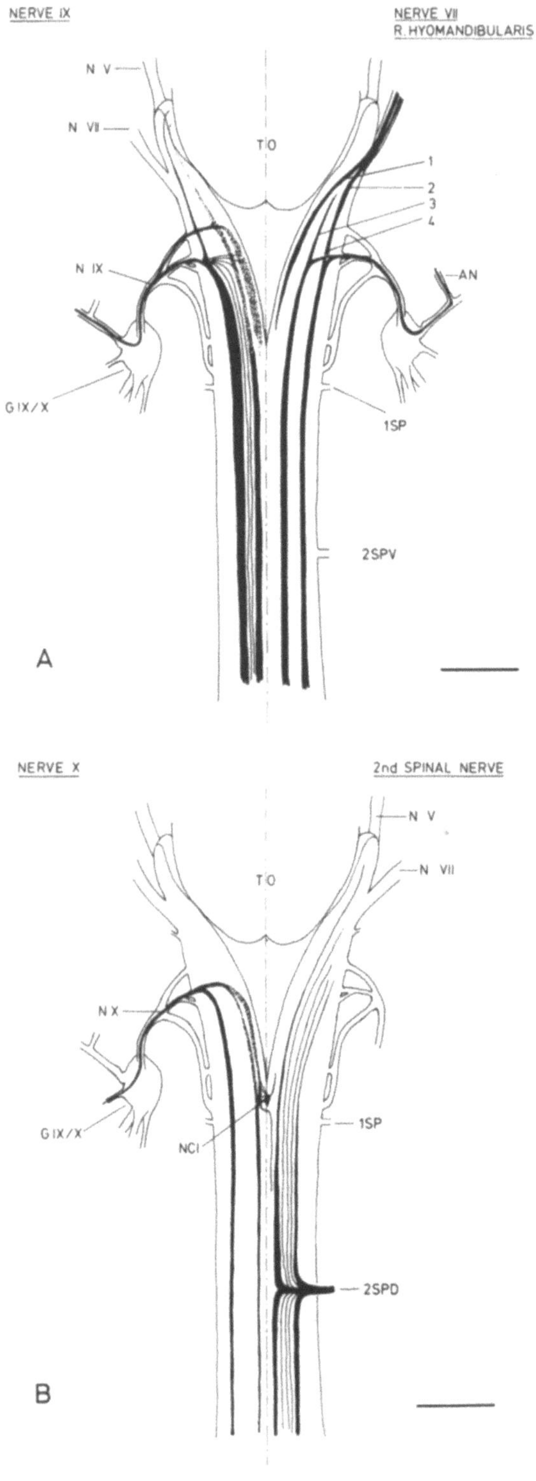

Fig. 4.8 A–B. Dorsal sensory projections of nerves related to feeding in plethodontid salamanders. **A** Nerve IX (*left*) and nerve VII ramus hyomandibularis (*right*). **B** Nerve X (*left*) and second spinal nerve (*right*). *Stippled bands*: Fasciculus solitarius (visceral sensory); *solid bands*: superficial (somatosensory) tracts. *Numbers* in **A** indicate different sensory pathways of r. hyomandibularis. Bar = 500 μm. (From Roth and Wake 1985 b)

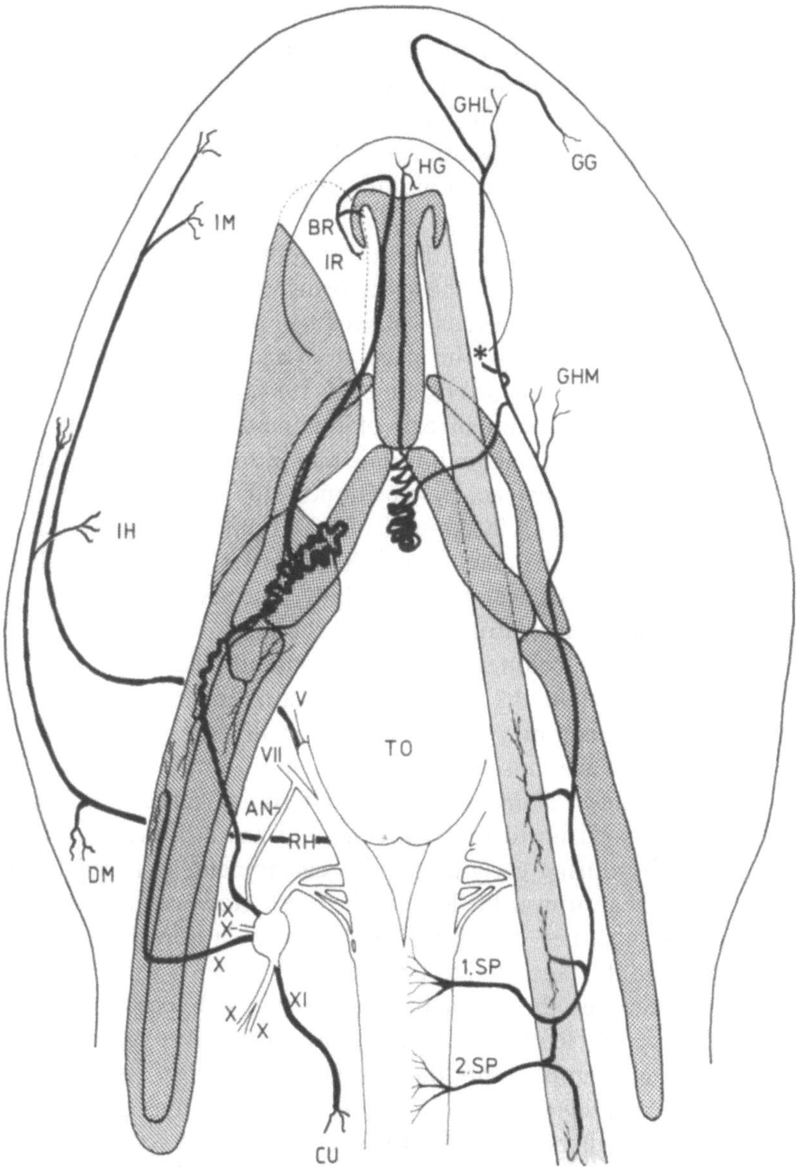

Fig. 4.9. Pattern of innervation of the tongue and associated musculature in *Batrachoseps attenuatus*. The figure is composed of ventral and dorsal views of the hyobranchial apparatus and the brain. It shows the lower jaw with the tongue skeleton (*stippled area*), the tongue protractor SAR (*hatched area*) and the tongue retractor RCP (*parallel-lined band*), the course of the nerves supplying SAR (*left side*) and RCP (*right side*), and the outlines of the tongue pad and the brainstem and cervical spinal cord. For abbreviations, see list

Wake 1985b). The first large root of this complex belongs exclusively to the ninth, glossopharyngeus, nerve. It has dorsal, visceral sensory, and ventral, visceral motor, components. The second root of this complex is as large as the first and belongs to the ninth and tenth (vagus) and partly to the r. hyomandibularis of the seventh nerve. It is a somatosensory root. The third, smaller root lies in close proximity to the second. It contains motor fibers of the vagus.

The fourth root, often with several rootlets, of this complex is either occupied by motor fibers of the accessorius vagi or "bulbar accessorius", as part of the vagus system, or, as – at least – in plethodontids and ambystomatids, by the motor fibers of the eleventh or accessorius spinalis which has its motor cells in the spinal cord (Roth et al. 1984, Roth and Wake 1985b, Wake et al. in prep.). The point of exit of this nerve varies greatly. Whereas in Ambystomatidae and Salamandridae, and in some plethodontid species like *Desmognathus* and *Eurycea*, it is positioned only slightly caudal to the second and third root of the IX/X complex, in the Bolitoglossini and Plethodontini of plethodontid salamanders it exits far posteriorly in the medulla oblongata at the obex region, or even more caudally. The seventh nerve has a separate motor root situated ventrally between the main seventh root and the first ninth root, which belongs to the r. hyomandibularis of the facial nerve. Between the level of the caudal facial roots and the rostral roots of the ninth-tenth root complex, and situated medially, the roots of the sixth nerve, abducens (third eye muscle nerve) are found.

In most adult salamanders, the first spinal nerve is represented by a ventral root, with up to four rootlets, and has a dorsal root and a dorsal ganglion only during early stages of development. It is strictly motor and exits the brain at or near the level of the obex. However, at least in *Dicamptodon ensatus*, a rather "primitive" salamander, a dorsal ganglion is present even in the adult stage (Wake et al. in prep.). There is still a debate about the homology between the first spinal nerve and the hypoglossal nerve (twelfth cranial nerve, cf. Roth et al. 1984, Roth and Wake 1985b, Wake et al. in prep.). In most non-plethodontid salamanders, e.g. *Ambystoma* (Herrick 1948) and *Salamandra* (Francis 1934) the ramus hypoglossus seems to be constituted by fibers from both the first and second spinal nerves which are connected by an anastomosis, although Norris (1908) claimed that in *Amphiuma means* this anastomosis is absent and the ramus hypoglossus is constituted by the first spinal nerve alone. In most plethodontids, especially in the Bolitoglossini, despite the presence of an anastomosis between the first and second spinal nerves, the ramus hypoglossus is clearly constituted by the first spinal alone. According to this fact and because it is not a typical spinal nerve, it should be considered in these salamanders to be homologous to the hypoglossal nerve of amniotes (Roth et al. 1984, Wake et al. in prep.). Thus, in most salamanders the first genuine spinal nerve, therefore, possessing both dorsal and ventral roots and a dorsal ganglion, is the "second" spinal nerve.

The dorsal sensory fiber system of the medulla oblongata is divided into two subsystems (Fig. 4.8). The first is a superficial one, constituted by the system of the tractus descendens nervi trigemini, to which all somatosensory roots contribute with both ascending and, more massively, with long descending fibers. The second is the system of the fasciculus solitarius situated somewhat below the surface and embedded in the dorsal gray matter. It extends from the rostral pole of the medulla oblongata to the obex. In it the fibers of special viscerosensory function of nerves V, VII, IX and X, are collected with ascending and descending fibers.

In aquatic larvae and in permanently or temporarily aquatic adult salamanders the lateral line nerve fibers enter the brain together with the eighth and ninth nerve, then bifurcate into ascending and descending branches consisting of two parallel bundles. The ascending branch extends to the cerebellar region, whereas the descending ones extend almost to the obex. In the adjacent dorsal gray matter of the medulla oblongata the area dorsalis octavo-lateralis is situated. It consists of three longitudinal cell masses. The dorsalmost is the n. dorsalis areae octavo-lateralis (Opdam and Nieuwenhuys 1976, Fritzsch et al. 1983). It extends from the entrance of the fifth nerve to the entrance of the ninth nerve or is shorter. It receives afferents predominantly from the ampullary electroreceptive organs (Fritzsch 1981 a). Below this nucleus the n. intermedius areae octavo-lateralis is situated. It extends further caudally, to the entrance of the tenth nerve or to the obex. It receives afferents mostly from mechanoreceptive neuromasts (Fritzsch 1981 b). The most ventrally situated cell mass is the n. ventralis areae octavo-lateralis. It extends as caudal as the n. intermedius or is shorter. It receives afferents from the inner ear (McCormick 1982).

Immediately caudomedial to the obex, in a very superficial position, the nucleus commissuralis Cajal is situated, in which many viscerosensory ascending and descending fibers decussate contralaterally.

The basal plate of the medulla oblongata is constituted by a medial somatomotor area and a lateral visceromotor area, divided by the sulcus intermedius ventralis. The ventral motor area comprises the rostral part of the spinal motor column and the nucleus of the sixth (abducens) nerve, the rhombencephalic part of the medial reticular formation and part of the nucleus interpeduncularis.

In the lateral motor area the nuclei of the visceromotor components of nerves V, VII, IX and X are found (Fig. 4.7). According to Opdam and Nieuwenhuys (1976), in *Ambystoma* these nuclei constitute an almost continuous column. The motor nucleus of the IX and X nerves, glossopharyngeus und vagus, in that species constitute together a single cell mass, situated in the lateral part of the basal plate and extending to the level of emergence of the abducens (VI) roots into the caudalmost part of the rhombencephalon. In plethodontids, the situation is somewhat different: the glossopharyngeus and the portion of the vagus that innervates the tongue-protractor muscle, as the entire glossopharyngeus does, have a more or less identical, oval-shaped nucleus that extends from the level of the first to the level of the second root of the IX/X complex (or slightly beyond it). The whole motor nucleus of the vagus, including parasympathetic functions, includes this tongue nucleus as its anterior portion, but extends further caudally to the level of the obex (Roth and Wake, 1985b).

In all salamanders, the motor nucleus of the facial nerve extends from the level of the motor root of the seventh nerve (or slightly beyond it) to the second root of the IX/X complex or to the obex. It completely overlaps the motor nucleus of the ninth and tenth nerves innervating the m. subarcualis rectus (Roth et al. in prep.) nucleus, and innervates the mm. depressor mandibulae (mouth opening) and subhyoideus (floor of the mouth). The fibers leave the motor nucleus in a conspicuous fashion forming the well-known "facialis knee": first, they extend medially almost to the midline of the brain, then run forward, curve around the medially situated nucleus of the sixth nerve (abducens), run laterally and exit through the motor root in the vicinity of the main facialis root.

The motor nucleus of the trigeminal nerve extends from the level of the root of the trigeminus to the mediolateral fibers of the facialis knee, i.e., at the level or shortly beyond the main facialis root. It never overlaps with the motor nucleus of the seventh nerve (Roth et al. in prep.). It innervates the m. levator mandibulae (mouth closing) and the m. intermandibularis which constitutes, together with the m. subhyoideus, the floor of the mouth (Roth et al. in prep.).

The position of the nucleus of the first spinal, or hypoglossal, nerve varies greatly among urodeles. In bolitoglossine plethodontids it is a rather narrow band of pear-shaped and medially situated neurons, with a striking absence of the lateral motor column (Fig. 4.10B). It extends rostrally almost to the level of the third root of the ninth-tenth root complex, and caudally to the midpoint between the ventral roots of the first and second spinal nerves. In other plethodontids as well as in all other salamanders studied so far it is shorter in its rostro-caudal extension but broader (Roth and Wake 1985b, Wake et al. in prep.). Here, as in all lower vertebrates studied so far, it consists of two rows of cells, a medial row of small, pear-shaped cells, constituting the medial motor column, and a lateral row of large, spindle-shaped cells, constituting the lateral motor column (Fig. 4.10E). The nucleus of the first spinal nerve innervates the tongue-retractor muscle m. rectus cervicis profundus as well as the m. genioglossus (if present), the m. geniohyoideus medialis and lateralis, the m. suprapeduncularis (all involved in the control of the direction of tongue projection), and some tongue pad muscles.

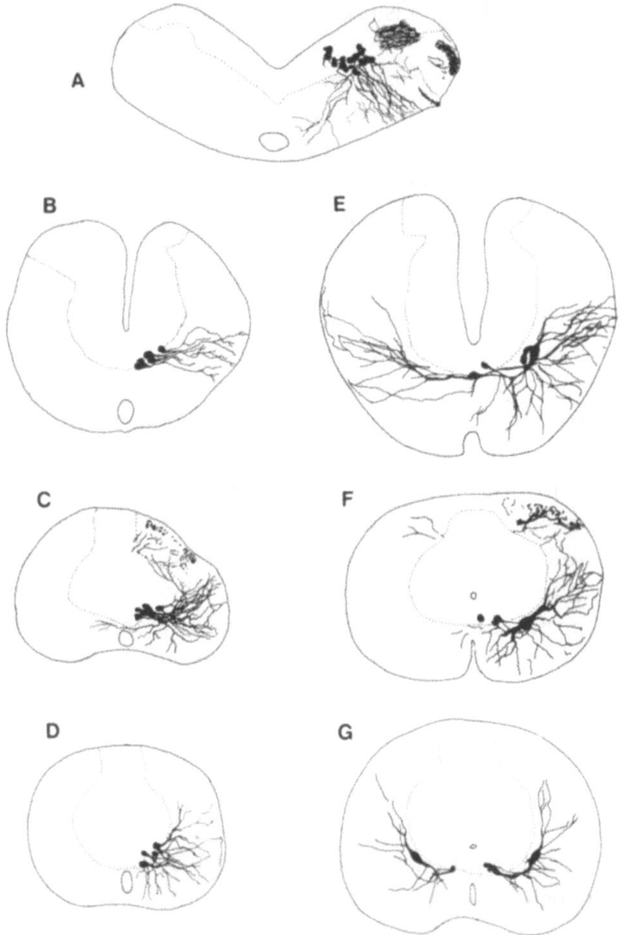

Fig. 4.10 A–G. Reconstruction of motor neurons related to feeding. Transverse sections through the respective nuclei. **A–D** *Batrachoseps attenuatus*; **E–G** *Plethodon jordani*. **A** Nucleus VII r.h./IX/X. The fasciculus solitarius is shown dorsally. **B, E** Nucleus of first spinal (hypoglossal) nerve. **C, F** Nucleus of second spinal nerve. Dorsally, somatosensory pathways are shown. **D, G** Nucleus of accessory spinal nerve. Note the differences in cytoarchitecture of spinal motor nuclei between *Plethodon* and *Batrachoseps*. (From Roth and Wake 1985 b)

In the Bolitoglossini, the motor nucleus of the second spinal nerve overlaps only slightly with that of the first spinal nerve and extends backward to the midpoint between the ventral roots of the second and third spinal nerves. This nucleus also comprises the motor units of the spinal accessory nerve, the fibers of which leave the brain far forward through the last root of the ninth-tenth root complex. There are the same differences in the cytoarchitecture of this nucleus between Bolitoglossini and non-bolitoglossine salamanders as those found in the nucleus of the first spinal nerve, namely with regard to the absence of a lateral motor column in the Bolitoglossini (Fig. 4.10 C, F). The nucleus of the second spinal nerve mainly innervates the m. rectus cervicis profundus (and superficialis), whereas the accessory spinal nerve innervates the neck musculature (in plethodontids, the m. cucullaris).

4.2 Visual and Visuomotor Projections and Pathways

Whereas the system of visual afferents and central visual pathways of anurans has been studied in great detail by many authors (e.g., Scalia 1976, Fite and Scalia 1976, Lázár 1978, Neary 1975, Wilczynski and Northcutt 1977, Neary and Northcutt 1983, Montgomery et al. 1981, 1985) and by means of modern tracing methods (e.g. HRP and autoradiography), the visual pathways of salamanders have been investigated with these methods only recently (e.g., Fritzsch 1980, Rettig and Roth 1982, 1986, Cowie and Baisden 1982, Finkenstädt et al. 1983, Rettig 1984, Wicht 1984, Naujoks-Manteuffel and Manteuffel 1986, Compoint and Clairambault 1986). Earlier reports were based on the Golgi method, as the still fundamental work of Herrick (e.g., Herrick 1942) or the degeneration technique (Riss et al. 1963, Jakway and Riss 1972, Gruberg 1972), which however, are not able to give as precise and reliable a picture of projections comparable to the HRP method. The following description is based on the HRP studies, with the study of Fritzsch (1980) on *Triturus alpestris* and *Salamandra salamandra*, and of Rettig and Roth (1982, 1986) on a larger number of plethodontid species concerning primary visual afferents as the most complete and detailed studies. It will be completed by recent studies on the cytoarchitecture of the visual centers and the secondary visual pathways.

4.2.1 Visual Afferents to the Brain

The optic nerve is constituted by the axons of the retina ganglion cells which leave the eye through the optic disk. Generally, salamanders seem to have fewer fibers in the optic nerve than anurans. After Gruberg (1972), *Ambystoma tigrinum* possesses 50,000 fibers which would be about ten times less than in the frog (Maturana 1959, Fite and Scalia 1976, Grüsser and Grüsser-Cornehls 1976). In *Notophthalmus viridescens* Ball and Dickson (1983) counted 73–80,000 optic nerve fibers, of which 2.4% were myelinated. In *S. salamandra*, Fritzsch (pers. comm.) counted 52, 970 fibers of which 49,250 were fine, unmyelinated fibers and 3,720 coarse, myelinated ones. This is a ratio of about 93% to 7% which is the same ratio found by Maturana (1959) in the frog. Our own counts of the fibers of the optic nerve in *Bolitoglossa subpalmata* revealed nearly 30,000 fibers of which about 850 (3.5%) were myelinated. Surprisingly, *Batrachoseps attenuatus*, a miniaturized plethodontid, either had very few (30–40) or no myelinated fibers at all (Linke and Roth in prep.).

According to Herrick (1942), in *Ambystoma tigrinum* the number of fibers of the optic nerve increases about tenfold at the time of metamorphosis. The adult fibers are arranged in 25–50 or even more fascicles. Throughout the length of the nerve to the chiasma opticum the separate fascicles shift their relative position, and individual fibers pass from one fascicle to another. At the entrance of the optic nerve into the brain, the fibers are arranged in four bundles. The ventral bundles 1 and 2 contain most of the fine fibers, the anterodorsal bundles 3 and 4 the coarsest. Most of the finer fibers of all bundles enter the tractus opticus marginalis (marginal optic tract, MaOT), the coarsest of the bundles 3 and 4 the tractus opticus axialis (axial optic tract, AOT) (Herrick 1942).

The fibers of the optic nerve cross in the chiasma opticum at the ventral side of the diencephalon to the other side of the brain. Part of the fibers cross below the nucleus praeopticus, whereas more dorsally running fibers cross behind it. All fibers seem to send a minute amount of collaterals to this nucleus (Fritzsch 1980, Cowie and Baisden 1982). A certain part of the fibers, mainly from the rostrally located ones, turns back to the ipsilateral side of the brain.

Behind the chiasma, the main fiber pathway runs dorsally within the white matter of the diencephalon. A minor number of coarser fibers separates early forming a distinct fascicle, the basal optic tract (BOT) which extends caudally over the surface of the infundibulum of the hypothalamus. Rostral to the root of the third cranial nerve, oculomotorius, these fibers form a dense terminal field, the basal optic neuropil (BON).

The dorsally running main pathway of retinofugal fibers forms several subtracts: (1) the marginal optic tract (MaOT) which comprises the majority of the fibers and runs along the rostrolateral margin of the thalamus directly under the surface; (2) the axial optic tract (AOT) which is constituted mainly by those fibers crossing behind the nucleus praeopticus, and which runs deeper inside the thalamic white matter; however, more dorsally, it joins the MaOT and is then no longer separable from it. In their study on plethodontids, Rettig and Roth (1986), in contrast to Fritzsch (1980), did not distinguish between MaOT and AOT, since according to their observations the fibers of the AOT fuse continuously with the MaOT withouth sharp boundaries. (3) The medial optic tract (MeOT) which comprises part of the fibers that enter the chiasma rostrodorsally and runs dorsally together with the MaOT, but being clearly separated from this tract in the dorsal thalamus, and then extends medially to the praetectum and optic tectum.

According to Fritzsch (1980), in Salamandridae, the following situation is found (Fig. 4.11). Three visual terminal fields can be distinguished within the rostral thalamus: the corpus geniculatum thalamicum (C, CGT), a large, egg-shaped neuropil which is situated superficially within the ventral thalamus. Immediately dorsomedial to the CGT, the neuropil Bellonci, pars lateralis (NBl), is situated. It extends parallel to the CGT in a ventrocaudal direction from the surface into the white matter. It is smaller than the CGT but more densely stained in HRP experiments. Mediodorsal to the NBl, a

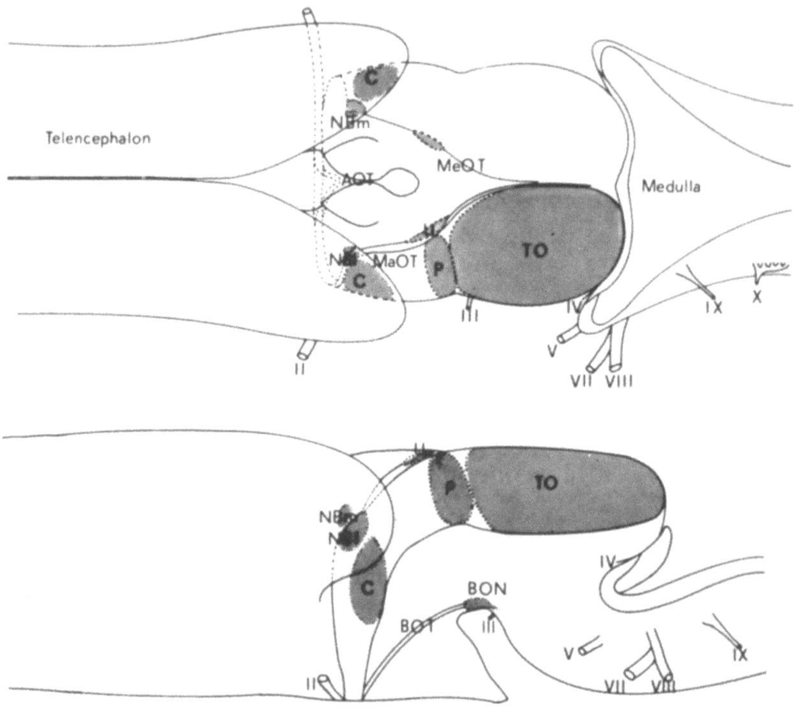

Fig. 4.11. Dorsal and lateral aspect of the brain of *Triturus alpestris* showing the size and position of the retinofugal terminal fields (*hatched areas*) and the course of the tracts supplying the fields. For abbreviations, see list. (From Fritzsch 1980)

third thalamic field, the neuropil Bellonci, pars medialis (NBm), can clearly be separated. It was first described by Fritzsch (1980). It has only half the size of the NBl and is located rostrolaterally and parallel to it. The NBm does not extend as far caudally as the NBl. The fibers supplying the NBm derive from the MeOT and from the NBl. Fritzsch (1980) mentioned fibers running from the rostral part of the NBm to the periventricular gray matter surrounding some cells at the position of Herrick's "nucleus Bellonci". This could be interpreted as further corroboration of the existence of such a nucleus in the urodele diencephalon.

Dorsocaudal to these terminal fields, the MaOT and the MeOT are clearly separated, the former turning laterally caudalward whereas the latter ascends most dorsomedially and then runs caudalward. In the praetectum region the MaOT forms the laterally oriented neuropil posterior thalami (or better neuropil praetectalis) (P), whereas the MeOT supplies a small sickle-shaped neuropil, the area uncinata or "uncinate field" (U, UF) which lies directly medial to the P, but lateral to the commissura posterior. The neuropil area between the NBl and the P, the lateral neuropil, is free of visual terminals.

These thalamic and pretectal projection sites are present both contra- and ipsilaterally. However, whereas the NBm and UF are equally labeled in HRP stainings both contra- and ipsilaterally, ipsilaterally, the NBl (Fritzsch, 1980) and the P (Naujoks-Manteuffel and Manteuffel 1986) show only an outer shell leaving the core empty, and the projection to the CGT is poor and patchy. Therefore, the ipsilateral projections to the diencephalon and praetectum are mostly constituted by the MeOT.

In densely HRP-stained brains there is no sharp border between the pretectal neuropil and the rostral tectum. After long survival times of the experimental animals, however, when the fibers have completely disappeared and only the terminal fields are visible, the tectum is clearly separated from the posterior thalamic/pretectal neuropil by a staining-free zone.

Behind the praetectum, the contralateral MaOT separates into two fascicles, or brachia, a dorsomedial and a lateral one which both extend around the optic tectum giving collaterals to the tectum. The MeOT enters the optic tectum most medially. Contralaterally, it runs caudally along the dorsomedial fascicle of the MaOT almost to the caudal tectal border, whereas ipsilaterally it reaches only the rostral tectum.

In the contralateral tectum, the incoming fibers are arranged in several layers. In the rostral part of the tectum which corresponds to the binocular visual field, four densely stained layers are found, separated by three layers of less intense staining. The deepest layer is constituted by the neuropil of the area uncinata continuing into the optic tectum. In the caudal part of the tectum only two intensely stained layers occur. The fibers entering the tectum through the MeOT seem to terminate mostly in the innermost layer. According to Fritzsch (1980), from this layer thin fiber collaterals run deeper into the white matter, sometimes reaching the periventricular cells.

In *S. salamandra* the MeOT and its related fields seem to be less distinct and smaller compared to *Triturus alpestris* (Fritzsch 1980). In *Ambystoma tigrinum* only two tracts, the MaOT and the AOT have been described on the basis of degeneration technique studies (Jakway and Riss 1972, Gruberg 1972). However, the MeOT seems to be present, too, at least in *Ambystoma mexicanum* (Fritzsch and Himstedt 1980). Correspondingly, the existence of the NBm in Ambystomatidae is still uncertain, but this projection field should be present. Again in *Ambystoma mexicanum*, Guillery and Updyke (1976), using ^{3}H-proline as autoradiographic tracer, found the densest ipsilateral staining in the NB and in the praetectum. They also found stained fibers throughout the rostrocaudal extent of the ipsilateral tectum. Due to the possibility of transsynaptic transport of ^{3}H-proline the authors could not exclude that these results are based on indirect ipsilateral staining via mesencephalic (tectal or tegmental) commissures. Therefore, the amount of direct ipsilateral input into the tectum of *Ambystoma* remains to be determined exactly. In *Necturus maculosus*, with the same method, Caldwell and Bermann (1977) found an ipsilateral projection to the tectum only in the anterior part and only in the deepest fiber layer, whereas the caudal two-thirds of the tectum were free of autoradiographic labels. This corresponds to the situation described in *Triturus* and *Salamandra*.

The situation is different in Plethodontidae, mostly with regard to the ipsilateral retinofugal projections (Rettig and Roth 1982, 1986, Figs. 4.12 and 4.13a–r).

The contralateral projections to the diencephalon are similar to the situation in salamandrids with the following exception. Similar to the situation found in *S. salamandra*, in the Desmognathinae, Hemidactyliini, and Plethodontini the NBm is mostly supplied by fibers coming from the NBl (which, in turn, is mostly supplied by the MeOT). In the Bolitoglossini the NBm has direct supply through the MeOT. Rettig and Roth (1986) described a fourth thalamic projection field in plethodontids, the rostroventral neuropil (RVN) which is situated rostral to the NBm and is innervated by the MeOT. It is well observable in the bolitoglossines *Batrachoseps attenuatus*, *Bolitoglossa subpalmata, and Chiropterotriton lavae*. Small-sized CGT, NBl, and NBm are found in the genera *Plethodon*, *Desmognathus*, and *Hydromantes*; large-sized ones in *Eurycea* and all Bolitoglossini except *Hydromantes*. The by far largest thalamic projection fields, relative to brain size, are found in the miniaturized bolitoglossines *Batrachoseps* and *Thorius*. Here, these projection sites constitute about 20% of all visual projections.

The tractus opticus basalis (BOT) runs in a broad band along the infundibulum into the basal area of the tegmentum where it forms the basal optic neuropil (BON). Some fibers surround the BON or traverse it in a caudal direction. However, they could never be followed beyond the tegmentum isthmi. In addition to the fibers forming the BOT, there are many fibers branching throughout the MaOT, especially at the pretectal level, joining the BON. In the tribe Bolitoglossini, the BON shows a shift from a lateral to a more ventral position.

As already mentioned, major differences exist between plethodontids and all other urodele families, as well as among the different groups of plethodontids, with regard to the amount of ipsilateral projections. In all plethodontids the NBm is the most heavily stained ipsilateral neuropil and shows nearly the same intensity as the contralateral NBm. The ipsilateral NBl is less intensely stained compared to the contralateral one in the Desmognathinae, Hemidactyliini, and Plethodontini, but almost equally stained in the Bolitoglossini. In the miniaturized bolitoglossines *Batrachoseps* and *Thorius* it shows the same staining intensity contra- and ipsilaterally.

The CGT, supplied by the ipsilateral MaOT, is only weakly stained in *Ensatina*, *Plethodon*, *Aneides*, *Eurycea*, and *Desmognathus*. Again in the Bolitoglossini, with the exception of *Hydromantes* (which is similar to *Desmognathus*), we find a massive staining of the CGT.

In the praetectum, the ipsilateral UF, supplied by the MeOT, has more or less the same size as the contralateral one, although it is less densely

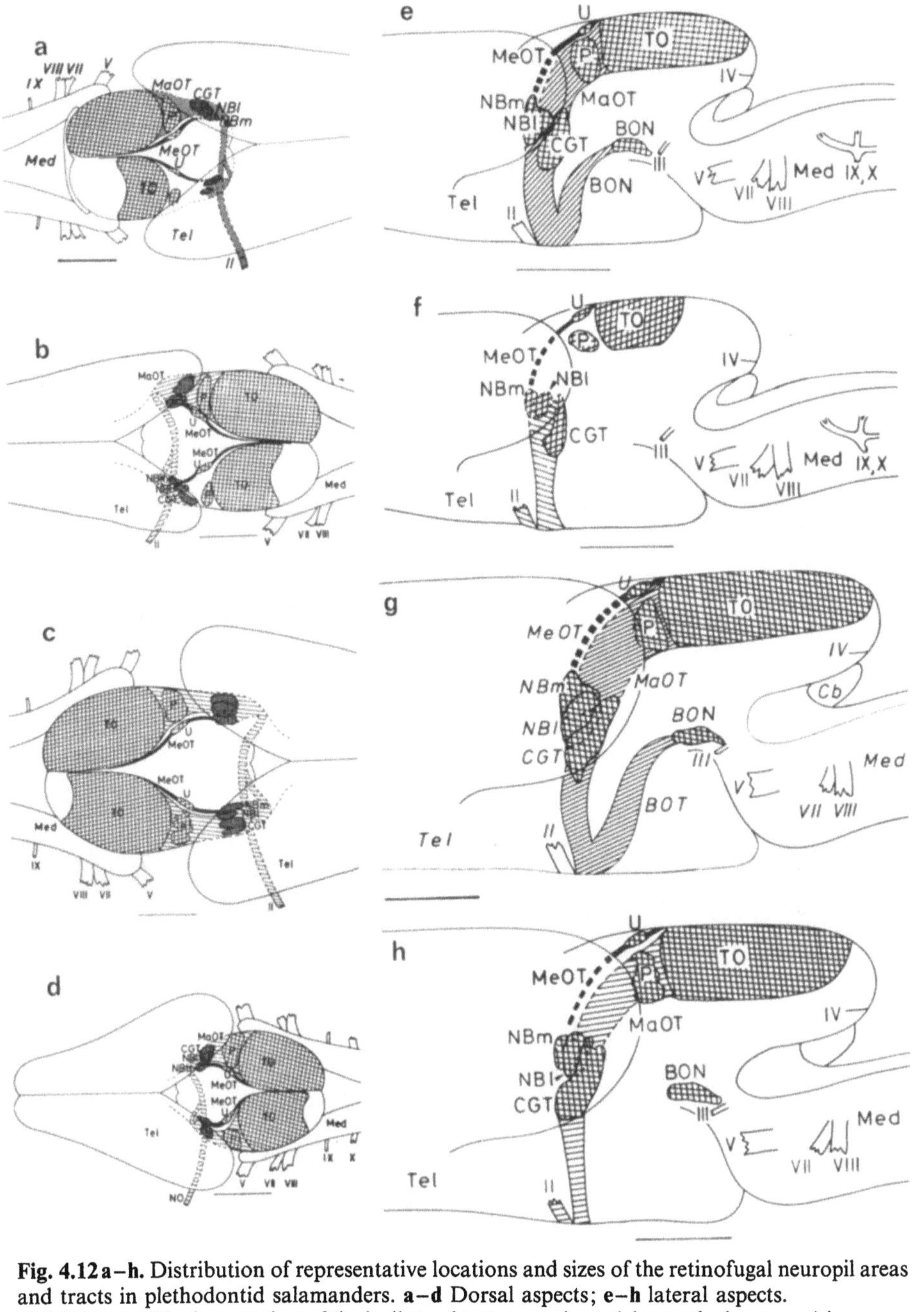

Fig. 4.12 a–h. Distribution of representative locations and sizes of the retinofugal neuropil areas and tracts in plethodontid salamanders. **a–d** Dorsal aspects; **e–h** lateral aspects.
Left side: **a–d** The innervation of the ipsilateral tectum opticum (always the lower one) increases from *P. cinereus* (**a**) over *E. bislineata* (**b**), *B. subpalmata* (**c**) to *T. narisovalis* (**d**).
Right side: **e–h** Contralateral (**e, g**) and ipsilateral (**f, h**) views of the brains of *P. cinereus* (**e, f**) and *B. subpalmata* (**g, h**). The ipsilateral terminal areas of *P. cinereus* are smaller than the contralateral ones, but a well-developed tractus opticus medialis is visible. In contrast, the ipsilateral terminal areas of *B. subpalmata* are nearly identical in shape to those contralateral. *Bar* = 500 μm. (From Rettig and Roth 1986)

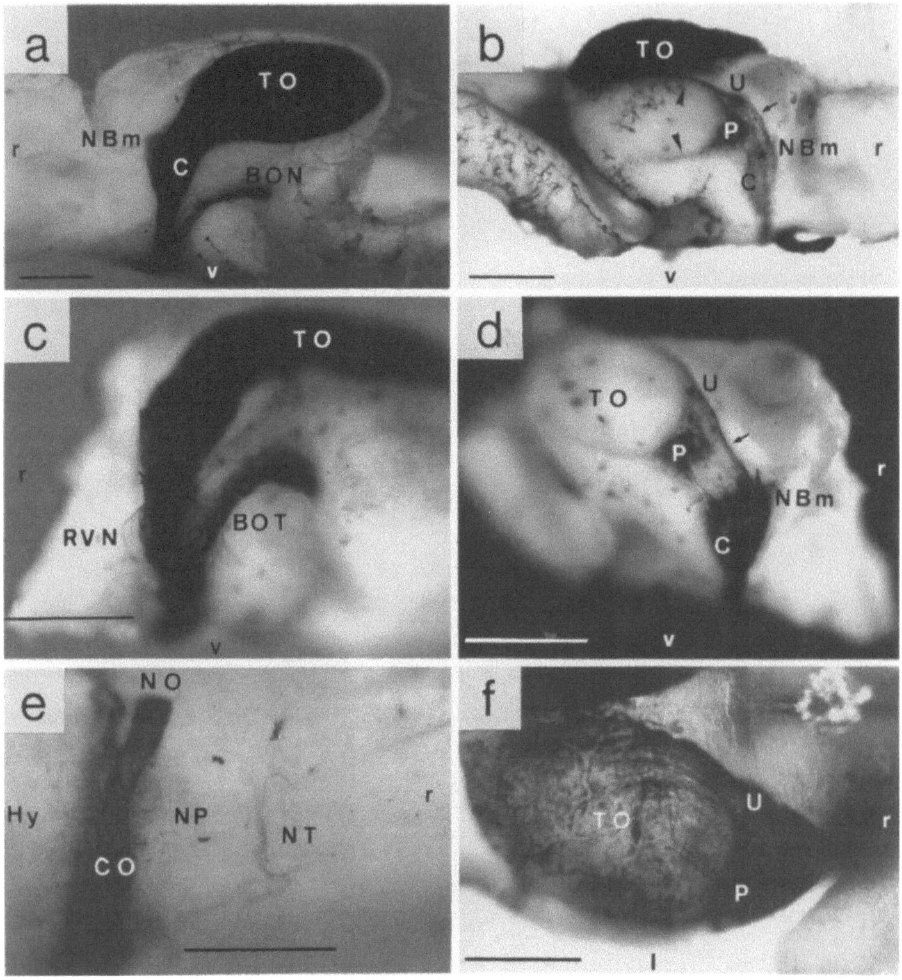

Fig. 4.13a–f. Photomicrographs from retinofugal projections in plethodontid salamanders.
a Lateral view of a brain of *H. genei*, contralateral aspect.
b Dorsolateral view of a brain of *E. bislineata*, ipsilateral aspect. Note the fasciculated projection of the tractus opticus marginalis in the dorsal thalamus. *Asterisk* indicates the neuropil Bellonci pars lateralis, *arrow* the tractus opticus medialis. The *arrowheads* show the two branches of the tractus opticus marginalis indicating the extension of the ipsilateral projection into the tectum opticum.
c Whole mount preparation of a brain of *B. subpalmata*, rostrolateral view, ipsilateral aspect with the forebrain cut off. The neuropil rostroventralis (*RVN*) is weakly stained and the broad projection of the tractus opticus basalis (*BON*) is visible.
d Whole mount preparation of a brain of *B. subpalmata*, rostrolateral view, ipsilateral aspect with the forebrain cut off. The neuropil Bellonci pars medialis (*NBm*) and the neuropil posterior thalami (*P*) are patchily stained. The tractus opticus marginalis connecting these areas is organized in fascicles (*arrow*).
e Ventral view of a brain of *B. subpalmata*. Note the fibers branching from the chiasma opticum (*CO*) into the nucleus praeopticus (*NP*) and the hypothalamus (*Hy*).
f Dorsal view of a brain of *C. lavae*, ipsilateral tectum opticum, cleared in methyl salicylate. Note the fibers deep inside the white matter of the tectum opticum. *Bar* = 400 μm (**a–f**)

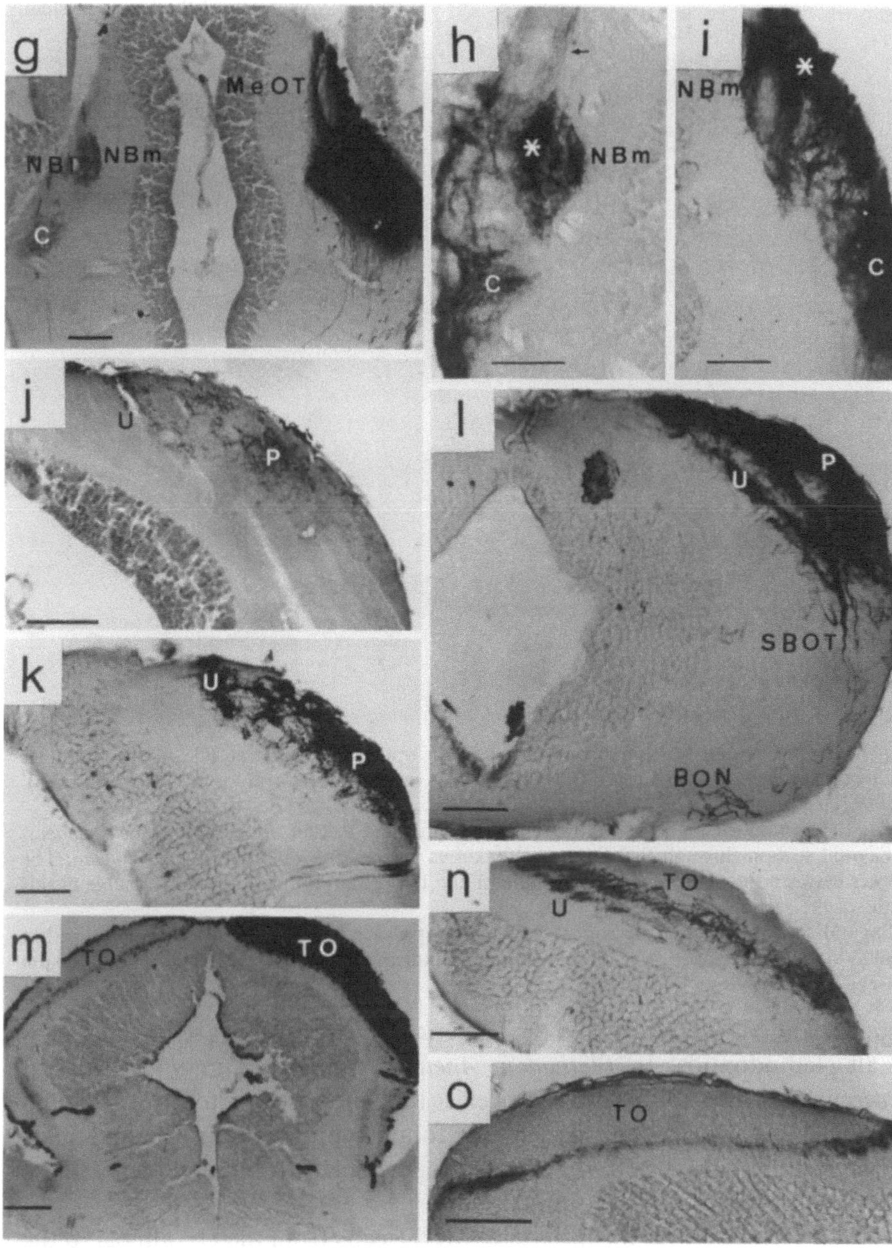

Fig. 4.13g–m. Coronal sections at the level of the rostral thalamus, ipsilateral and contralateral sides of the brain of *D. ochrophaeus* (**g**) and ipsilateral sides of the brain of *E. bislineata* (**h**) and *B. subpalmata*, respectively (**i**). The ipsilateral terminal areas of *D. ochrophaeus* are more weakly and more homogeneously stained compared to *E. bislineata* and *B. subpalmata*. *Asterisk* indicates the neuropil Bellonci pars lateralis, *arrow* the tractus opticus medialis.

j–l Coronal sections at the pretectal level, ipsilateral side of the brains of *P. cinereus* (**j**), *B. subpalmata* (**k**) and *C. lavae* (**l**). The projection of the neuropil posterior thalami in *P. cinereus* lies laterally, whereas in *B. subpalmata* the entire area is unevenly stained. The projection in *C. lavae* shows the heaviest staining of all plethodontid species studied so far. Note the projection of the fasciculus superior tractus optici basalis (*SBOT*) and the relatively well-stained neuropil opticus basalis (*BON*) of *C. lavae*.

m Coronal section at the caudal-tectal level of a brain of *Ps. cephalica*. Ipsilateral is to the *left*. Note that the ipsi- and contralateral projections show the same distance to the periventricular gray matter. (Legend to Fig. 4.13n, o see page 160)

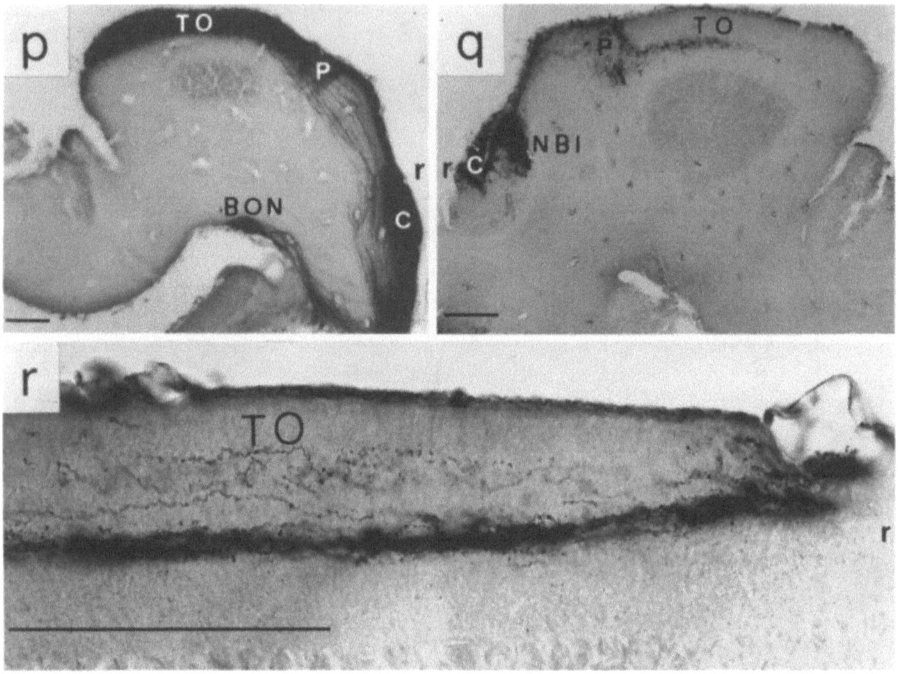

Fig. 4.13n–r. Coronal section through the tectum opticum, ipsilateral side of the brain, at midtectal level of *T. narisovalis* (**n**) and at the rostral-tectal level of *B. subpalmata* (**o**). *Bar* = 100 μm (**g–o**).
p Sagittal section through the contralateral thalamus and tectum opticum of *D. ochrophaeus*.
q Sagittal section through the ipsilateral thalamus and tectum opticum of *E. bislineata*. The border between the neuropil posterior thalami and the tectum opticum is readily visible in this plane of the section.
r Sagittal section through the rostral half of the ipsilateral tectum of *Ba. attenuatus* with a heavily stained innermost projection layer and an indistinct layer more dorsally. *Bar* = 100 μm (**p–r**)

stained. The MaOT running to the posterior thalamus shows several fascicles in all plethodontids. In *Plethodon*, *Aneides*, *Desmognathus*, *Eurycea*, and *Hydromantes*, two to three fascicles exist that support two to three terminal fields. In *Bolitoglossa*, four fascicles are found supplying four terminal fields in the P underlying the four fascicles. In the smallest bolitoglossines these fascicles are not clearly separable.

In all plethodontids the ipsilateral MeOT runs from the UF almost to the caudal margin of the tectum. The lateral projection of the UF extends into the deepest retinofugal projection and is clearly distinguishable from the deepest projection layer of the MaOT. The ipsilateral MaOT shows two brachia as does the contralateral one.

In the Plethodontini, the ipsilateral retinofugal projections cover mostly the rostral tectum, forming two layers. The deep layer is continuous through the whole width of the tectum, whereas the superficial layer covers only the mediorostral part. In the genera *Desmognathus*, *Eurycea*, and *Hydromantes* both layers extend further caudally. In the Bolitoglossini (except *Hydromantes*) the deep layer extends almost to the caudal margin of the tectum. The superficial layer extends much further caudally than in nonbolitoglossines, but is relatively indistinct in the caudal optic tectum. The strongest ipsilateral projections to the tectum are found in *Chiropterotriton lavae*. The deep layer of ipsilateral fibers is found at about the same vertical level as the lowest layer of the contralateral projections. A clear separation between ipsilateral and contralateral fibers in the same side of the brain could not be observed in the thalamus, the praetectum, and the tectum.

The ipsilateral BOT consists of fewer fibers than the contralateral one, which, however innervate the whole BON. The tribe Plethodontini again shows the weakest ipsilateral innervation of the BON, whereas in the Bolitoglossini the BOT and the BON are always well stained with a mirror-symmetric position to the contralateral side.

4.2.2 Comparison with Anuran Retinofugal Projections

On the basis of the Fink-Heimer degeneration technique, in *Rana esculenta* Lázár and Székely (1969) found the following contralateral, retinofugal projection sites: in the thalamus the nucleus (= neuropil) Bellonci (NB) and the lateral geniculate body (LGB) (= corpus geniculatum thalamicum, CGT), and in the praetectum, a dorsolateral projection site which, due to its curved form, in horizontal sections appears as two neuropils, a small medial and a larger lateral area. Later, other authors, by means of HRP studies, have established evidence for the existence of two independent pretectal neuropils, the posterior thalamic neuropil and the uncinate field (Fite and Scalia 1976).

In the tectum, Lázár and Székely found four laminae of retinofugal fibers: lamina 1 and 2, situated immediately below the surface and consisting mostly of thin, unmyelinated fibers; lamina 3 of myelinated fibers, situated above layer 8; and lamina 4, consisting of few, thick unmyelinated fibers, situated in layer 8 and beneath it (cf. Székely and Lázár 1976, and Fig. 4.21 for nomenclature of tectal layers in anurans). In the ipsilateral thalamus the nucleus Bellonci was stained only to one-third of its contralateral size and restricted to a half-moon shape covering the dorsocaudal part. The LGB was fully stained in its rostrocaudal portion and showed only a thin, stained surface in its caudal part. In the ipsilateral praetectum the caudomedial portion of the neuropil was stained. On the basis of HRP studies, Fite and Scalia (1976) reported ipsilateral projections in all diencephalic neuropils mentioned above. Lázár and Székely (1969), as well as Fite and Scalia (1976) found no evidence of direct ipsilateral fibers in the tectum of *Rana*.

In *Xenopus*, on the basis of ^{3}H-proline and -fucose experiments, Levine (1980) found three optic terminal sites in the contralateral thalamus: nucleus Bellonci, CGT, and the rostral visual nucleus, which is an equivalent of the rostral and of the posterior entopeduncular nucleus in *Rana* (Frontera 1952, Scalia 1976) and is located at the rostral edge of the MaOT beneath the lateral edge of the caudal pole of the telencephalic lobe. The nucleus Bellonci and the CGT have more or less the same position as in salamandrids and plethodontids. In the praetectum there is only one clearly defined projection site, the uncinate field. In addition there are two projection fields which with the technique applied are not clearly discernible: the thalamo-pretectal field, a horizontal band extending in the dorsomedial neuropil from midthalamus to the praetectum, and the praetectal field, a diffusely stained area immediately rostral and lateral to the uncinate field.

The nucleus Bellonci and the rostral visual nucleus as well as the uncinate field receive strong ipsilateral input, the CGT only a weak one. Both the thalamo-pretectal field and the pretectal field receive ipsilateral supply, too. The optic tectum receives a dense contralateral retinal input; however, in *Xenopus*, there is little evidence of laminar distribution compared to ranids and bufonids. In the ipsilateral tectum, there is only weak innervation, confined to the rostral tectum and consisting of two parts: one band of labeling which runs along the rostral edge of layer 8, and one restricted area in layer 9.

Of the three thalamic neuropils described above in salamanders, the corpus geniculatum thalamicum (CGT) and the neuropil Bellonci pars lateralis (NBl) seem to be homologous to the respective fields in anurans. There are great similarities with respect to position, shape, and fiber supply between the NBl in *Salamandra* and the nucleus Bellonci in *Rana* and *Xenopus*. According to Fritzsch (1980), the neuropil Bellonci pars medialis (NBm) has no obvious counterpart in anurans and, therefore, seems to have evolved independently. On the other hand, the rostral visual nucleus of *Xenopus* and the rostral end of the posterior entopeduncular nucleus of *Rana* have only a weakly stained correspondent part in salamanders. It is interesting, however, that the Australian lungfish possesses both an entopeduncular nucleus and a NBm-like neuropil (Northcutt 1980).

Whereas, according to Lázár and Székely (1969) and to Fite and Scalia (1976), only indirect ipsilateral tectal input exists in the frog, *Xenopus* is closer to the situation in urodeles in having a direct ipsilateral, two-layered input to the rostral tectum as is the case in plethodontid salamanders like *Desmognathus*. This might not be a sign of primitivity, but rather connected to the frontality of the eyes and the increased binocularity in both *Xenopus* and plethodontids.

4.2.3 Topic Organization of Retinal Projections to the Diencephalon and the Tectum

Modern HRP experiments have shown that the distribution of retinal afferents to the tectum is retinotopically arranged in the optic tracts ascend-

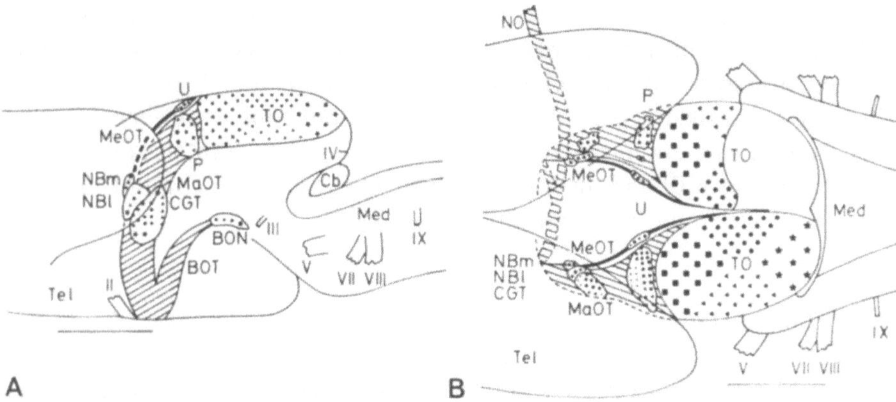

Fig. 4.14 A–B. Retinotopic map in the diencephalon and midbrain of *Triturus alpestris* showing the terminal areas of the four retinal quadrants. **A** Contralateral aspect; **B** dorsal view; ipsilateral projections *above*; contralateral projections *below*. *Squares*: Projections from the temporal quadrant of the retina; *black dots*: projections from the ventral quadrant; *triangles*: projections from the dorsal quadrant; *stars*: projections from the nasal quadrant. *Bar* = 400 µm. (From Rettig et al. submitted)

ing contralaterally over the surface of the diencephalon. This arrangement strongly corresponds to the retinotopic organization of the retinal terminals in the optic tectum.

Rettig (1984) and Rettig et al. (in prep.) have studied the retinotopic organization of visual projections to the visual centers in *Triturus alpestris*, *Salamandra*, and in plethodontid salamanders. The results from *Triturus alpestris* are given here with respect to the four retinal quadrants (Figs. 4.14 A–B):

4.2.3.1 Nasal Quadrant

Fibers from the nasal quadrant run in the contralateral MaOT to the diencephalic neuropils CGT, NBl and P which, however, show only light staining. The MeOT and the neuropils NBm and UF are not stained. In the optic tectum fibers from the MaOT enter the lateral as well as the medial brachium and terminate in the caudal tectum. Some thick fibers run to the BON. Only very few ipsilateral fibers from the nasal quadrant are found in the anterior thalamus. The tectum is free of ipsilateral fibers from the nasal quadrant.

4.2.3.2 Ventral Quadrant

Contralateral fibers from the ventral quadrant project via the MaOT to the CGT and via the MaOT and MeOT to the NBl and terminate in their medial parts. Only a few fibers run to the UF via the MeOT. The P shows weak staining in its medial part. In the tectum, fibers from the ventral quadrant are seen almost exclusively in the medial branch of the MaOT projecting to the medial part of the tectum. Few, but consistent fibers project to the BON. Ipsilateral projections from the ventral quadrant are detectable in the NBm and NBl and, in small number, in the CGT, UF, and P. In the tectum they are found in its medial part in layer 4.

4.2.3.3 Dorsal Quadrant

Contralateral projections from the dorsal quadrant are found in the lateral parts of the NBl and NBm, the CGT, and the P, in the whole UF and, via the lateral branch of the MaOT, in the lateral tectum. Some coarse fibers consistently project to the BON. Ipsilateral fibers from the dorsal quadrant are always heavier than those from the nasal and ventral quadrant. They project to all termination fields. In the ipsilateral tectum, layer 4 is stained in its lateral part.

4.2.3.4 Temporal Quadrant

Fibers from the temporal quadrant enter the chiasma opticum in two bundles, a larger caudal and a smaller rostral one, from which the MeOT derives.

Contralateral projections are found in the caudo-axial parts of the CGT, NBl, and P, and nontopically in the UF and NBm. In the tectum they project to the rostral part, entering the tectum directly between the two brachia. There is a well-developed lamination of the projections: The contralateral BON is well-stained.

Ipsilateral fibers from the temporal quadrant are stained massively in the MeOT and diffusely in the MaOT. The ipsilateral NBl, NBm, and UF are more, the ipsilateral CGT is less intensely stained than the contralateral ones. They do not show any topic arrangement, whereas the P does. In the ipsilateral tectum fibers are seen in the rostral part of the tectum. The ipsilateral BON is stained in most cases.

In summary, the temporal quadrant corresponding to the binocular visual field shows the strongest projections. The temporal quadrant is followed by the dorsal and then by the ventral one in strength. The nasal quadrant shows the fewest projections. This order is true for both contralateral and ipsilateral projections.

A detailed electrophysiological analysis of the retinotopic organization of the tectum was done by Himstedt and Fischerleitner (1975) in the newt

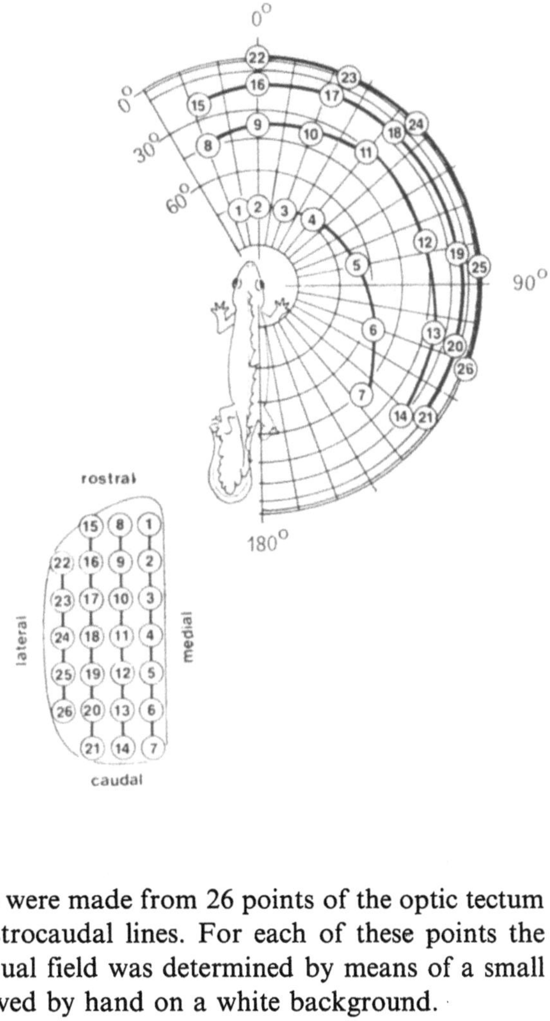

Fig. 4.15. Retinotectal projection in *Triturus cristatus* as revealed by microelectrode recordings from the tectal surface. The visual field of the right eye above the horizon (= *0° circle*) is shown as viewed from above. The *concentric circles* are "degrees of latitude", the *radial lines* are "meridians" of the visual field. *Left*: Dorsal part of the left tectum with 26 electrode positions. The same *numbers* in the visual field indicate from what position a visual stimulus elicited a response in a certain point on the tectal surface. (From Himstedt and Fischerleitner 1975)

Triturus cristatus. Recordings were made from 26 points of the optic tectum arranged in four parallel rostrocaudal lines. For each of these points the corresponding area in the visual field was determined by means of a small piece of black cardboard moved by hand on a white background.

In Fig. 4.15, the tectal points 22–26 represent the visual horizon, whereas line 2–9–16–22 corresponds to the frontal "meridian". The point 22, therefore, corresponds to the 0/0 coordinate of the visual field. Whereas the most lateral rostrocaudal projection line of the tectum (22–26) strictly follows the horizon, lines in the visual field corresponding to the more medially situated rostrocaudal tectal lines (15–21, 8–14, 1–7) deviate increasingly from the parallels of "latitude" toward the horizon as soon as they reach the lateral and caudal parts of the visual field. The upper visual field of the contralateral eye in *Triturus cristatus* extends from 20° of the left visual field to 130–140° temporalwards. The binocular field, therefore, is about 40° large.

It is known that in anurans the visual afferents to the diencephalic neuropils are retinotopically organized. Fite and Scalia (1976) reported that in the NB, CGT, the posterior thalamic neuropil and the UF of *Rana pipiens* the temporal and dorsal quadrants of the retina project to the posterior halves of the neuropils, and the ventral and nasal quadrants to the anterior halves. Furthermore, the ventral and temporal quadrants project to the dorsal halves and the nasal and dorsal quadrants project to the ventral halves. This means that with respect to the retino-tectal map, the nasotemporal axis of the retino-diencephalic map is reversed, but that the representation of the dorso-ventral axis is unaltered. The ipsilateral projections to the diencephalic neuropils stem almost entirely from the temporal quadrant of the retina, only a small portion comes from the dorsal and ventral quadrants. The situation in *Rana* is, therefore, very similar to that found in urodeles.

4.2.4 Development of Retinofugal Projections

The development of retinofugal fibers and of the corresponding diencephalic and tectal termination areas has been studied by Rettig et al. (1981) and by Rettig (1984) in the newt *Triturus alpestris* by means of the HRP-staining method (Figs. 4.16 and 4.17).

The first visual fibers, belonging to the MaOT, reach the brain in the embryonic stage 32 (stage of "early swimmers" after Herrick, 1941) (Fig. 4.16). They run to the tectum without branching in the diencephalon. These fibers are of large caliber and follow the pathway of adult fibers. The first fibers of the basal optic tract are recognizable at stage 34 ("strong swimmers"). They are also thick and show no branching on their way to the midbrain tegmentum. These large caliber fibers of the MaOT and BOT seem to exert a guiding function for the following fibers, at least in normal development, and can, thus, be called "pioneer fibers" (Rettig 1984).

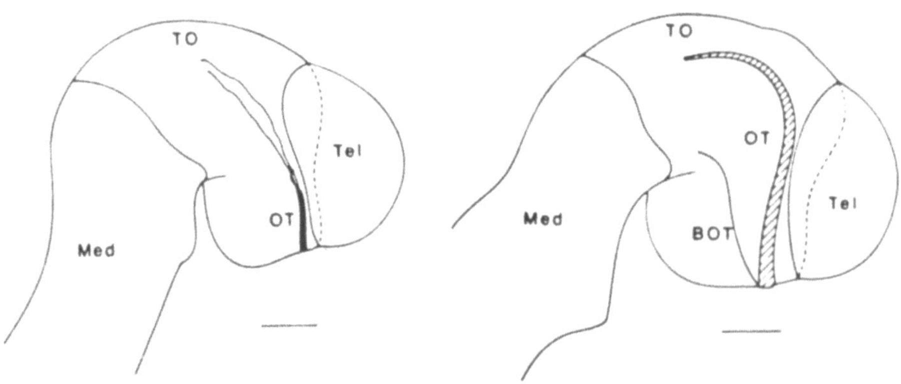

Fig. 4.16. Lateral aspect of embryonic brains of *Triturus alpestris* showing pioneering axons of the optic tract (*OT*) to the tectum opticum (*TO*) and along the basal optic tract (*BOT*). *Left*: Stage 32; *right*: stage 34. (From Rettig et al. submitted)

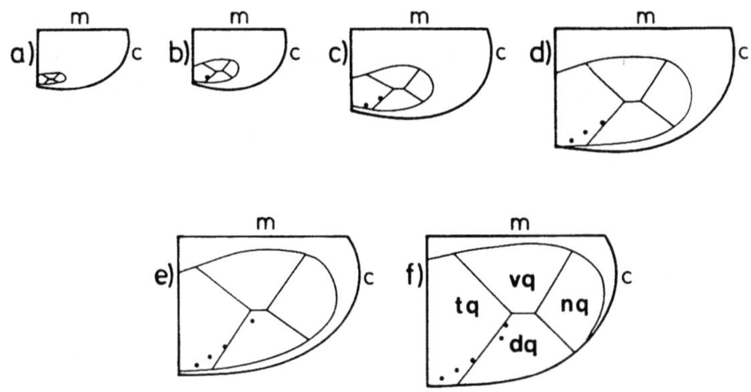

Fig. 4.17 a–f. Left tectal hemisphere showing the growth of retinofugal projections and the tectum in *Triturus alpestris*. *Dots* indicate the shift of the center of retinofugal projections from embryonic stages to 3-month-old juveniles. a Embryonic stage 37; b early larval stage 41; c midlarval stage 46; d late larval stage 53; e 1-month-old juvenile; f 3-month-old juvenile. *m* Medial; *c* caudal; *tq*, *dq*, *vq*, *nq* projections from the temporal, dorsal, ventral and nasal retina. *Bar* = 100 μm. (From Rettig et al. submitted)

The formation of the MeOT is distinguishable at stage 37 and runs axomedially. For this tract no "pioneer fibers" are detectable. At hatching (stage 39) the projections to the optic tectum and praetectum are restricted to about one quarter of the tectum, covering its rostrolateral part. At this stage, the immature neurons of the medial and caudal tectum reach the surface without forming an intervening visual neuropil.

During subsequent development the area of visual projections extends continuously from rostrolateral to caudomedial. Three months after metamorphosis the tectal surface is almost completely covered by visual neuropil (Fig. 4.17). The two brachia, the medial and the lateral one, are first visible at about stage 50 (24 days after hatching). From stage 51 on (25 days after hatching) the visual tectal neuropil shows some lamination which becomes stronger during further development. The adult pattern, however, is still not present 3 months after metamorphosis.

In the diencephalon, the first projection sites are detectable at stage 39, shortly before hatching. They are situated in the ventral thalamus and form the CGT and the NBl. These projection sites, as well as the projections to the optic tectum, therefore, develop in the embryo before any structured visual experience occurred. The moment of appearance as well as the position of the first projections in the tectum and the thalamus correspond, with minor delay, with the differentiation sites of the central neurons.

The next diencephalic projection field, after CGT and NB, is formed at the midlarval stage (stage 46, about 19 days after hatching) as the area uncinata (UF) which, as already mentioned, is mainly supplied by the MeOT. The last diencephalic projection and the last of the contralateral projections to appear, is recognizable in late larval stage (around stage 52, 30 days after hatching) in the rostral thalamus where the differentiation of the NB into a medial and a lateral part takes place. At stage 52 all contralateral, retinofugal, thalamic neuropil areas are present. They develop into the adult pattern by increase of field size.

The formation of massive projections to the mesencephalic tegmentum takes place with some delay to the first occurrence of fibers in the basal optic tract (BOT). The basal optic neuropil (BON) is first stained massively at stage 42. This is correlated with the onset of first eye movements and the first stained neurons of the oculomotor nerve (third cranial nerve) (Herrick 1941). The dendrites of these neurons project massively to the BON, but this relative amount of oculomotor dendrites within the BON diminishes during further development. Since for adult amphibians an involvement of the BON in the optomotor nystagmus is well established (Lázár 1973, Manteuffel et al. 1983), it may be assumed that early eye movement is elicited by the strong visual input onto the neurons of the n. oculomotorius which is later superimposed by other inputs.

The first ipsilateral projections are identifiable at the embryonic stage. Their number slightly increases during development until the midlarval stage (around stage 48) and then decreases afterwards (after stage 49) until the onset of metamorphosis. These fibers are not clearly correlated with contralateral projection sites. During metamorphosis the number of ipsilateral fibers strongly increases, and all ipsilateral projection fields known from adults are qualitatively present. However, they do not reach their final size within the first 3 months after metamorphosis.

The results concerning the development of the ipsilateral fibers seem to corroborate the hypothesis of Lázár (1971) that the ipsilateral projection is associated with the development of binocular vision. However, this assumption is somewhat inconsistent with the finding that ipsilateral tectal afferents develop much later than the end of the movement of the eyes to the more frontally oriented, adult position.

The data on the development of contralateral projections demonstrate that the development of the thalamic visual projection fields are delayed in comparison to the tectal projections. This is consistent with findings in mammals (Frost et al. 1979). Apparently, the MeOT and the MaOT show a different time schedule with respect to their rostrocaudal development. It is of interest that the formation of substantial rostrothalamic projection coincides with the beginning of visually guided feeding behavior.

Further insight into the induction of ipsilateral tectal fibers is provided by experiments of Stirling and Brändle (1982) in which expansion of the visual tectal projections in the neotenic *Ambystoma mexicanum* ("axolotl") was achieved by induction of metamorphosis. As known, axolotls normally do not metamorphose and are, therefore, permanent larvae and reproduce as larvae. Metamorphosis, however, can be induced in some axolotls by administration of thyroxine.

In larval *Ambystoma mexicanum* of any size, the visual projections were found to be confined to the rostral part of the contralateral tectum, covering only 60% of the tectal surface. The caudal part of the tectum did not respond to stimulation by any modality. There was no visual input to the ipsilateral tectum.

During induced metamorphosis the ipsilateral projection developed, but no expansion of the contralateral projection occurred. Such an expansion, however, took place between the second and third week after the completion of metamorphosis, after the animal had left the water. Now, the retinofugal projections covered 80–90% of the contralateral tectal surface. Interestingly, this process of expansion was always the same independent of the size of the larva in which metamorphosis was induced.

During metamorphosis, in the axolotl the eyes protrude from the head. However, the angle between the visual axes of the eyes does not change; therefore, expansion of the retinotectal map was not correlated with an increase in binocularity. Furthermore, the expansion seemed to involve mainly the temporal visual field or nasal retinal quadrant projecting to the caudal tectum. The projection to the rostral pole of the tectum, corresponding with the central, binocular visual field (or temporal retina) was not significantly different before or after metamorphosis.

However, there seems to be an inhibitory interaction between the expansion of the contralateral projection and the growth of ipsilateral projections: expansion of the first seems to inhibit growth of the latter. Stirling and Brändle found that during expansion of the contralateral projection the ipsilateral visual responses became very difficult to record.

In *Rana pipiens* the contralateral fibers of the retino-diencephalic projections are formed during the earliest larval stage, but ipsilateral fibers begin to develop near the onset of the metamorphic climax (stage 20) (Currie and Cowan 1974a, b). The eyes begin to migrate from a lateral to a more dorsofrontal position in later larval stage (19–20).

Preliminary studies in *Bolitoglossa subpalmata* (Rettig et al. submitted) show that in these animals, and probably in all Bolitoglossini, the development of retinofugal projections is different from that of salamanders possessing an aquatic larval stage. In *Bolitoglossa subpalmata* the whole development within the egg until hatching takes 6 to 8 months. About 4 weeks before hatching the embryos show the complete ipsilateral and contralateral adult distribution of retinofugal projections in the diencephalon and mesencephalon, as described above. Additionally, the embryos show much more frontal eyes than other salamander embryos or larvae.

4.3 Intrinsic Organization and Cytoarchitecture of the Optic Tectum and the Diencephalon

4.3.1 Tectum Opticum

It has been repeatedly stated by eminent comparative neuroanatomists that the optic tectum of salamanders shows a rather primitive, undifferentiated organization compared to that of other lower vertebrates. In anurans and teleosts and even in cyclostomes, sharks, and rays, the tectum opticum is multistratified with alternating fiber and cellular layers. Furthermore, tectal cells of these vertebrates show a morphological differentiation which range from at least four different cell types in lampreys (Kennedy and Rubinson 1984) to 15 in teleosts (Meek 1983, Vanegas et al. 1984). In contrast, the salamander tectum, like that of caecilians (pers. observation) and lungfishes (Northcutt 1977), shows only a two-layered structure, as described above. Furthermore, the morphology of tectal neurons seems to be equally undifferentiated. Herrick, who devoted most of his lifetime to the neuroanatomy of the salamander brain, stated somewhat resignatively: "The tectal neurons are of various forms ... Attempts to classify these types of cells in terms of their regional localization or of their fibrous connections have yielded little that seems significant" (Herrick 1948, p. 249). Similarly, Leghissa 1962, in his comparative study on the structure of the optic tectum of lower vertebrates, stated that urodeles represent the most primitive state among vertebrates with respect to the organization of the tectum. With regard to the morphology of tectal cells, he recognized the following types:

1. Large piriform neurons of the deep cell layer with long shafts and extensive ramification within the fiber layer. Their axons leave the tectum via the fiber layer which he calls the "deep layer of unmyelinated fibers" within the periventricular cellular layer.
2. Large piriform neurons of the outer cellular layer with dendrites that extend to the tectal surface. Their axons leave the tectum via the main efferent fiber layer.
3. Small piriform neurons that send their axons outside the tectum via the deep layer of unmyelinated fibers.
4. Fusiform neurons with two dendritic trees, i.e., one tree extending toward the tectal surface and one extending toward the deep fiber layer.
5. Multipolar cells with several main dendrites originating from the soma.
6. Mesencephalic trigeminal neurons.

Leghissa held that all tectal neurons are projection neurons in the sense that their axons always leave the tectum. Thus, he denied the existence of intrinsic tectal cells or interneurons.

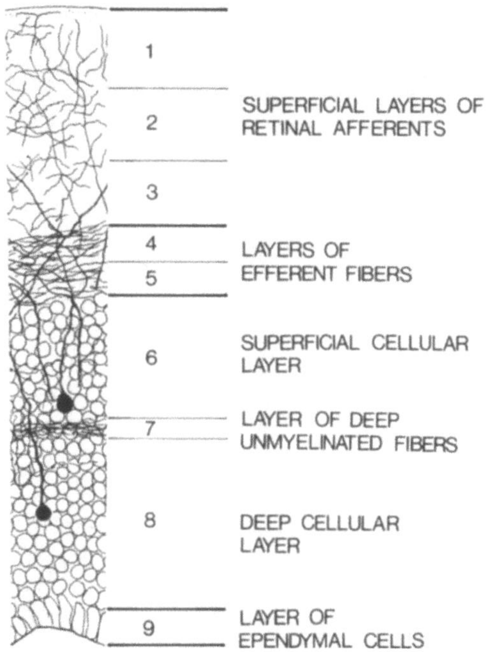

Fig. 4.18. Lamination of the salamander tectum as revealed by Golgi and HRP studies

Roth et al. (in prep.) have conducted extensive studies on the morphology of tectal cells in a larger number of salamander species, using Golgi as well as HRP techniques. The species studied were *Desmognathus ochrophaeus*, *Thorius narisovalis*, *Bolitoglossa subpalmata*, *Hydromantes genei*, *H. italicus*, *Batrachoseps attenuatus*, *Plethodon jordani*, and *Ambystoma tigrinum*. The tecta of *D. ochrophaeus*, *Ba. attenuatus*, *Th. narisovalis*, and *P. jordani* were the most extensively studied.

On the basis of the Golgi study a classification of tectal cells was made, based primarily on the following criteria: (1) position and shape of the soma; (2) angle of dendritic arborization; (3) structure of the dendritic tree and site of main dendritic arborization.

As shown in Fig. 4.18, the tectum was divided into the following layers: layer 1–3: superficial layers of retinal afferents, layers 4–5: layers of efferent fibers; layer 6: superficial cellular layers; layer 7: layer of deep unmyelinated fibers; layer 8: deep cellular layer; layer 9: layer of ependymal cells. This division differs from that of Herrick presented above.

The following main cell types were found according to the criteria mentioned above (Fig. 4.19 1–3):

Type-1 Cells (Large Ganglionic Cells). Neurons of this type have very broad and flat dendritic trees (130–180°) that extend exclusively within layers 4–5 of efferent fibers, except for a few fine dendritic processes that extend

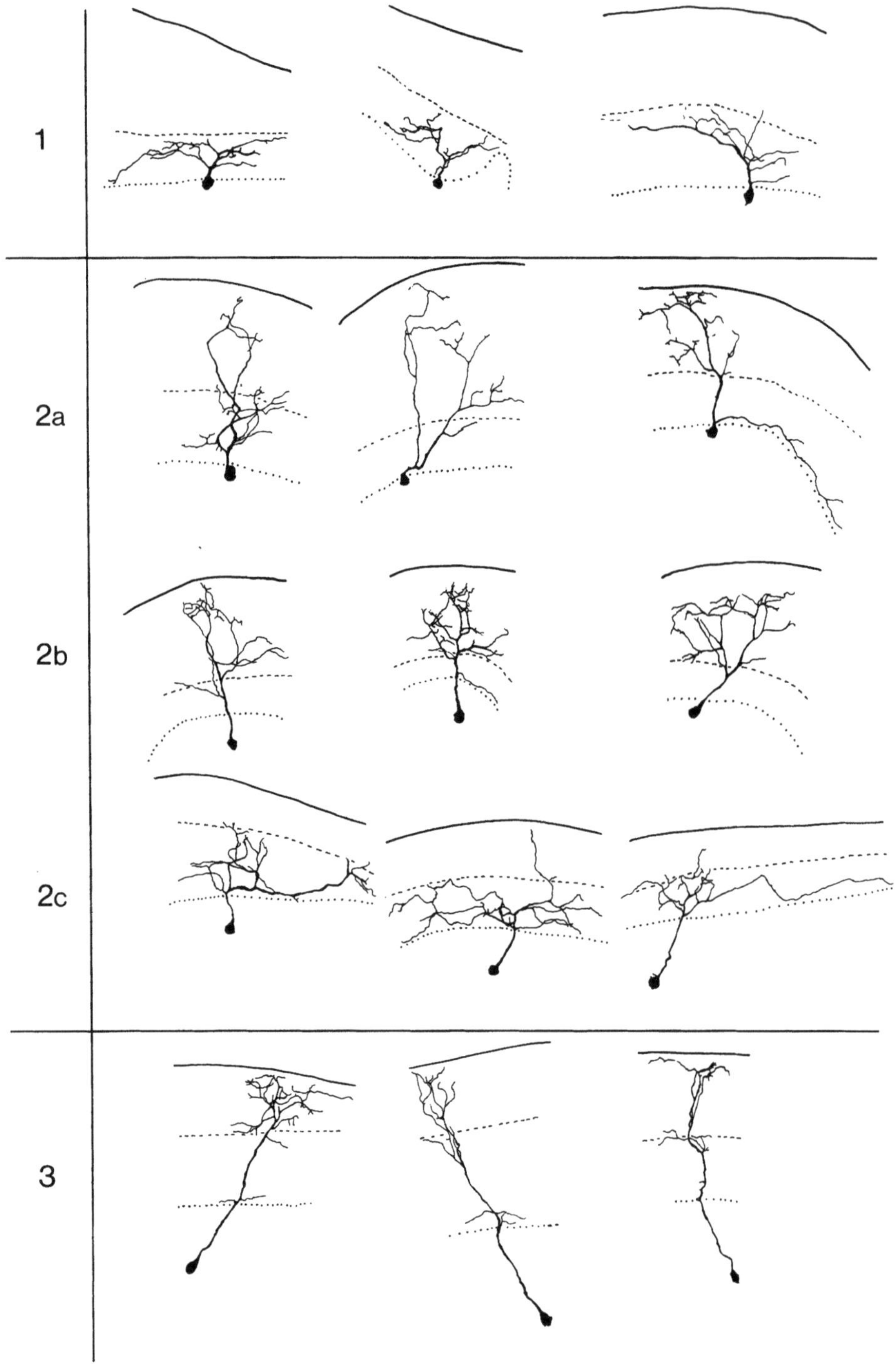

Fig. 4.19. Morphological classification of neurons found in the tectum of *Desmognathus ochrophaeus*. 1 Type-1 cells (large ganglionic cells); 2 a–c type-2 cells (large piriform and pyramidal cells); 3 type-3 neurons (small piriform cells). *Upper solid line* indicates surface of the tectum; *broken line* indicates the border between fiber layers 3 and 4 (see Fig. 4.19); *dotted line* indicates border between layers 5 and 6 (border between fiber and cellular layers in the tectum). For further explanation see text

toward the tectal surface. The somata of these neurons are round or oval with 11–14 μm diameter (in *Desmognathus*) and are found in most cases in the most superficial part of layer 6, directly beneath the efferent fiber layers. Rarely, they are found more deeply, but never beneath layer 6. The main dendrite is mostly very short and always bifurcates at the border between layers 5 and 6. The axons originate mostly from thick secondary dendrites immediately above layer 6 and leave the tectum via layers 4–5.

Two subtypes can be distinguished: 1a-cells with symmetric dendritic trees in which the main apical dendrite quickly bifurcates into two major branches that arborize widely within layers 4–5; 1b-cells with strongly asymmetric dendritic trees in which the main apical dendrite immediately bends laterally and extends parallel to the surface of layer 6 for a certain distance, giving off small secondary dendrites extending vertically or obliquely into layers 4–5.

Sometimes 1a-neurons with double dendritic trees arborizing at two different levels within the layers 4–5 are found.

Type-2 Neurons (Large Piriform and Pyramidal Cells). These neurons are different from the previous ones in that the dendritic trees are narrower (35–120°) and always extend into the superficial fiber layers 1–3 where the retinal afferents enter the tectum. There are several subgroups with regard to the angle of arborization and the site of the somata:

a) 2a-Neurons: Here, the pear-shaped somata (10–14 μm in diameter in *Desmognathus*) are situated at the dorsal edge of layer 6 as in neurons of type 1. The apical dendrite is very short and bifurcates quickly into two secondary dendrites. However, the angle of arborization is rather small (35–60°) and part of the dendrites always extends into layers 1–3, where they arborize extensively. Other secondary and tertiary dendrites also arborize in layers 4 and 5.

b) 2b-Neurons: The somata of these cells (13 μm in diameter) are located in the upper half of layer 6. The apical dendritic shaft is thus longer, but the bifurcation point is always within layers 4 and 5. The dendritic tree is wider than in 2a-neurons, arborizing widely with secondary and tertiary branches in layers 2 and 3 or deeper, but not in layer 1.

c) 2c-Neurons: The oval- or pyramid-shaped somata of these cells (12 μm in diameter) are situated deeper in layer 6 and often show extensive basal dendrites extending into layer 7. The arborization angle is very wide (90–180°). The first point of bifurcation is again at the dorsal border of layer 6 or in layer 5. The dendritic tree arborizes mainly in layers 4 and 5.

Type-3 Neurons (Small Piriform Cells). The dendritic trees of these cells are narrow (25–60°), and their dendrites always extend into layers 1 and 2 where they ramify extensively with fine branches extending parallel to the tectal surface. In part of the neurons these horizontal branches are confined to layer 1 or 2. Axons are difficult to identify. When present they originate from secondary and tertiary dendritic branches and leave the tectum via layer 4 or 5. Three subtypes can be distinguished:

a) 3a-Neurons: The somata of these neurons are situated deep, mostly in layer 8. The first bifurcation point of the dendritic tree is located at the upper border of layer 4 or more superficially.

b) 3b-Neurons: Their somata are again situated in the deep tectum, but the point of first dendritic bifurcation is located in layer 5.

c) 3c-Neurons: These rather rare neurons have somata situated within the upper part of layer 6, with points of first dendritic bifurcation in layer 5, as in 3b-neurons.

Type-4 Neurons (Migrated Cells). The salamander tectum contains a number of neurons that are found within the fiber layer. These migrated cells do not form a continuous layer, but are dispersed within the stratum album, with a certain concentration immediately above layer 4.

The morphology of these cells, studied by means of the Golgi and HRP techniques (Roth et al. in prep.) shows great diversity. Some of the neurons show a morphology similar to that of non-migrated tectal neurons, i.e. they have a pear-shaped perikaryon and a small to medium wide dendritic tree that extends more or less vertically toward the tectal surface. Other cells show perikarya that are polygonal in shape, i.e. several major dendrites originate from the soma extending vertically to the tectal surface. Many migrated cells, however, have horizontally oriented, narrow dendritic trees, either with one main dendrite or with two major dendrites extending in opposite directions. Some neurons are found that show an upside-down orientation, i.e., their dendritic tree is inverted and points to the stratum griseum.

4.3.1.1 Neurons of the Nucleus Mesencephalicus Nervi Trigemini

The neurons of the nucleus mesencephalicus nervi trigemini are prominent in the tectum in possessing by far the largest cell somata of all tectal cells (Roth et al. in prep.). Most of these cells are located along the midline of the tectum, almost immediately below the tectal surface. Others are located more laterally, but almost always within the superficial part of layer 6. Rarely, migrated mesencephalic V cells are found. All these cells are pseudo-unipolar cells, i.e., they possess a thick process that does not ramify within the tectum, but leaves it within fiber layer 4 and extends to the medulla oblongata where it bifurcates at the level of the root of the trigeminal nerve. The peripheral branch exits through the root and projects to the jaw muscles, the central branches descend further downward within the medulla; at least some of them make contact with the dendrites of the motor nucleus of the trigeminus (Roth et al. in prep.).

4.3.1.2 Functional Classification of Tectal Cells

A functional classification of tectal cells was attempted with regard to the projection site of the neurons (Roth et al. in prep.). HRP was applied in crystalline form or injected by pressure or iontophoretically in the following sites of the brain of *H. italicus* and *B. subpalmata* (Fig. 4.20 A):

(1) Ventrolateral medulla oblongata, between the roots of the seventh and ninth nerve, into the region of the tractus tecto-bulbaris rectus and cruciatus; (2) level of the commissura tuberculi posterioris ("peduncle commissure"); (3) commissura postoptica; (4) contralateral optic tectum.

After injection of HRP into the medulla oblongata, neurons were stained retrogradely in the ipsilateral and contralateral tectum (Fig. 4.20 B). Ipsilaterally, the somata of these neurons were always situated in the upper half of layer 6. These neurons belonged to type-1 described above, more rarely to type-2a. Contralaterally, the somata of the labeled neurons were located only at the dorsal edge of layer 6, thus directly underneath the large efferent fiber layer (layer 5). They belonged exclusively to type-1 and showed always very flat, mostly asymmetric dendritic trees.

After injection into the tegmentum lateral to the commissura tuberculi posterioris the same situation was found contralaterally. Ipsilaterally, in addition to the neurons described above, cells with somata situated more deeply within layer 6 were found.

Injections of HRP into the commissura postoptica revealed neurons in the ipsilateral as well as the contralateral tectum with somata situated mostly in the ventral part of layer 6 and dorsal part of layer 8. Somata in the ventral part of layer 8 were rare. All labeled cells had small dendritic trees extending into layers 1–3, thus belonging to type-3. In addition, the nucleus isthmi was always stained.

After injection of HRP into the tectum only a few cells were labeled contralaterally which either belonged to type-2 or to type-3.

Thus, type-1 cells can be considered as efferent neurons constituting the descending tecto-bulbar tracts. The crossed tecto-bulbar tract seems to originate only from cells with extremely flat dendritic trees and with somata situated in the very superficial gray matter, whereas the uncrossed tecto-bulbar tract may also involve neurons of type-2a. In contrast, neurons of type-3 seem to be the main origin of ascending tecto-thalamic tracts, the

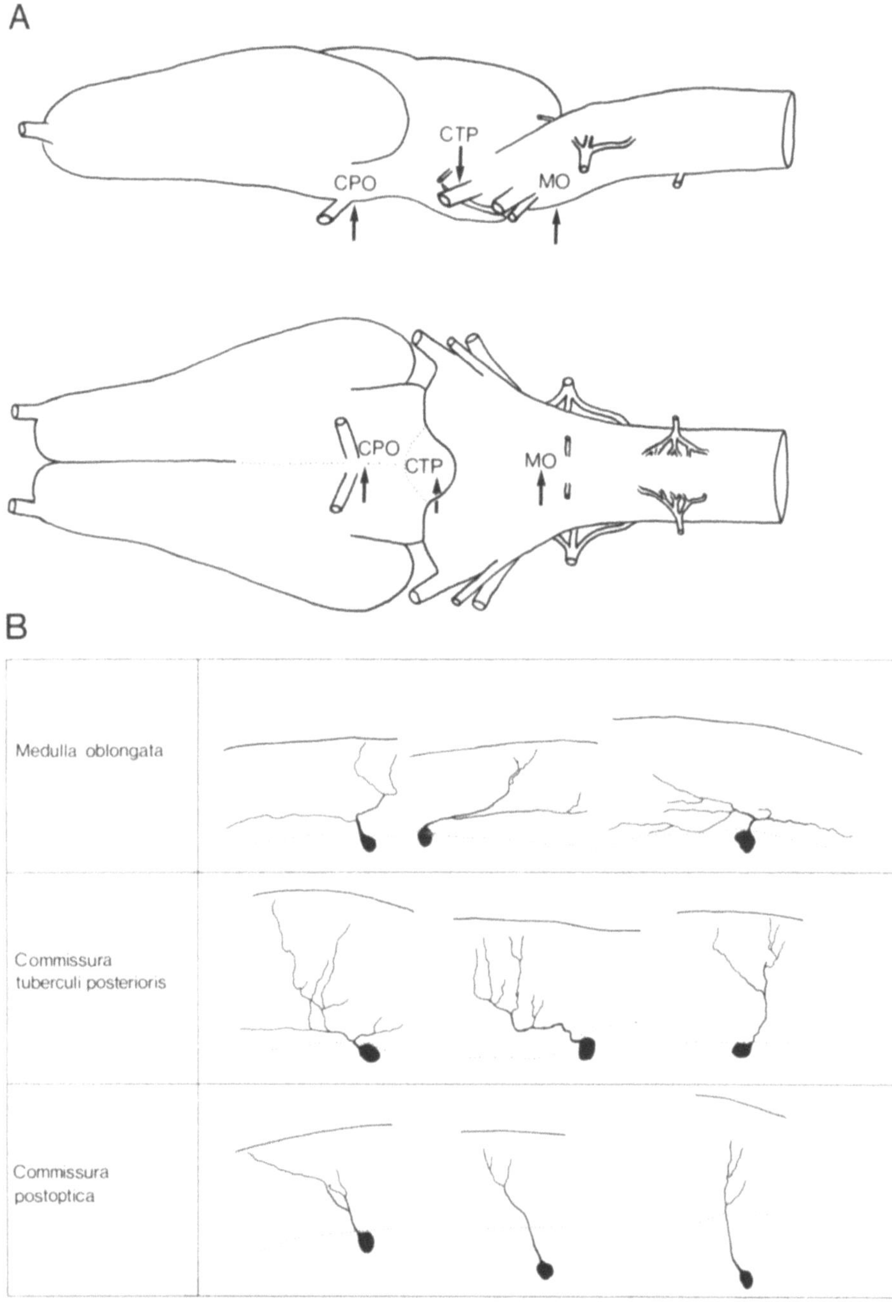

Fig. 4.20 A, B. Functional classification of neurons found in the tectum of *Hydromantes italicus* and *Bolitoglossa subpalmata*. **A** Lateral and ventral view of the brain of *B. subpalmata*, indicating the sites of HRP injection (*arrows*). *CPO* Commissura postoptica; *CTP* commissura tuberculi posterioris; *MO* medulla oblongata. **B** Representative neurons labeled retrogradely from the medulla oblongata, the commissura tuberculi posterioris, and the commissura postoptica. For further explanation, see text

tecto-tectal connections and the tecto-isthmic tract, and, furthermore, constitute the mass of intrinsic neurons. i.e. those with no projections outside the tectum.

4.3.1.3 Comparison with Tectal Cytoarchitecture of Anurans

On the basis of Golgi studies, Székely and Lázár (Lázár and Székely 1967, Székely and Lázár 1976, Lázár 1984) described the following major types of neurons in the tectum opticum of *Rana esculenta* (Fig. 4.21):

1. Large piriform cells. These are the main elements of layers 2, 4, and 6 of the frog tectum. They have round or oval perikarya. The most numerous of them have rather small dendritic trees which arborize only in the upper laminae of layer 9. Other cells of this type start arborizing above layer 8. Their axons originate from the dendritic trunk, ascend to layer 9, and terminate in the neighborhood of the neuron. The somata show only a few tiny basal dendrites. Another group of large piriform cells has widely arborizing dendritic trees which start branching in layer 7. Some of them arborize only in one lamina, others in more than one. The basal dendrites are well developed and extend for some distance. They are the main constituents of deep fiber layers 3 and 5.
2. Small piriform cells. These cells have smaller perikarya and narrower dendritic trees. They are exclusively found in layers 8 and 9. Lâzâr et al.

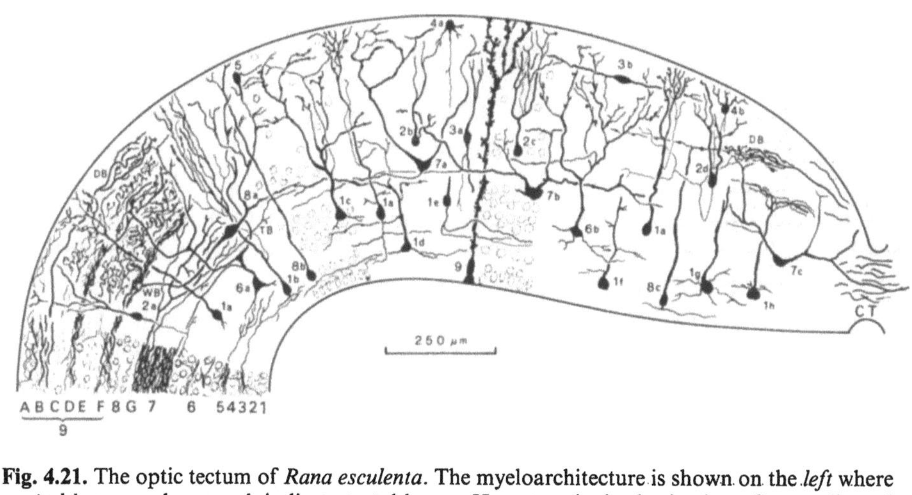

Fig. 4.21. The optic tectum of *Rana esculenta*. The myeloarchitecture is shown on the *left* where *capital letters* and *numerals* indicate tectal layers. Here, terminal arborization of unmyelinated retinal fibers can be seen in laminae *C* and *E*. Cellular layers (*1, 2, 4, 6, 8*) are also shown in two columns in the middle of the drawing. *1a–h* Large piriform cells; *2a–d* small piriform cells; *3a–b* fusiform cells; *4a–b* stellate cells; *5* amacrine cell; *6a–b* pyramidal cells; *7a–c* large ganglionic cells; *8a–c* unclassifiable cells; *9* ependymoglial cell. The *finest lines* represent axons. *DB, WB, TB* Optic terminals; *CT* commissura tecti. (From Lázár 1984)

(1983), on the basis of cobaltic lysine studies, described two subtypes. Subtype-1 arborizes in lamina B of myelinated retinal afferents, subtype-2 in all laminae of retinal afferents except lamina D. The axon of small piriform cells originates either from the perikaryon and descends to layer 6 (subtype-2) or from a thicker dendrite and arborizes within the dendritic tree (subtype-1).

3. Pyramidal cells. These cells are located in deeper cellular layers, mainly in layer 6. The perikarya have a pyramidal or oval shape. The thick main dendrite diverges into two to four secondary branches in layer 7. They run horizontally for a certain distance, then extend to the tectal surface forming a wide dendritic tree. The few tertiary branches terminate mainly in laminae B, C, and F. The axon originates from the main or from a secondary dendrite and enters layer 7. There are numerous cells of intermediate morphology between pyramidal and large piriform cells.

4. Large ganglionic cells. These cells are the largest cells of the frog's tectum (except for trigeminal cells). They are exclusively situated in layers 6 and 7. They have extremely wide dendritic arborization. Their perikarya are boat-shaped or spindle-like, because two to several thick dendrites originate directly from the soma and branch into thick secondary dendrites which extend obliquely to the surface. Smaller dendrites of subtype-1, with a soma located in layer 7, preferentially terminate in lamina F, less frequently in laminae B and D, whereas those of subtype-2, with somata located in layer 6, terminate in lamina B. The axons of these large ganglionic cells leave the tectum via layer 7.

In their HRP and cobaltic lysine study on tectal projecting neurons, Lázár et al. (1983) found that tectal projections to the diencephalon via the commissura postoptica originate mainly from small piriform cells situated in layer 8, less frequently from large piriform and pyramidal cells located in layers 4 and 6 and large ganglionic cells in layer 7. After injection of cobaltic lysine into the medulla oblongata, mainly large ganglionic cells were labeled via the descending tecto-bulbar tracts. Ipsilaterally, these cells were found mainly in layer 7, with dendrites arborizing predominantly in lamina F (myelinated retinal afferents). Contralaterally, labeled somata were found in layer 6 with dendrites terminating primarily in lamina B (unmyelinated retinal fibers).

If we compare the tectal cell morphology and projectivity of anurans and salamanders, we find a close resemblance between large ganglionic cells of anurans and type-1 cells of salamanders. In both groups of amphibians these cells are situated immediately below or even within the main efferent fiber layer (layer 7 in anurans, layers 4–5 in urodeles); they have very wide dendritic trees with a few thin processes extending to the surface, and their axons constitute the descending tecto-bulbar tracts. In both anurans and urodeles, the crossed and the uncrossed tecto-bulbar tracts originate from somewhat different types of these cells.

Type-2 cells of salamanders resemble in many ways large piriform and pyramidal cells of frogs. Both are found in deeper cellular layers of the tectum, their arborization is narrower than that of the previous type and varies from rather small to wide. They are mostly not involved in the descending, but rather in the ascending pathways.

Type-3 cells of urodeles and small piriform cells of anurans are comparable in the respect that both have narrow dendritic trees and give rise to the tecto-thalamic and tecto-isthmic tracts or represent intrinsic neurons. The main difference between anurans and urodeles consists in the position of the soma. Whereas small piriform cells of anurans are found in the uppermost cellular layers of the tectum (layers 8 and 9), in salamanders their somata are found mainly in deeper layers. A consideration of the ontogeny of these cells in anurans and urodeles, however, may help to explain these differences (Leghissa 1962, Jacobson 1978). In the tectum of frogs and salamanders the first cells to develop are the descending efferent cells, i.e. the large ganglionic cells in frogs and type-1 neurons in salamanders, which occupy a final position immediately below the main efferent layer. The next tectal neurons to appear are the large piriform/pyramidal cells in frogs and the type-2 neurons in salamanders which are situated at tectal cellular layers beneath that of the large descending efferent neurons. The latest developing cells are the small piriform cells in frogs and type-3 neurons in salamanders. At the time of invasion of retinal fibers into the tectum, in the frog these cells migrate through all previously established cellular layers toward the surface and constitute the uppermost cellular layers 8 and 9. In salamanders, these cells migrate only to a minor degree within the periventricular gray matter or remain in the deepest layers, near their point of origin.

Thus, the main difference in tectal morphology of anurans and urodeles seems to concern mainly the presence or absence of migration of the small piriform neurons. It is still a completely open question why migration in the salamander tectum cell does not take place similar to anurans. In vertebrates we generally find a strict inverse correlation between the degree of lamination of the tectum (as well as other parts of the brain) and cell differentiation, on the one hand, and genome size and cell size on the other. Those vertebrates with the largest genome and largest cells (including nerve cells), i.e., lungfishes and salamanders (cf. Olmo 1983), have the least laminated tectum, whereas those with the smallest genome and the smallest cells, i.e., teleosts and birds, show the highest degree of lamination and cell differentiation in the tectum. Anurans, on the average, have a genome size ten times smaller and neuron sizes five to eight times smaller than salamanders (Roth et al. in prep.), but their genome size is larger than in teleosts, reptiles, and birds, and the degree of tectal lamination and cell differentiation in anurans lies between salamanders and these latter vertebrate groups. Thus, an increase in genome size and a slow-down of the differentiation rate of neurons as a consequence, may also affect neuronal migration and the formation of nuclei.

4.3.2 Diencephalon

The diencephalon is perhaps the least understood region of the urodele brain, both with respect to its morphology and its functional organization. On the one hand, there is a remarkable complexity of at least five different visual termination fields (neuropils) which show clear differences among each other, both with respect to ipsilateral and contralateral retinal afferents, as has been described in the preceding chapters. With respect to this organization of visual termination fields, there is almost no difference between anurans and salamanders. This situation is in striking contrast to the organization of the cellular layer. As has already been discussed, in the urodele diencephalon we find an almost continuous formation of periventricular gray substance whereas in the anuran diencephalon a larger number of nuclei has been formed. The only major boundaries in the urodele diencephalon are the longitudinal sulci.

The intrinsic organization of the urodele diencephalon has been studied recently by Wicht (1984) by means of the HRP technique. Wicht applied HRP iontophoretically to different sites within the stratum album and the stratum griseum in the diencephalon of the newt *Triturus alpestris* (Fig. 4.22).

Wicht found that despite rather small injection sites in different parts of the diencephalon, the distribution of the cell somata was always rather diffuse, extending mostly along the rostrocaudal line. As shown in Fig. 4.22, an injection of HRP into a very restricted region in the medial part of the thalamus produces the staining of many neurons throughout the whole rostrocaudal extension of the dorsal thalamus. These neurons are more numerous near the injection site, but even far away a considerable number of stained neurons can be found. The neurons are, with few exceptions, restricted to the region of the dorsal thalamus; only a few stained neurons are found that were clearly situated ventrally to the sulcus medialis. These neurons did not show axons.

In another case, where the injection was made in the vicinity of the CGT, i.e. in the ventral thalamus, only neurons in the ventral thalamus were stained, but as in the previous case they extended in a horizontal band rostrocaudally throughout the whole diencephalon. Surprisingly few cells were found in the gray matter directly below the injection site. In all cases the stained neurons did not form a continuous band, but showed a rather "scattered" arrangement; i.e., there were many neurons in between that obviously had not been filled.

The fact that despite rather small injection sites, horizontal bands of retrogradely stained neurons were always found throughout the diencephalon, leads to the conclusion that there must be an enormous overlap of diencephalic neurons with regard to their dendritic arborization. This overlap seems to be restricted to the horizontal bands formed by the horizontal sulci

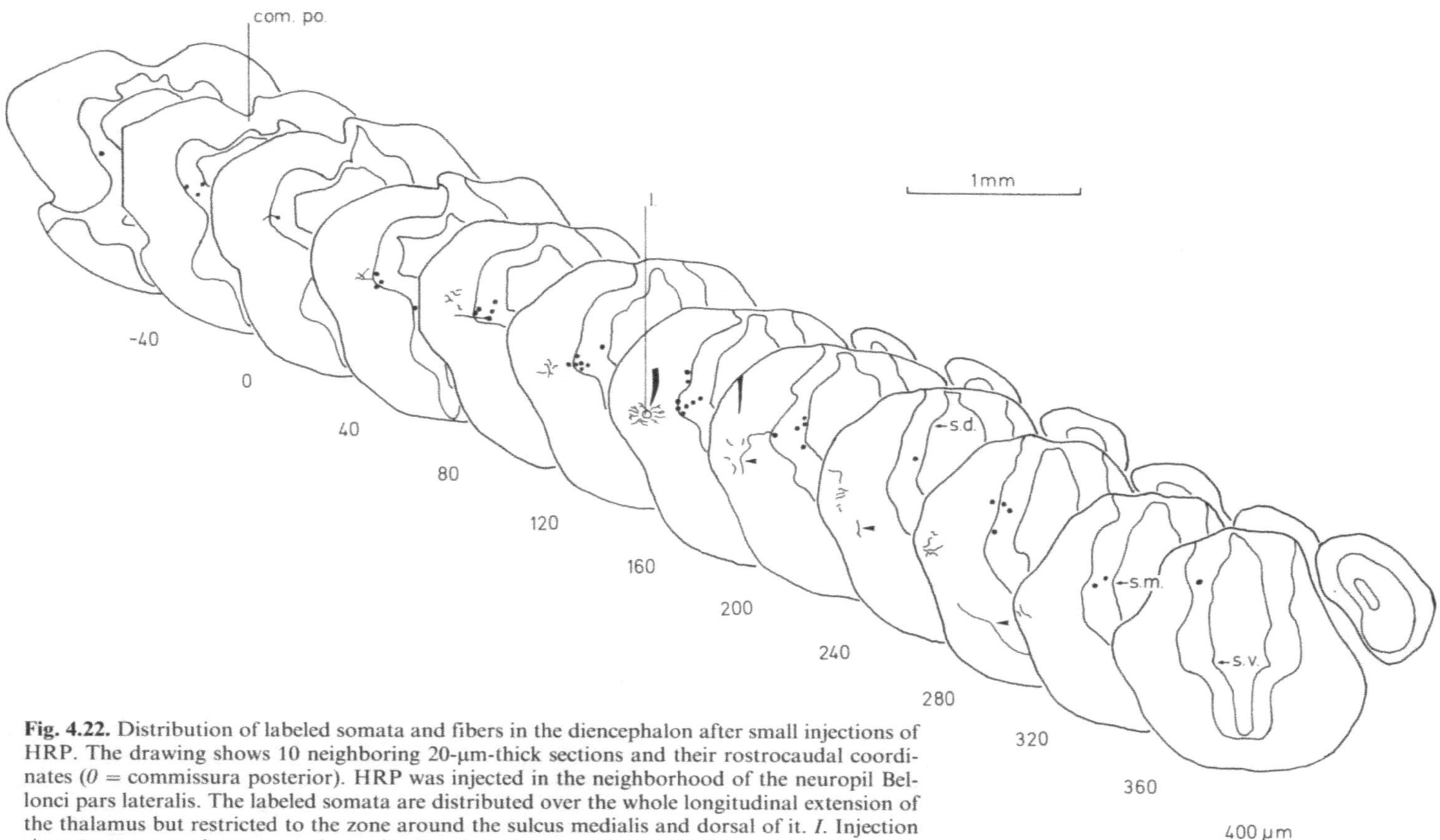

Fig. 4.22. Distribution of labeled somata and fibers in the diencephalon after small injections of HRP. The drawing shows 10 neighboring 20-µm-thick sections and their rostrocaudal coordinates (*0* = commissura posterior). HRP was injected in the neighborhood of the neuropil Bellonci pars lateralis. The labeled somata are distributed over the whole longitudinal extension of the thalamus but restricted to the zone around the sulcus medialis and dorsal of it. *I.* Injection site; *com.po.* commissura posterior; *s.d.* sulcus dorsalis; *s.m.* sulcus medialis; *s.v.* sulcus ventralis. (From Wicht 1984)

(sulcus dorsalis, s. medialis, s. ventralis). Some neurons had a dendritic extension of 480 µm.

The conclusion from these studies may be that there is only a dorsoventral functional separation of the periventricular gray matter, but no sharp rostrocaudal one. This means that the clear-cut organization of the diencephalic visual neuropils finds no analog in a similar nucleization of the periventricular gray matter. Many neurons from very different parts of one horizontal band seem to make contact with one neuropil. Considering the large horizontally extended dendritic arborization of thalamic neurons one is inclined to assume that one neuron has contact with more than one neuropil. Wicht stressed the fact that most neurons showed the morphological character of a pseudo-unipolar cell which is a cell type often found in nonnucleized and nonlaminated parts of the brain and in which the axon usually arises from the base of the dendritic arborization. Unfortunately, experiments in the anuran diencephalon with comparably small injections of HRP are lacking, so that it is difficult to decide whether there are indeed fundamental differences between anurans and salamanders with regard to the structural organization of the connectivity between visual neuropil and secondary visual neurons.

4.4 Central Visual and Visuomotor Pathways

4.4.1 Tectal Afferents and Efferents

The afferents and efferents of the optic tectum in salamanders have been revealed recently by means of the HRP method by Finkenstädt et al. (1983) in *S. salamandra*, by Rettig (1984) in *S. salamandra* and *Bolitoglossa subpalmata*, and by Roth et al. (in prep.) in a larger number of plethodontids.

4.4.1.1 Afferents (Figs. 4.23, 4.24, and 4.25)

After unilateral injection of HRP, in the optic tectum retrogradely stained neurons were found in the following parts of the brain:

1. Telencephalon (Figs. 4.23b and 4.24b): Retrogradely labeled cells of small diameter (6–9 µm) are found in the rostrolateral amygdala and striatum after large injections into the rostral tectum (Rettig 1984). Finkenstädt et al. (1983) also reported stained neurons in the ventral part of the lateral pallium. The labeled neurons in the telencephalon send their axons to the tectum via the lateral forebrain bundle.

2. Diencephalon (Figs. 4.23c–f, and 4.24c–e): All studies report three groups of retrogradely labeled neurons in the diencephalon:

a) Ventral thalamus, at the level of the sulcus ventralis thalami. This "nucleus" appears only ipsilateral to the injection site and is rather narrow in its dorsoventral extension and is more extended rostrocaudally. The stained neurons are of medium size (10–16 μm) with long apical shafts and narrow dendritic arborization. This region projects mostly to the rostral tectum. No labeling of neurons in this region occurs after injection of HRP into the caudal tectum.

b) Dorsal thalamus. Here a group of medium-sized cells (about 14 μm) is found ipsilaterally at the level of the sulcus medialis thalami. This region corresponds well to the "nucleus Bellonci" of Herrick. The cells again have long apical shafts with narrow dendritic trees. According to Rettig, this nucleus projects exclusively to the rostral tectum. Finkenstädt et al. found that its rostral part projects to the rostral tectum, whereas the more caudal part projects to the whole tectum.

c) Praetectum. Stained neurons are found both in the caudal praetectum below the rostral tectum and rostral to the optic tectum and immediately below the commissura posterior. Finkenstädt et al. reported that the caudal praetectum (which comprises the nucleus Darkschewitsch) shows ipsilateral projections to the tectum, whereas the rostral part (their "pretectal nucleus proper") projects bilaterally to the tectum. Rettig reported bilateral projections of the nucleus praetectalis mostly to the rostral, but also to the lateral and medial tectum, after injections penetrating deep tectal fiber layers. Many neurons are filled via dendrites; these neurons have axons coursing to the medulla oblongata. Other neurons are filled via axonic retrograde transport. These neurons are medium-sized (12–16 μm) and mostly possess narrow dendritic arborization. The n. Darkschewitsch shows a few stained cells after injection into the very rostral tectum.

In general, the praetectum projects mostly to the rostral tectum opticum, to a lesser degree to the lateral and medial tectum, and not at all to the caudal tectum.

No stained cells occurred in the n. praeopticus or in the ipsilateral hypothalamus.

3. Tectum (Figs. 4.23 g–h, and 4.24 f): Only a few labeled neurons are found in the contralateral tectum. Some of them are very large in diameter and are located near the midline of the tectum and predominantly in the caudal part. These are presumably cells of the n. mesencephalicus nervi trigemini. The remaining neurons had mostly short apical shafts and large dendritic arborization.

4. Tegmentum mesencephali (Figs. 4.23 g–h, and 4.24 f): Labeled cells were found exclusively in the dorsal tegmentum. These are medium- to large-sized neurons (16–22 μm) with short apical shafts and large dendritic arborization, mostly extended in a dorsolateral direction. Only a few neurons are found in the contralateral dorsal tegmentum which send their axons via the commissura tecti mesencephali.

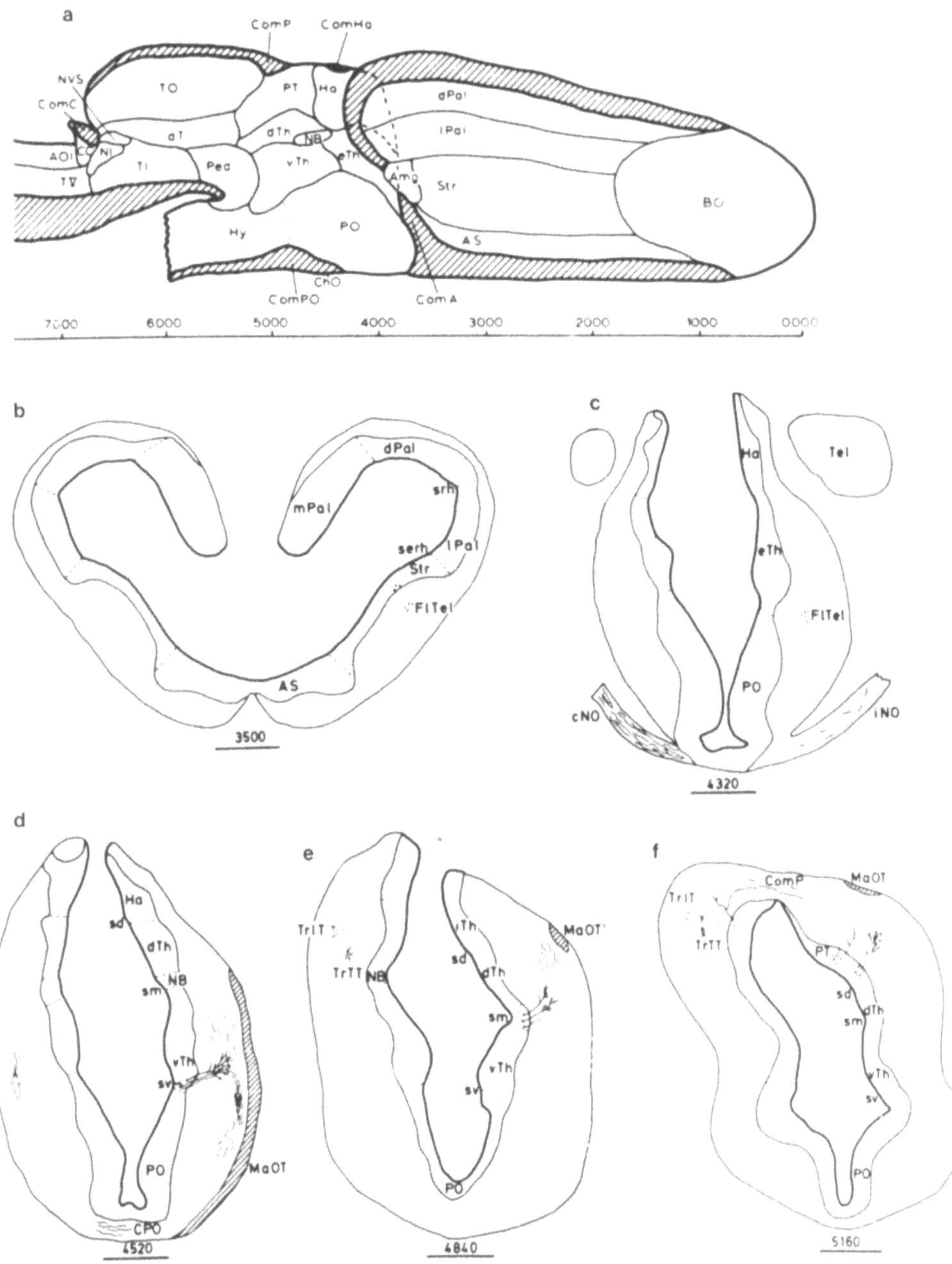

Fig. 4.23 a–f. Tectal afferents and efferents in *S. salamandra* as revealed by injection of HRP into the tectum (injection site: *hatched area* in **g** and **h**). **a** Schematic drawing of a sagittal section through the brain with rostrocaudal coordinates (μm). **b–j** Drawings from transverse sections through the brain at coordinates indicated below each drawing. **b** Caudal forebrain; **c** rostral diencephalon; **d** mid-diencephalon at the level of the neuropil Bellonci and the commissura postoptica; **e** mid-diencephalon at the level of the posterior end of the neuropil Bellonci; **f** caudal diencephalon/praetectum at the level of the commissura posterior

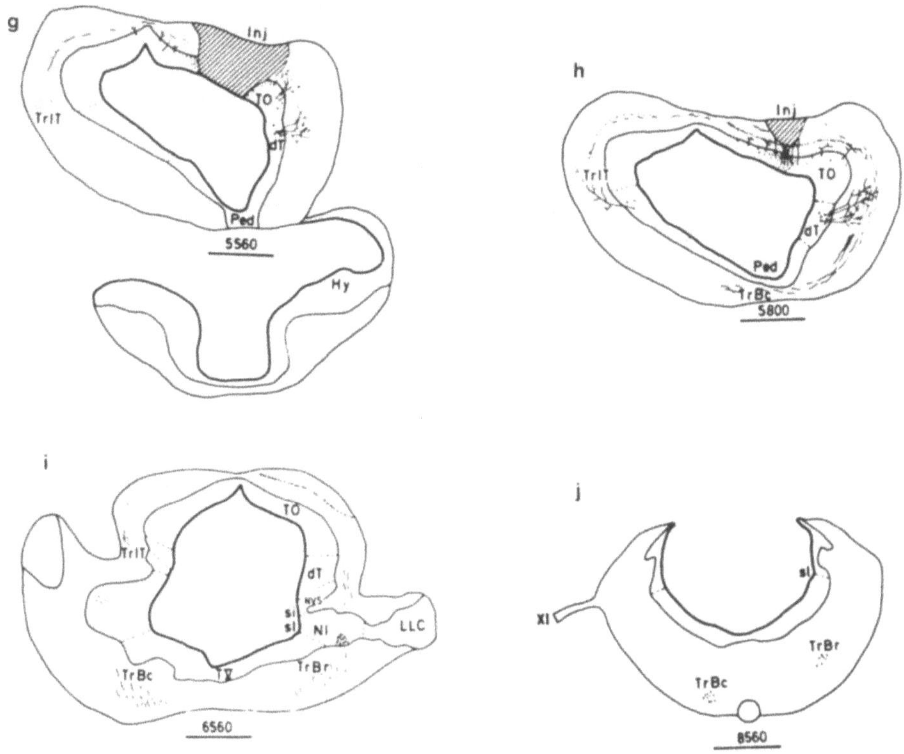

Fig. 4.23g–j. Rostral mesencephalon/hypothalamus; **h** mid-mesencephalon at the level of the commissura tuberculi posterioris; **i** caudal mesencephalon/rostral medulla oblongata at the level of the nucleus isthmi; **j** mid medulla oblongata at the level of the entrance of the 11th nerve. *Bar* = 400 μm. For abbreviations, see list. (From Rettig, submitted)

5. Tegmentum isthmi (Figs. 4.23i, 4.24g, and 4.29): Rettig found labeled cells only in the area ventrolateral to the nucleus cerebelli and rostrodorsal to the tegmentum trigemini. The nucleus visceralis secundarius and the rest of the tegmentum isthmi are unlabeled. The labeled region can be identified as the nucleus isthmi. Rettig found a somewhat "blurred" topic projection of this nucleus isthmi to the optic tectum: the ventrolateral part of the nucleus projects to the rostral tectum, whereas the dorsalmost part projects to the caudal tectum. The medially situated parts project to the medial and lateral tectum, with the tendency of the ventromedial part to project to the medial tectum and the dorsolateral part to the lateral tectum. Only the projections to the rostral and the caudal quadrants of the tectum show a clear separation. The contralateral nucleus isthmi shows stained cells only after injections into the rostral, lateral, and medial tectal quadrants, but not after injections into the caudal quadrant.

6. Medulla oblongata (Fig. 4.23j): In the medulla bilateral stainings of cells in the nucleus reticularis medius and in the nucleus vestibularis magno-

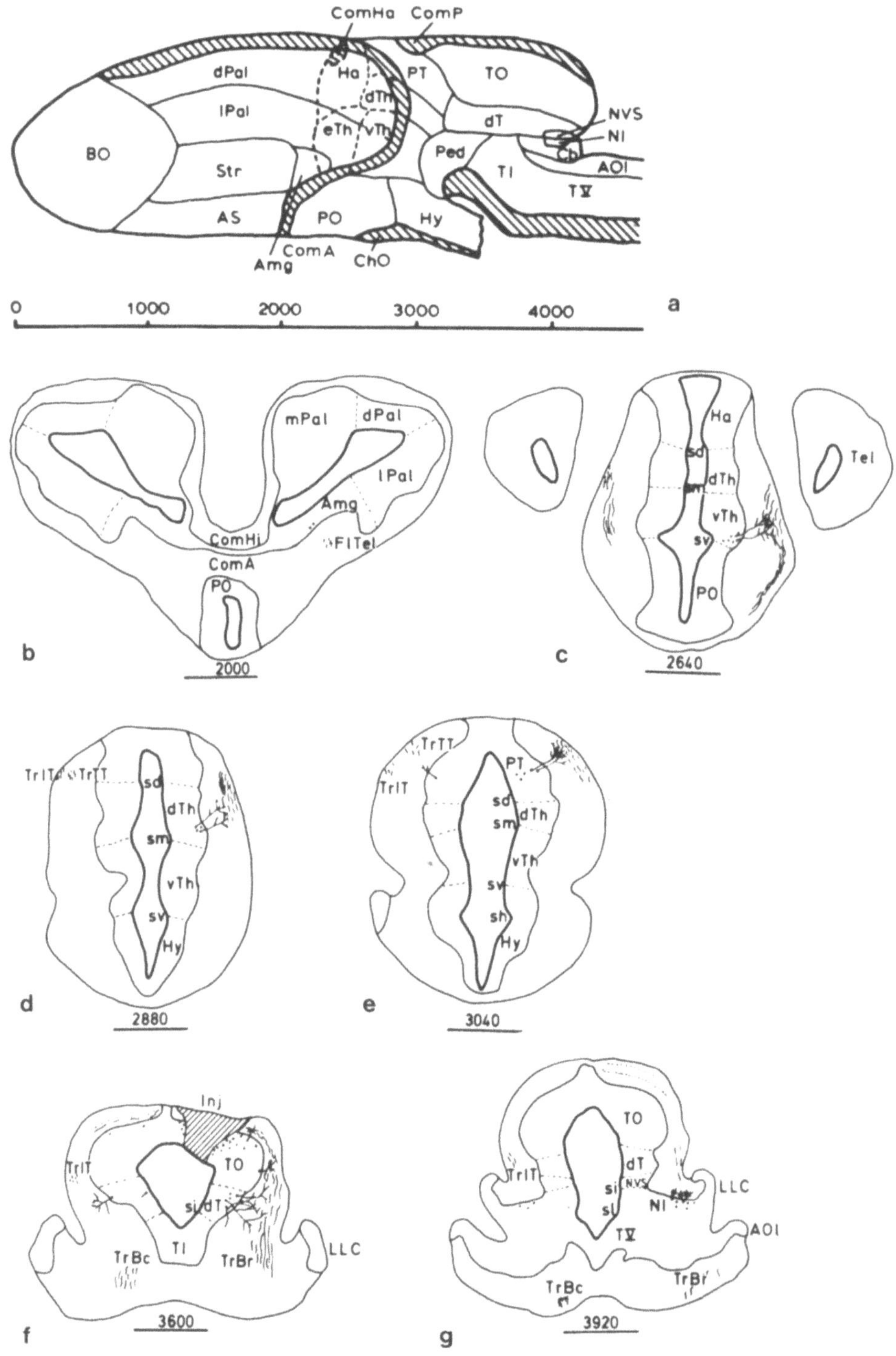

Fig. 4.24a–g. Tectal afferents and efferents in *Bolitoglossa subpalmata* as revealed by injection of HRP into the tectum. **a** Schematic drawing of sagittal section through the brain; **b–g** drawings from transverse sections through the brain at coordinates indicated below the drawing. **b** Caudal forebrain at the level of the commissura anterior (commissura hippocampi, c. amygdalae); **c** rostral diencephalon; **d** mid-diencephalon; **e** caudal diencephalon/praetectum; **f** mid-mesencephalon; **g** caudal mesencephalon/rostral medulla oblongata at the level of the nucleus isthmi. *Bar* = 400 μm. For abbreviations, see list. (From Rettig submitted)

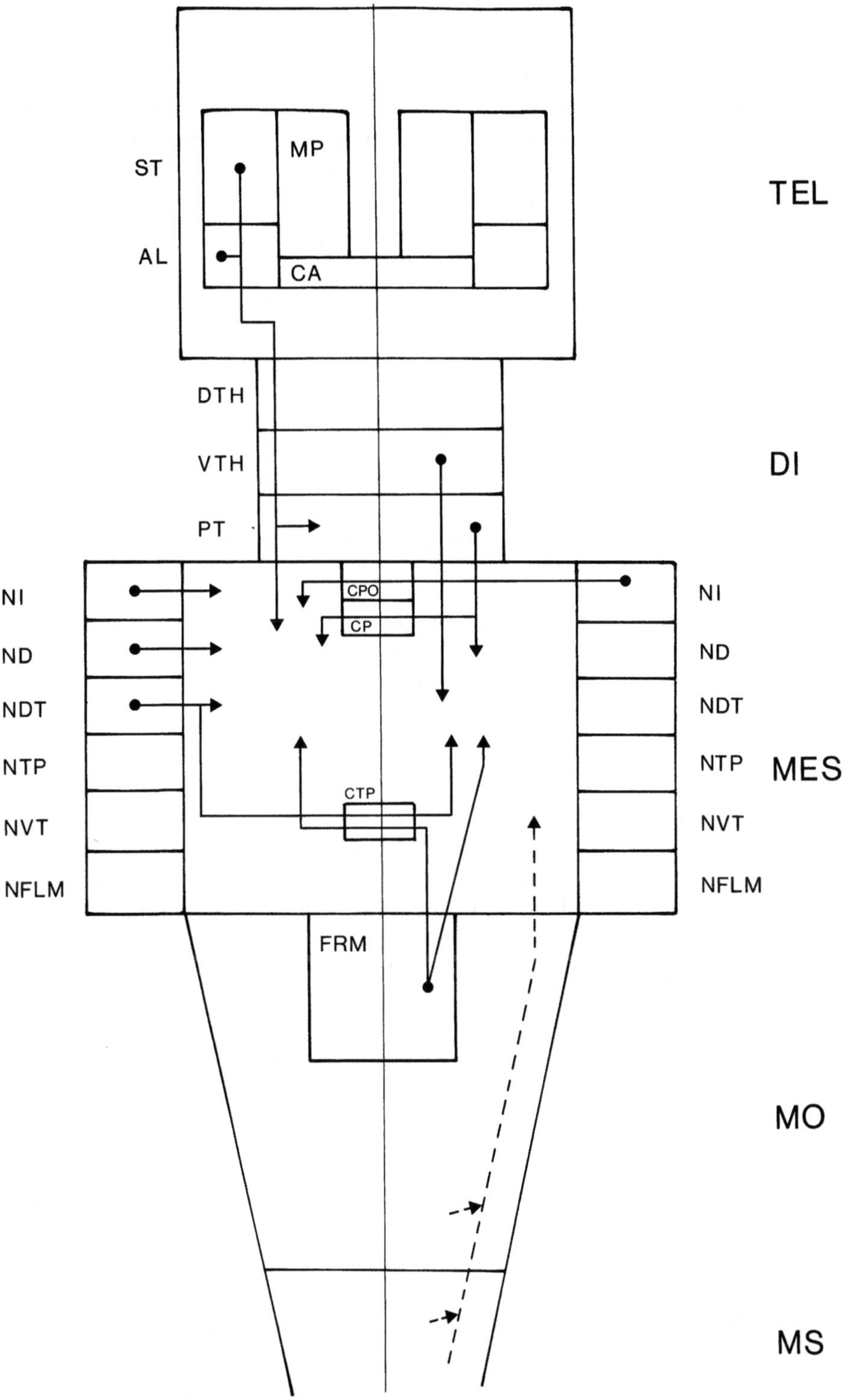

Fig. 4.25. Schematic drawings of tectal afferents in salamanders based on the studies discussed in the text. *Solid lines*: Pathways involved in vision; *dotted line*: lemniscus spinalis (somatosensory pathway). For abbreviations, see list

cellularis (after Opdam and Nieuwenhuys 1976) are found. Finkenstädt et al. as well as Roth et al. found bilateral, Rettig unilateral stainings in the area octavo-lateralis which occurred only after unilateral injection in the caudal part of the tectum. These cells are large (22–24 μm) with short apical dendrites and large dendritic arborization.

7. Spinal cord: Finkenstädt et al. reported stained cells in the dorsal gray matter of the ipsilateral spinal cord. Rettig found a few bilaterally stained cells predominantly located in the ventral horn of the gray matter.

No afferents from the cerebellum to the tectum seem to exist in urodeles or in anurans.

4.4.1.2 Efferents (Figs. 4.23h–j, 4.24f–g, and 4.26)

After unilateral injection into the tectum the following descending and ascending tracts are stained:

1. Tractus tecto-bulbaris (tecto-spinalis) cruciatus: This tract contains axons of neurons in the tectum and dorsal tegmentum. The fibers cross in the commissura tuberculi posterioris and then run rostroventrally to the lemniscus fibers to the medulla and extend along the midline. In well-stained brains this tract can be traced far caudally at least up to the dorsal root of the second spinal nerve and is, thus, also a tecto-spinal tract.

2. Tractus tecto-bulbaris rectus. This tract runs between the tractus tecto-bulbaris cruciatus and the lemniscus down to the obex region and sends collaterals to all tegmental nuclei and the formatio reticularis. It descends at least as far caudally as to the level of the root of the third spinal nerve. It is, thus, identical with the uncrossed tractus tecto-spinalis which other authors considered as a separate tract. Part of the fibers of the tecto-bulbar tract cross in the ventral commissure of the medulla and then ascend to the area acustico-lateralis. In the medulla oblongata, the tract is broad, with a concentration of fibers at the lateral edge. At the level of the root of seventh and eighth nerves, a small proportion of fibers deviates laterally and ascends obliquely in dorsocaudal direction to the level of the root of the tenth cranial nerve.

3. Tractus tecto-isthmicus. It terminates in the neuropil of the n. isthmi in a layered, topographic manner.

4. Tractus tecto-thalamicus rectus et cruciatus and tractus tecto-isthmicus cruciatus. The uncrossed tract shows numerous terminals in the dorsal thalamus, between the neuropil Bellonci and the posterior thalamic neuropil, fewer terminals in the praetectum and the ventral thalamus. The crossed tract, constituting the commissura postoptica, divides into a rostral and a caudal part. The fibers of the rostral part mostly belong to the tecto-tectal

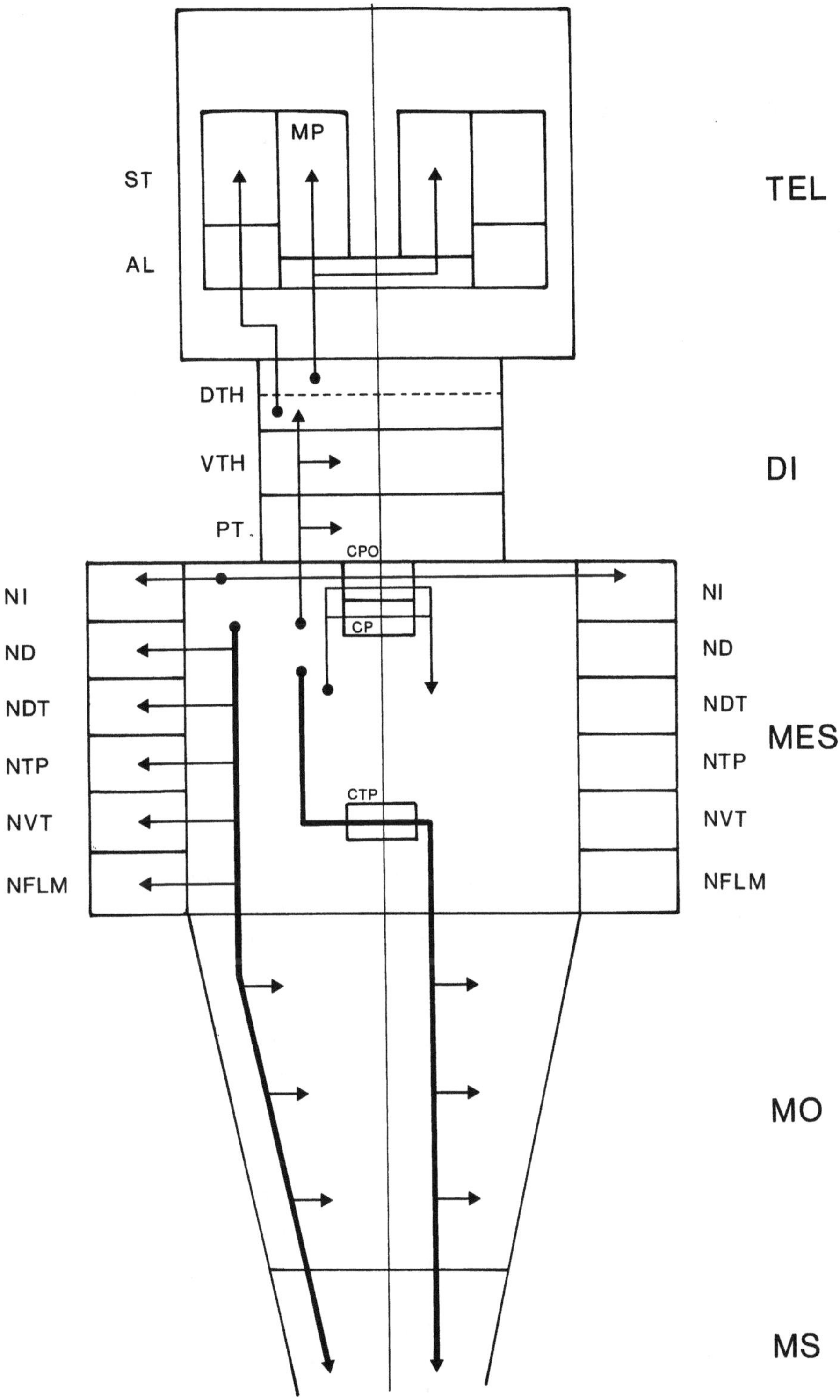

Fig. 4.26. Schematic drawing of tectal and dorsal-thalamic efferents. *Thick lines*: Crossed (medial) and uncrossed (lateral) descending tectal pathways; *thin lines*: other tectal and thalamic efferents; *thin lines* deviating from the thick ones represent collateral fibers of the descending tecto-bulbar/spinal tracts. For abbreviations, see list

tract; within the tectum they extend far caudally beneath the retinal afferents. Only a few fibers terminate in the contralateral thalamus. The posterior tract runs dorsocaudally to the contralateral n. isthmi, forming the tractus tecto-isthmicus cruciatus.

5. Tractus tecto-praetectalis et tecto-peduncularis: Fibers of this tract cross in the commissura posterior and run to the praetectum and the ventral tegmentum.

6. Tractus nuclei mesencephalici nervi trigemini: Thick fibers, originating from the tectal cells of the trigeminus descend ipsilaterally along the caudolateral edge of the tectum to the trigeminal root where they bifurcate and either leave the brain through the root or extend further caudally.

4.4.2 Afferent and Efferent Pathways of the Praetectum (Fig. 4.27)

Afferent and efferent pathways of the praetectum have been studied by Naujoks-Manteuffel and Manteuffel (1986). After injection of HRP into the tr. tecto-bulbaris at the level of the seventh cranial nerve, retrogradely filled neurons, constituting efferent pretectal pathways to the medulla oblongata, were found bilaterally in the nucleus praetectalis profundus, ipsilaterally in the n. praetectalis superficialis and in the n. Darkschewitsch which, according to its position, corresponds to the "caudal praetectum" of Finkenstädt et al. (1983).

The pretectal neuropil receives afferents from the nucleus Bellonci of Herrick (1948), the n. ventralis tegmenti, and the nucleus of the basal optic neuropil. This latter nucleus projects into the central core of the neuropil which is free of ipsilateral retinal afferents. Afferents via dendrites come, apart from the n. praetectalis profundus and superficialis, from the nucleus Darkschewitsch, the n. dorsalis tegmenti, and the n. fasciculi longitudinalis medialis. The same HRP experiments revealed that a rostrodorsal thalamic cell group, the nucleus praetectalis, the nucleus Darkschewitsch, the n. fasciculi longitudinalis medialis and the n. dorsalis and n. ventralis tegmenti project to their contralateral counterpart via the commissura posterior and via the commissura tecti diencephali in the case of the rostrodorsal thalamic cell group.

Naujoks-Manteuffel and Manteuffel (1986) also studied the connectivity of the migrated nucleus of the basal optic neuropil (nBON). This nucleus possesses strong reciprocal connections with the praetectum. It receives afferents from the n. ventralis tegmenti, n. Darkschewitsch, n. praetectalis profundus, n. Bellonci, a cell group within the ventral thalamus, and a cell group in the n. praeopticus. The nBON projects to the n. fasciculi longitudinalis medialis, the n. ventralis tegmenti, and the nuclei of the eye muscle nerves, i.e. the oculomotor and trochlear nuclei.

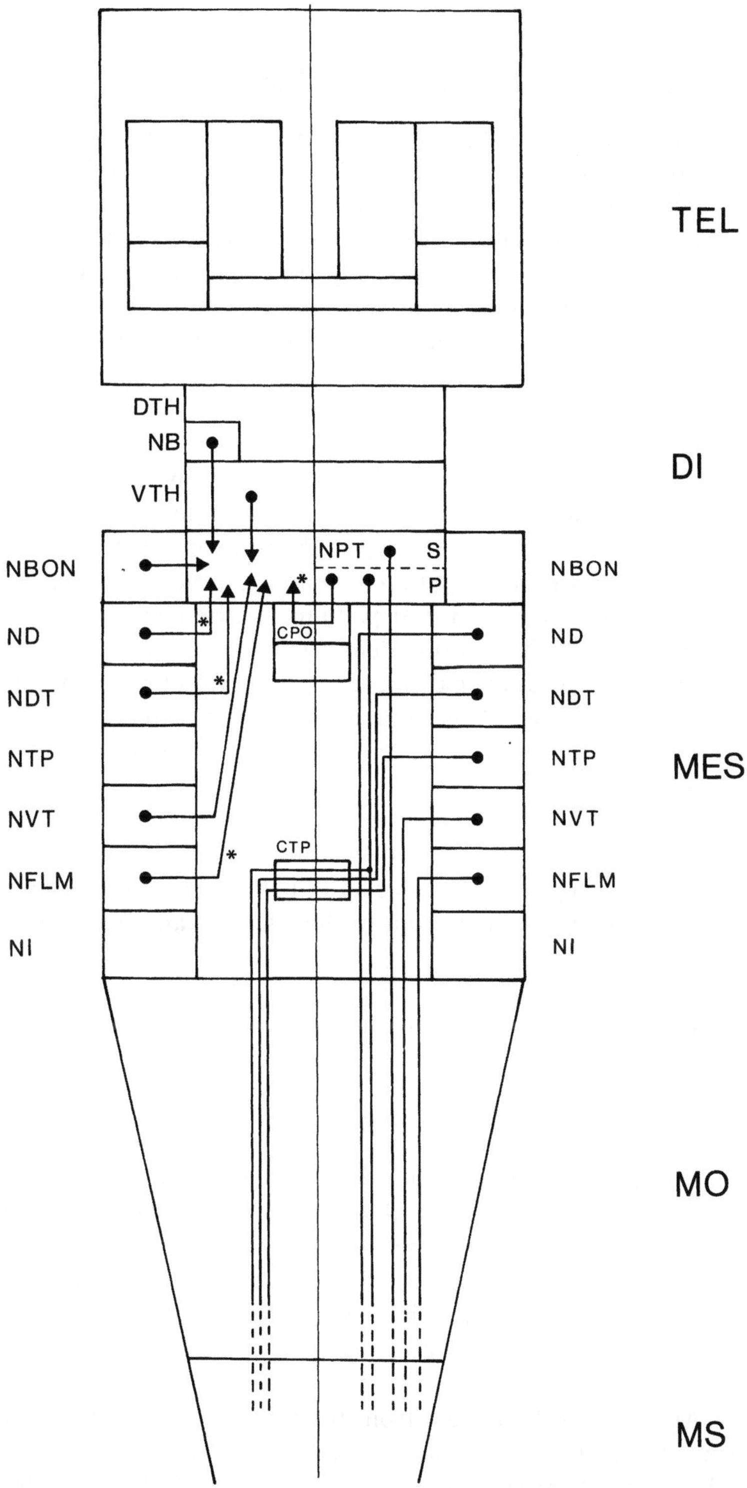

Fig. 4.27. Schematic drawing of afferents to the praetectum (*left side*) and efferents from the praetectum and mesencephalic tegmental nuclei (*right side*). *Asterisk* indicates dendritic afferents to the pretectal nucleus

4.4.3 Afferents and Efferents of the Thalamus

Wicht and Himstedt (1986) studied the connectivity of the thalamus in the newt *Triturus alpestris* (Fig. 4.28). After transsection of the medial forebrain bundle and application of HRP onto the disrupted fibers at the level of the mid-telencephalon, labeled somata were observed bilaterally in the anterior dorsal thalamus. The contralaterally projecting fibers crossed in the commissura anterior. The same bilateral thalamic labeling was observed after injection into the medial pallium and the medial part of the dorsal pallium. These regions are, thus, the termination sites of axons from the labeled thalamic neurons.

Injections of HRP into the striatum resulted in labeled somata in the ipsilateral, posterior dorsal thalamus. Correspondingly, injections of HRP into this region resulted in labeled fibers running rostrally in the lateral forebrain bundle and terminating in the ipsilateral neuropil of the striatum.

The anterior thalamic group of neurons, projecting to the medial pallium and medial part of the dorsal pallium via the medial forebrain bundle, is located medial to the neuropil Bellonci and may be identical with the nucleus Bellonci. The posterior thalamic cell group projecting to the striatum via the lateral forebrain bundle is situated beneath the lateral neuropil which does not receive direct afferents from the retina, but is the termination field of tectofugal fibers.

In summary, these experiments show the existence of two thalamo-telencephalic pathways, a retino-thalamo-telencephalic pathway, and a tecto-thalamo-telencephalic one. This latter pathway is of importance, because no direct projection of the tectum to the telencephalon seems to exist.

4.4.4 Summary and Comparison of Tectal and Diencephalic Afferents and Efferents

4.4.4.1 Tectal Afferents

The studies of Finkenstädt et al. (1983) and Rettig (1984 and in prep.) revealed a telencephalo-tectal connection in salamanders. It is unclear whether the tectum receives telencephalic input only from the amygdala and striatum, or from the lateral pallium, too. Similar to Rettig, Kokoros and Northcutt (1977) did not find pallio-tectal connections. Also, Wicht and Himstedt (1986) found no projection of the tectum or thalamus to the lateral pallium.

An input to the tectum from the forebrain is known from many vertebrate groups (e.g., Vanegas and Ebbesson 1973, in teleosts; Ebbesson and Schroeder 1971, in sharks; Butler and Ebbesson 1975, in reptiles), but has not been shown in anurans (Trachtenberg and Ingle 1974, Wilczynski and North-

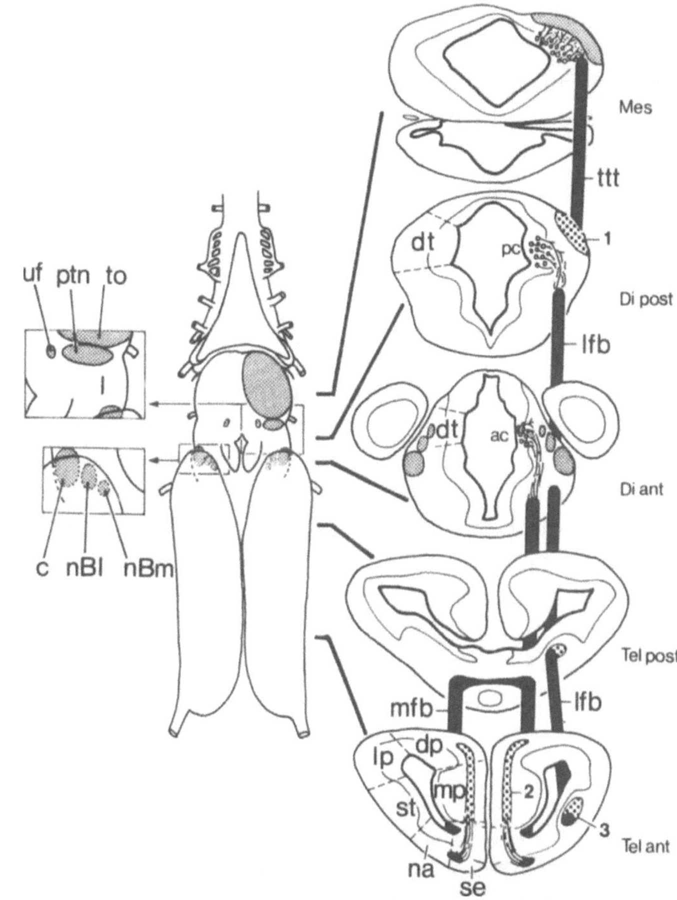

Fig. 4.28. Schematic diagram representing diencephalic connections and termination sites of retinofugal fibers in *Triturus alpestris*. *Left*: Dorsal view of the brain showing distribution and arrangement of projection sites of the right retina (*fine stippling*). *Right*: Transverse sections at various levels. Retinal projections are *stippled* as on the left, fiber tracts are represented by *black bars*. Termination sites of central ascending systems are depicted by *black dotted areas* (*1–3*). Populations of projection neurons are represented by *open circles* in the periventricular stratum griseum. *ac* Anterior cell group; *c* corpus geniculatum thalamicum; *Di ant* anterior diencephalon; *Di post* posterior diencephalon; *dp* dorsal pallium; *dt* dorsal thalamus; *l* lateral neuropil; *lfb* lateral forebrain bundle; *lp* lateral pallium; *Mes* mesencephalon; *mfb* medial forebrain bundle; *mp* medial pallium; *na* nucleus accumbens; *nBl*: neuropil Bellonci pars lateralis; *nBm* neuropil Bellonci pars medialis; *pc* posterior cell group; *ptn* posterior thalamic neuropil; *se* septum; *st* striatum; *Tel* telencephalon; *to* tectum opticum; *ttt* tectothalamic tract; *uf* uncinate field; *1* tectofugal projection site in the lateral neuropil; *2* thalamofugal projection site in *mp* and *dp*; *3* thalamofugal projection site in *st*. (From Wicht and Himstedt 1986)

cutt 1977). The existence of a telencephalo-tectal projection would be an important difference between the anuran and the urodele brain.

There is conformity about the projections of two visual thalamic centers to the tectum: one in the ventral and one in the dorsal thalamus, the "nucleus

Bellonci". Finkenstädt et al. (1983) reported a somewhat separate third center in the anterior dorsal thalamus, that projects exclusively to the rostral tectum. Similarly, the authors distinguished between two centers in the praetectum, the more rostrally situated "nucleus praetectalis proper", which shows bilateral projections to the tectum, and a more posterior pretectal field (which, according to Naujoks-Manteuffel and Manteuffel, 1986, is the nucleus Darkschewitsch) projecting ipsilaterally to the tectum.

The existence of a nucleus isthmi seems to be a common feature in most lower vertebrates (holosteans, Northcutt 1982; teleosteans, Grover and Sharma 1981, Fiebig et al. 1983; anurans, Gruberg and Lettvin 1980, Dann and Beazley 1982, Grobstein and Comer 1983; reptiles, Wang et al. 1981). Herrick named this region tegmentum isthmi, but denied the existence of a nucleus isthmi proper, mostly due to the absence of a clear anatomical differentiation. Larsell (1967) considered the nucleus isthmi to be part of the nucleus cerebelli, whereas according to Opdam and Nieuwenhuys (1976) it is identical with the nucleus visceralis secundarius. Gruberg and Harris (1981) described a nucleus isthmi found by means of HRP injection, and related it to the identification done by Opdam and Nieuwenhuys. However, the position of the nucleus described by Gruberg and Harris corresponds to the nucleus found by Rettig (cf. Fig. 4.29).

The anuran nucleus isthmi has been the subject of intense research (Gruberg and Lettvin 1980, Dann and Beazley 1982, Grobstein and Comer 1983, Udin 1985). Some authors see this nucleus as a relay station of tecto-tectal connections (Glasser and Ingle 1978, Grobstein and Comer 1983, Gruberg and Lettvin 1980, Gruberg and Udin 1978, Udin 1985). All authors found a topic arrangement of the projections from and to the tectum, i.e., the retinotectal and the isthmotectal maps are in register, which is also found in salamanders. This and the more or less similar relative position of the region identified in the isthmic tegmentum in salamanders justifies the assumption of the existence of a nucleus isthmi in salamanders. Furthermore, the nucleus isthmi has been clearly identified in salamanders by means of electrophysiological recordings and marking of the recorded cells (Wiggers and Roth in prep.).

Both Finkenstädt et al. and Rettig described the staining of two nuclei in the medulla oblongata projecting to the tectum, mostly after large injections into the caudal part of the tectum. The first nucleus is situated in the formatio reticularis and is bilaterally stained. It is described in most anamniotes (selachians, Smeets 1982; holosteans, Northcutt 1982; teleosteans, Fiebig et al. 1983; anurans, Wilczynski and Northcutt 1977). According to Gruberg and Harris (1981) in *Ambystoma* and Grover and Sharma (1981) in *Carassius* it can be doubted that this projection is related to vision; it is more likely to be somatosensory, as was shown by Gruberg and Harris by means of specific staining of serotonergic transmitters known to be typical for somatosensory

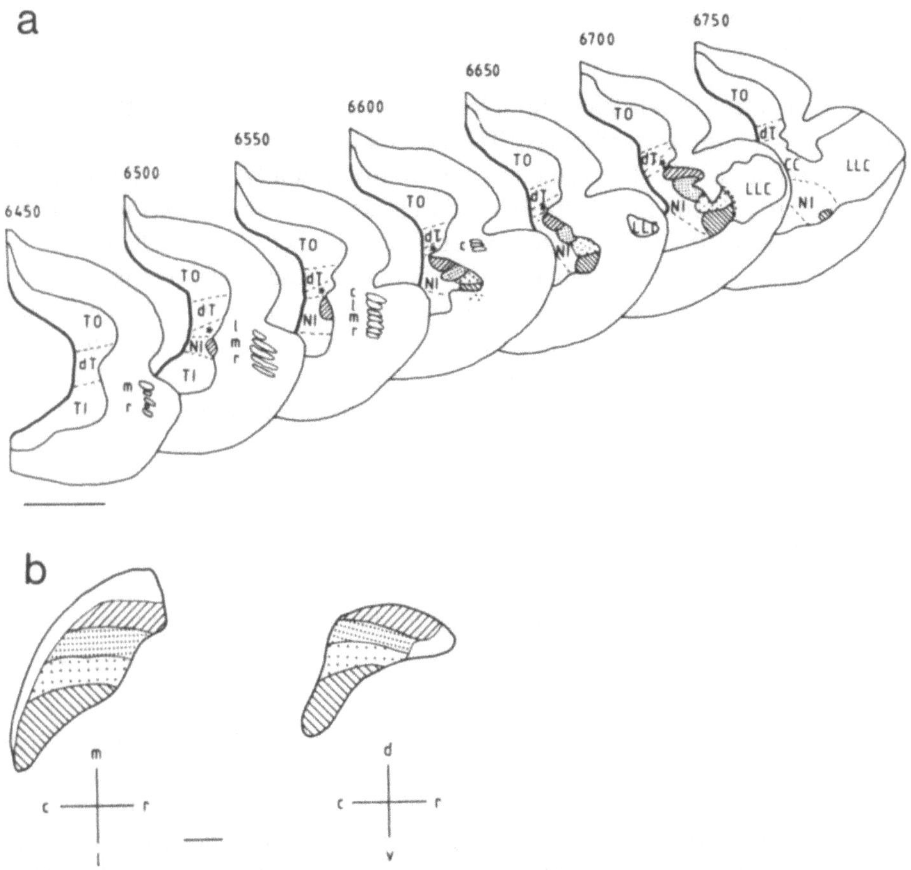

Fig. 4.29 a–b. Topic organization of the neuropil and nucleus isthmi in *S. salamandra* as revealed by injection of HRP into the tectum opticum. **a** Series of coronal sections through the midbrain at the level of the isthmic area connected to the four tectal quadrants. *Left-hatched*: Connection to the rostral tectum; *loose dotting*: connection with the medial tectum; *dense dotting*: connections with the lateral tectum; *right-hatched*: connections to the caudal tectum. Lateral to the nucleus isthmi, the isthmic neuropil is drawn showing a layered structure corresponding to connections to the caudal (*c*), lateral (*l*), medial (*m*) and rostral (*r*) tectum. Star indicates position of the nucleus visceralis secundarius. *Bar* = 400 µm. **b** Schematic dorsal (*left*) and lateral (*right*) aspect of the nucleus isthmi. *Symbols* are the same as in **a**. *Bar* = 300 µm. (From Rettig submitted)

fibers. According to these authors the fibers project immediately below visual units in the superficial layer of the tectum. Our experiments show a clear association of this nucleus with the ascending lemniscus fibers (Roth et al. in prep.). The same seems to be true for the second nucleus within the area octavo-lateralis which, according to Rettig, is stained only contralaterally. This nucleus is also found in other anamniotes.

4.4.4.2 Tectal Efferents

A tractus tecto-thalamicus (et hypothalamicus) rectus et cruciatus seems to exist in all anamniotes (Smeets 1982, Grover and Sharma 1979, Wilczynski and Northcutt 1977). The same is true for the caudally descending tractus tecto-bulbaris rectus and tecto-bulbaris cruciatus. In anurans the crossed tract is stronger than the uncrossed one (Wilczynski and Northcutt 1977, Gruberg and Harris 1981) whereas in salamander and fishes the opposite is the case (Smeets 1982, Grover and Sharma 1979, Luiten 1981, Roth et al. in prep.). There seems to be no direct tecto-telencephalic connection. The connection is mediated in both anurans and urodeles by the posterior thalamic cell group described by Wicht and Himstedt (1986) for salamanders and the lateral thalamic nucleus of the frog (see below).

4.4.4.3 Connections of the Diencephalon in Anurans and Comparison to Urodeles

Neary and Northcutt (1983) have studied the nuclear organization of the diencephalon in the bullfrog *Rana catesbeiana*, and the authors discussed the afferent and efferent connections of this part of the brain. According to this study, within the dorsal thalamus the nucleus anterior receives afferents from the medial pallium, the ventral hypothalamus, the posterior entopeduncular nucleus, the posterior thalamic nucleus, and the obex region, and indirect visual input from the retina via the nucleus Bellonci; it projects to the dorsal, especially medial pallium, the n. posterior thalami, the ventral hypothalamus, and the optic tectum.

The connectivity of the anterior thalamic nucleus of salamanders, as described by Wicht and Himstedt (1986), corresponds exactly to the nucleus anterior of the dorsal thalamus in *Rana catesbeiana* (cf. also Neary 1984).

The central nucleus receives afferents from the torus semicircularis and projects to the striatal complex, whereas the lateral nucleus receives afferents from the optic tectum and projects back to the optic tectum and to the torus. As the authors stated, the middle zone nuclei are "important elements in pathways carrying auditory and visual information to the ventrolateral telencephalic wall". Furthermore, this zone seems to be involved in the pathways transmitting somatosensory information to the striatal complex. Again, the tecto-thalamo-striatal pathway involving the lateral nucleus corresponds to that described by Wicht and Himstedt involving the posterior thalamic cell group.

The posterior nucleus does not project to the telencephalon. It receives afferents from the ventral hypothalamus and the obex region. It projects to the anterior nucleus and the dorsal part of the ventrolateral nucleus, the tectum opticum, the torus semicircularis, and possibly to the brain stem and the spinal cord. The posterodorsal division of the lateral nucleus receives afferents, mostly together with the posterior nucleus from the anterior entopeduncular nucleus, the anterior preoptic area, the dorsal part of the ventrolateral nucleus and the nucleus Bellonci. The posterodorsal part of the lateral nucleus receives afferents from the striatal complex and the tectum. It projects to the optic tectum and the brain stem.

Within the ventral thalamus, the afferents and efferents are less well known and the definition of nuclear masses is much less well established. The ventrolateral nucleus has no projections to the telencephalon, but has reciprocal connections with the posterior thalamic zone.

The nucleus Bellonci, placed by the authors into the ventral thalamus and assumed to be incomparable with the mammalian dorsolateral geniculate nucleus, receives dense retinal input via the neuropil Bellonci. Again, it has no connections with the telencephalon. It projects to the posterior thalamic zone and the optic tectum.

The ventromedial nucleus and the ventral part of the ventrolateral nucleus receive afferents from the obex region and the retina. They project to the optic tectum, and the ventromedial nucleus projects also to the spinal cord. The superficial ventral nucleus receives retinal input.

Montgomery et al. (1985) have recently described the afferents and efferents of the pretectal nucleus lentiformis mesencephali. Afferents come from the contra- and ipsilateral retina, leaving free the central core of the neuropil, from the nucleus of the basal optic neuropil, from the ipsilateral tectum, the nucleus rotundus, the mesencephalic pretectal gray matter, and the contralateral nucleus fasciculi longitudinalis medialis. The nucleus lentiformis mesencephali sends efferents to the BON, the tectum opticum, and to the medulla oblongata via the tractus tecto-bulbaris.

Recently, Tóth et al. (1985) examined the descending pathways in *Rana esculenta* using the cobalt-lysine method. They found *bilateral* descending pathways originating from the striatum ventrale, the area praeoptica, the n. ventralis hypothalami, the n. posterior thalami, the optic tectum, the n. anterodorsalis and anteroventralis mesencephali in the midbrain tegmentum, the n. vestibularis lateralis, and the reticular nuclei of the rhombencephalon. *Ipsilaterally*, uncrossed descending pathways were found originating in the n. ventrolateralis thalami, the n. fasciculi longitudinalis medialis, the n. profundus and the n. posteroventralis in the midbrain tegmentum, and in the n. tractus spinalis nervi trigemini. *Contralaterally*, crossed pathways were found originating from the nucleus ruber in the midbrain tegmentum, the n. cerebelli, the n. tractus solitarii and the n. vestibularis descendens in the rhombencephalon.

The authors found that the mesencephalic tegmental nuclei receive afferents from the contralateral tegmentum, the ipsilateral striatum, and bilateral medial-telencephalic input via the medial forebrain bundle.

All these findings in anurans correspond more or less exactly with those obtained from salamanders. It is, thus, remarkable that despite the striking differences in the gross morphology, for example the absence of clear lamination of the optic tectum and of the formation of nuclei in the diencephalon (except the n. praetectalis superficialis and the n. of the basal optic tract), the primary as well as secondary and tertiary visual pathways are essentially identical in anurans and urodeles. This again raises the question of the functional significance of anatomical features such as lamination and nucleization.

5 Neurophysiology of Visually Guided Behavior

5.1 Electrophysiological Methods

The neurophysiological data presented in this chapter were obtained by single cell recordings. The procedure of these experiments is more or less the following. Before the experiments the salamanders are immobilized by intraperitoneal injection of a 0.2–1% solution of succinyl-choline that blocks the transmission of nerve impulses onto cross-striped muscles. Since succinylcholine also blocks the respiration muscles, lung-breathing salamanders have to be covered with a thin layer of wet cellulose paper in order to facilitate oxygen uptake through the skin. Lungless salamanders have proved to be better with regard to the maintenance of oxygen supply during recording, due to a much higher vascularization of the skin. Here, too, a covering of the surface with wet cellulose paper is of advantage in order to avoid desiccation of the skin.

After complete immobilization of the animals (after 10–30 min) the skin above the parts of the brain from which recordings are to be made is removed under local anesthesia. In salamanders with bony skulls, the skull is opened by means of a dentist's drill. In salamanders that possess a cartilaginous skull, the skull can be opened using a microknife and microscissors. Then, the meninges are disrupted.

For the presentation of visual stimuli during the recording experiments, a so-called perimeter apparatus is usually used. As shown in Fig. 5.1, such a perimeter may consist of a platform on which the animal is placed and which carries the microelectrode holder. Around the platform a screen is moved three-dimensionally. On this screen visual stimuli are usually presented on a movable plastic tape on which the stimuli are fixed. The tape is driven by means of an electronically controlled electromotor and can be moved at different velocities as well as with different movement patterns. Usually the screen can be moved not only along the x-, y- and z-axes but can also be rotated around the z-axis in order to move the stimulus in different directions. An alternative type of perimeter apparatus consists of a large screen on which stimuli are projected and moved by means of movable mirrors that are electronically controlled. Here, the projection screen is moved only vertically and along the z-axis, the movement within the horizontal plane is achieved by rotating the platform on which the salamander is placed.

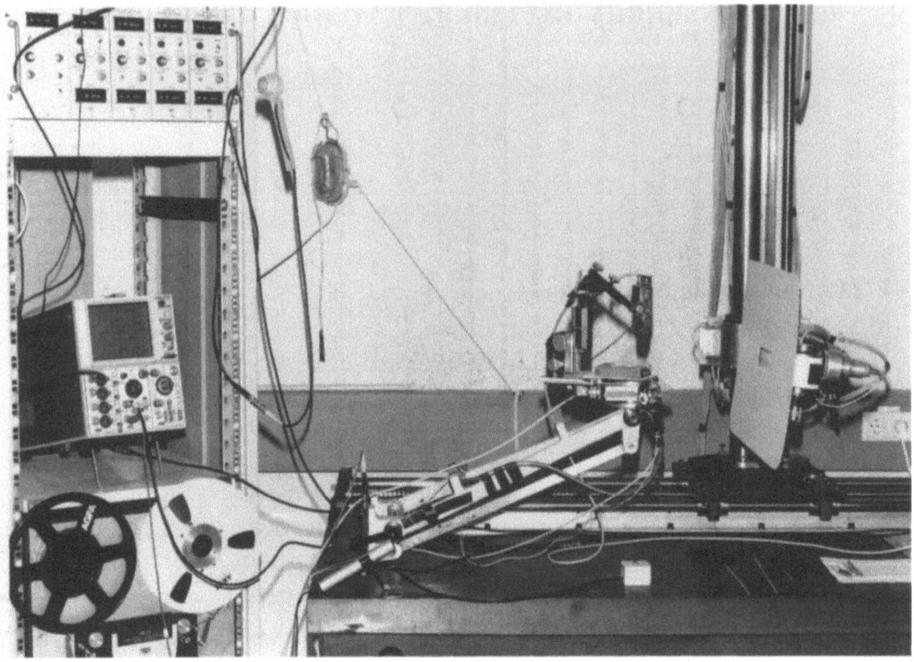

Fig. 5.1. Perimeter apparatus for electrophysiological studies of the visual system. The animal sits on a platform above a heavy metal table looking at a screen on which moving stimuli are shown. In order to bring the stimuli into the receptive field of a recorded cell, the screen is moved three-dimensionally over a spherical surface relative to the animal. Relative movements in the horizontal plane are achieved by rotation of the platform together with the animal. For variation of the stimulus distance, the screen is moved vertically and horizontally. For variation of the direction of stimulus movement, the screen can be rotated

For single-cell recordings different types of microelectrodes are used. In some of the experiments reported here, metal (mostly steel or tungsten) microelectrodes were used that were sharpened electrolytically and laquer-coated leaving the very tip of the electrode uncoated (tip diameter 1–2 μm). Metal electrodes can be used for extracellular recordings only. In other experiments glass micropipettes were used that were produced with a micropipette puller. The pipettes were filled with a 1.5–3 M KCl or NaCl solution. For extracellular recordings the tip of the micropipette was broken so that the stump had an inner diameter of 1–2 μm. In order to determine the exact position of the recorded cells, the micropipettes were filled with an NaCl-alcian blue solution which was applied iontophoretically (about 30 s at 500 nA) and resulted in small marking points (6–10 μm i.d.) (Harnischfeger 1979). In some extracellular recording experiments, HRP-filled electrodes (10% HRP in 3 M NaCl) were used. HRP was applied iontophoretically (500 nA, 1–5 s) which often resulted in a filling of only one cell (presumably the recorded one) (Manteuffel et al. 1983). For intracellular recordings the micropipettes were pulled to a tip diameter equal to or smaller than 0.5 μm and filled with a 1.5 M KCl solution (impedance 80–150 MΩ). For anatomical characterization of the recorded cell the micropipette was filled with a 10% solution of HRP desolved in 1.5 M KCl (injection time between 30 s and 10 min).

After insertion into the brain the microelectrode is moved through the neural tissue either by hand via a micromanipulator or by means of an electronically controlled "stepper" until a single neuron is found the activity of which can be sufficiently distinguished from the activity of the surrounding cells. Then, the perimeter screen is moved in such a way that its center coincides with the receptive field of the isolated cell and the stimuli can be moved exactly through the receptive field.

The major problem of recording experiments in amphibians is the insufficient stability of the electrode position in relation to the recording site. This is especially true for intracellular recordings in which the electrode tip usually remains inside the cell for only a few minutes. Extracellular recordings, which are stable for 15–30 min can be obtained much more easily. To date, it is still somewhat unclear which the exact sites of a cell are from which recordings are made. It seems most probable that with metal electrodes recordings are mostly made from synaptic structures, e.g., from glomerular structures in the upper tectum. With glass microelectrodes extracellular recordings are most probably made from the soma or from the axon hillock of the neuron.

The neural activity of the recorded cells is usually evaluated by calculation of the average discharge (impulses per second) between the first and the last impulse. As was shown by Schürg-Pfeiffer and Ewert (1981), different methods for evaluating neuronal activity elicited by stimuli used in the experiments presented here do not lead to significantly different results in comparison to the calculation of the average impulse frequency.

For further technical details the reader is referred to the original papers cited in this chapter.

5.2 Response Properties of Retina Ganglion Cells to Preylike Stimuli

The first recordings of the activity of retina ganglion cells of urodeles were done by Hartline (1938) from optic nerve fibers of *Necturus*. Hartline found: (1) fibers that showed a brief response to changing light intensity ("light on/off"); (2) fibers that showed a prolonged response (up to 30 s) to "light on"; (3) fibers that were activated only by "light off". These latter "off" responses were very weak and could be obtained only after at least 10 s of illumination.

During recording from the optic nerve and from the superficial layers of the optic tectum of *Triturus vulgaris*, Cronly-Dillon and Galand (1966) found

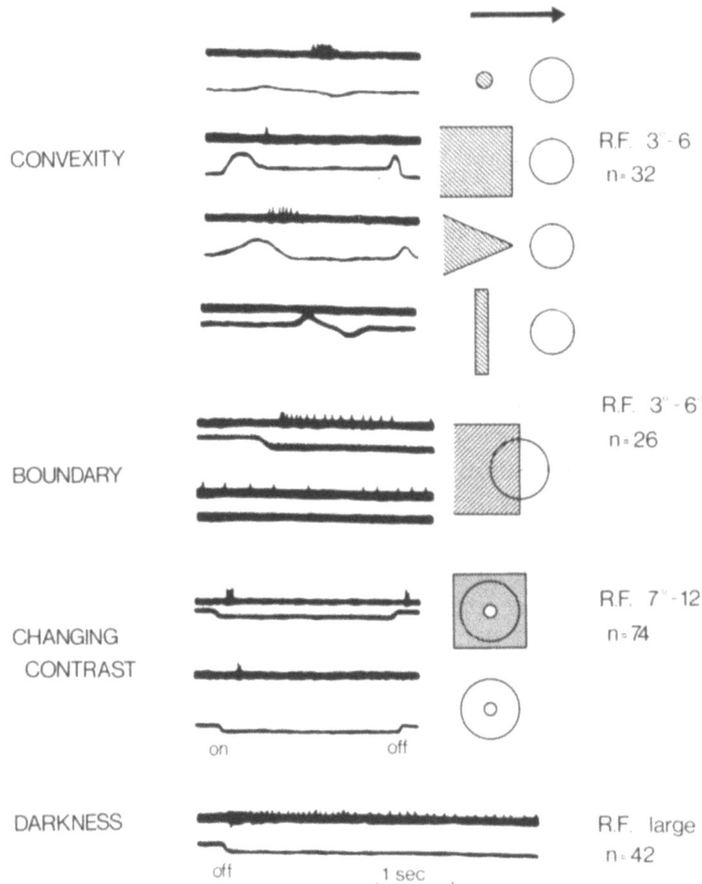

Fig. 5.2. Types of single units found in the optic nerve of the newt *Triturus vulgaris*. *Left*: Recording traces. *Upper trace* of each record represents activity of the retina ganglion cell; *lower trace* photocell response. The following four types types are found:
1. *Convexity units* which respond to a small, black dot stimulus (first stimulus from above, right) or to a convex-pointed edge (third stimulus from above) moved through the receptive field. These units do not respond to movement of stimuli with a straight edge (second and fourth stimuli from above).
2. *Boundary units* which respond with a long-lasting tonic discharge to a sharp stationary edge within the receptive field.
3. *Contrast units* which respond with a brief high-frequency burst to any moving boundary that passes through the receptive field. The example shows an on-off response when the center of the receptive field alone was stimulated with a small 1° spot of light flashed briefly on and off. *Lower record trace* shows the diminished response when the periphery of the receptive field was illuminated with a background light. Thus, an antagonism exists between center and periphery of the receptive field.
4. *Darkness units* are inhibited by light and respond with a long tonic discharge in the dark. (After Cronly-Dillon 1968)

four different classes of neurons which were described according to the classification of retina ganglion cells of *Rana* made by Lettvin et al. (1959) (Fig. 5.2):

1. Convexity detectors: Neurons of this class responded to movement of a small, dark object on a bright background through the excitatory receptive field (ERF) of the retina ganglion cells recorded. The diameter of the ERF was 3–6°. Since a stimulus larger than the ERF size elicited no response, one can assume that the ERF is surrounded by an inhibitory receptive field (IRF).

2. Boundary detectors: These neurons responded with prolonged activation to a stationary light-dark boundary running through the ERF. The ERF size was again 3–6°.

3. Contrast units: These cells responded with a short, high-frequency burst to any moving, light-dark boundary within the ERF. The ERF size varied between 7 and 12°.

4. Darkness units: These cells were inhibited by light and showed prolonged activation after "light off". They had large ERFs. Their responses to moving dark stimuli were somewhat direction-dependent.

The responses of retina ganglion cells of *S. salamandra* were investigated by Grüsser-Cornehls and Himstedt (1973, 1976) and by Himstedt and Roth (1978 unpublished data). Grüsser-Cornehls and Himstedt found three layers of optic nerve fibers in the superficial part of the white matter of the optic tectum.

Layer-1 units had relatively large excitatory receptive fields (12–16°) and did not respond to diffuse light on off. They showed a relatively short activation period (100–200 ms) after a change in the position of a moving stimulus. These neurons were to some degree similar to class-2 retina ganglion cells of anurans in showing a strong inhibitory surrounding and being activated only by small moving objects (Fig. 5.5).

Layer-2 units had ERFs of 6–9° and responded with a short on-off activation to a change of diffuse light. These units behaved similarly to class-3 rgc of anurans, and their surrounding inhibition was relatively weak (Figs. 5.3 and 5.5) Thus, they optimally responded to larger stimulus sizes than layer-1 neurons. With regard to stimulus velocity, layer-2 neurons did not respond to velocities below 2°/s. However, with increasing velocity the increase in discharge rate, was rather steep (curve A in Fig. 5.4).

Layer-3 units showed ERFs of 10–20°. They exhibited tonic on-off responses to a change of diffuse light. Similar to class-4 rgc in anurans, they showed weak surrounding inhibition (Fig. 5.5). They optimally responded to still larger stimulus sizes than layer-2 neurons. Furthermore, they responded to rather low stimulus velocities down to 0.05°/s, but with increasing velocity the increase in discharge rate was less strong compared to layer-2 neurons (curve B in Fig. 5.4).

The terminals of the optic nerve fibers within the optic tectum of *Salamandra salamandra* were demonstated by means of the Fink-Heimer degener-

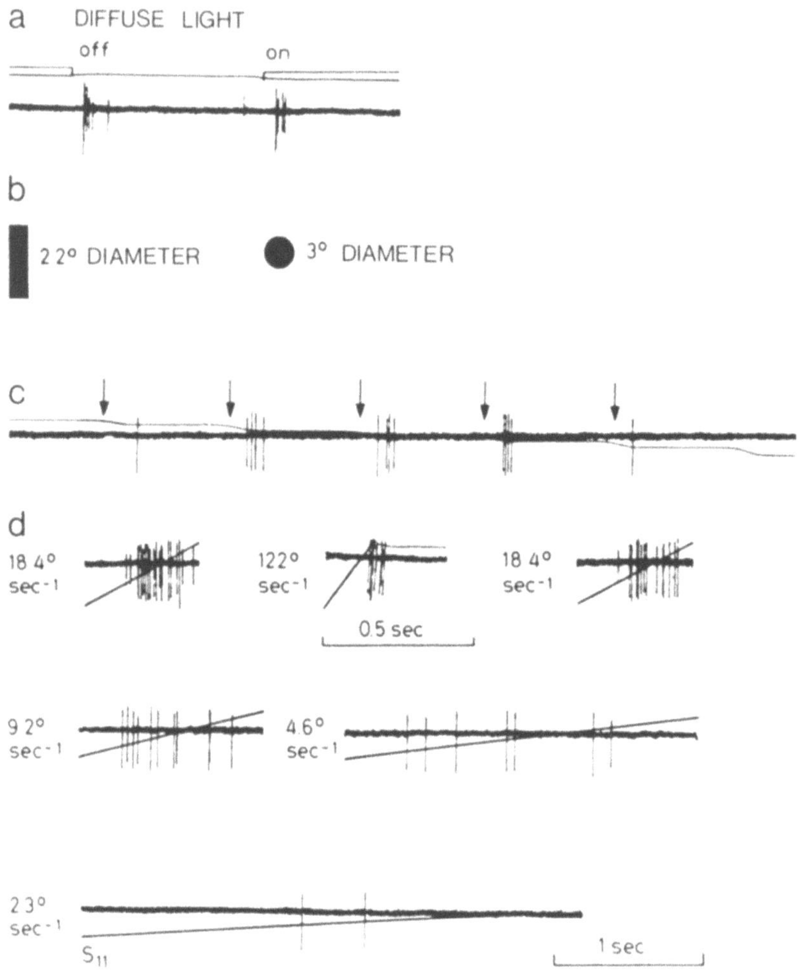

Fig. 5.3a–d. Response of retina layer-2 neurons in *S. salamandra*. **a** Response of a neuron to diffuse light turned off and on. **b** Stimuli used in experiments **c** and **d**: a vertically oriented *black bar* measuring 2.2 × 20° and a *black spot* with a diameter of 3°. **c** Response of the same neuron as in **a** to the spot stimulus moved in steps through the receptive field. *Arrows* indicate the onset of stimulus movement. **d** Response of the same neuron to the vertical *bar* and the *spot* moved at different velocities through the receptive field. The example shows a decrease in the discharge rate with decreasing velocity. (After Grüsser-Cornehls and Himstedt 1973)

ation method and the HRP technique (Fig. 5.5, lower right). On the basis of marking the recording points by means of stainless steel electrodes and estimation of penetration depth during recording, it was assumed that the functional types are localized within the following anatomical structures: layer-1 neurons form no continuous layer within the tectum. Some terminals are scattered directly below the surface which can probably be ascribed to this

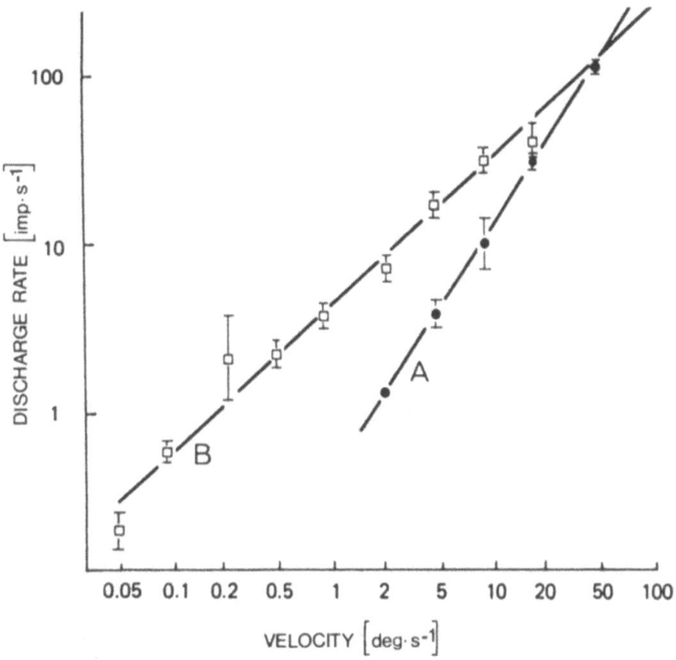

Fig. 5.4. Relationship between impulse frequency and stimulus velocity in layer-2 and layer-3 neurons in *S. salamandra*. *Curve A*: Responses of 11 layer-2 neurons. As stimuli, a 3.1° black spot, a 5 × 5° black square, and a 2.4 × 20° vertical bar were used. Equation of regression line: $R = 0.5v^{1.4}$. *Curve B*: Responses of 5 layer-3 neurons. As stimulus, a 3.1° black spot was used. Equation of regression line: $R = 4.2v^{0.85}$. *Bars* indicate standard deviations. (From Grüsser-Cornehls and Himstedt 1976)

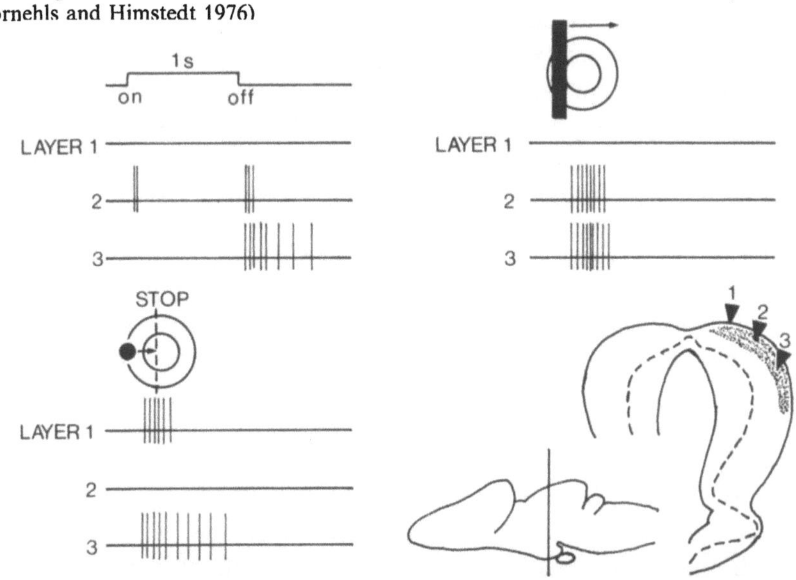

Fig. 5.5. Responses of layer-1, layer-2, and layer-3 neurons in the optic tectum of *S. salamandra*. Stimuli were light on-off (*upper left*), a small moving object stopping in the excitatory receptive field (*lower left*), a large vertical bar passing through the excitatory and inhibitory receptive field (*upper right*). Drawing (*lower right*) shows the three layers of terminals of retina ganglion cells. (From Himstedt and Roth unpublished)

type. Somewhat deeper there follows a dense layer measuring 45–60 μm in depth. From this region, responses of type layer-2 can be obtained. Below this region, there is a layer measuring 20–30 μm in depth in which the fibers of type layer-3 probably terminate. These two regions converge in the caudal half of the tectum. In general, the different types of fibers do not seem to be separated as clearly as in anurans. Sometimes responses of layer-2 neurons can be obtained directly from the surface and responses from layer-1 neurons can be recorded more deeply than usual.

Himstedt and Roth (unpublished) studied the response properties of retina ganglion cells in *S. salamandra* to stimuli of different configuration and velocity (Fig. 5.6). In these experiments, the following black stimuli on white background were presented: a square measuring 8 × 8° visual angle, and rectangles measuring 2 × 8° oriented either parallel or perpendicular to the direction of movement. The patterns were moved horizontally at velocities of 2°, 6°, and 20°/s through the receptive field of the neuron under investigation (cf. Fig. 5.10). The distance between the stimulus and the eye of the subject was 28.6 cm.

On the basis of their relative ERF diameters and the differences in their inhibitory surrounds, the three types of retinal neurons previously described by Grüsser-Cornehls and Himstedt (1973) respond differently to the presented stimuli. Figure 5.6A shows that layer-1 neurons respond more strongly to a horizontally oriented rectangle measuring 2 × 8° than to a vertically oriented rectangle of equal dimensions and to a square measuring 8 × 8°. During its passage through the ERF a horizontally oriented rectangle stimulates a greater portion of the excitatory center than of the inhibitory surround and, thus, leads to good activation of the neuron. A rectangle oriented vertically and a square, on the other hand, stimulate a considerably larger area of the IRF so that the response level of the neuron is lower. The stimuli of 8° extension, used in these experiments, however, did not suppress the response completely.

The fact that layer-1 neurons respond best to small objects or small vertical edges of 2–3° in diameter or length, indicates that that the ERF size of 12–16° found by Grüsser-Cornehls and Himstedt is too large to explain the stimulus preference found by Himstedt and Roth. Actually, it might be close to 6° which would also correspond better to the results obtained by Cronly-Dillon and Galand (1966) in *Triturus* and to the range of ERF of class-2 rgc in anurans. In general, it proved to be difficult in *Salamandra* to obtain clear single-unit recordings from retinal afferents within the tectum using metal electrodes, so that the ERF may appear larger in size due to multi-unit recordings. Himstedt and Roth, using glass micropipettes, found ERFs of layer-1 units within a range of 8–12° which again is rather large. In *Batrachoseps attenuatus* Wiggers and Roth (in prep.) found minimum ERF sizes of rgc of 5°.

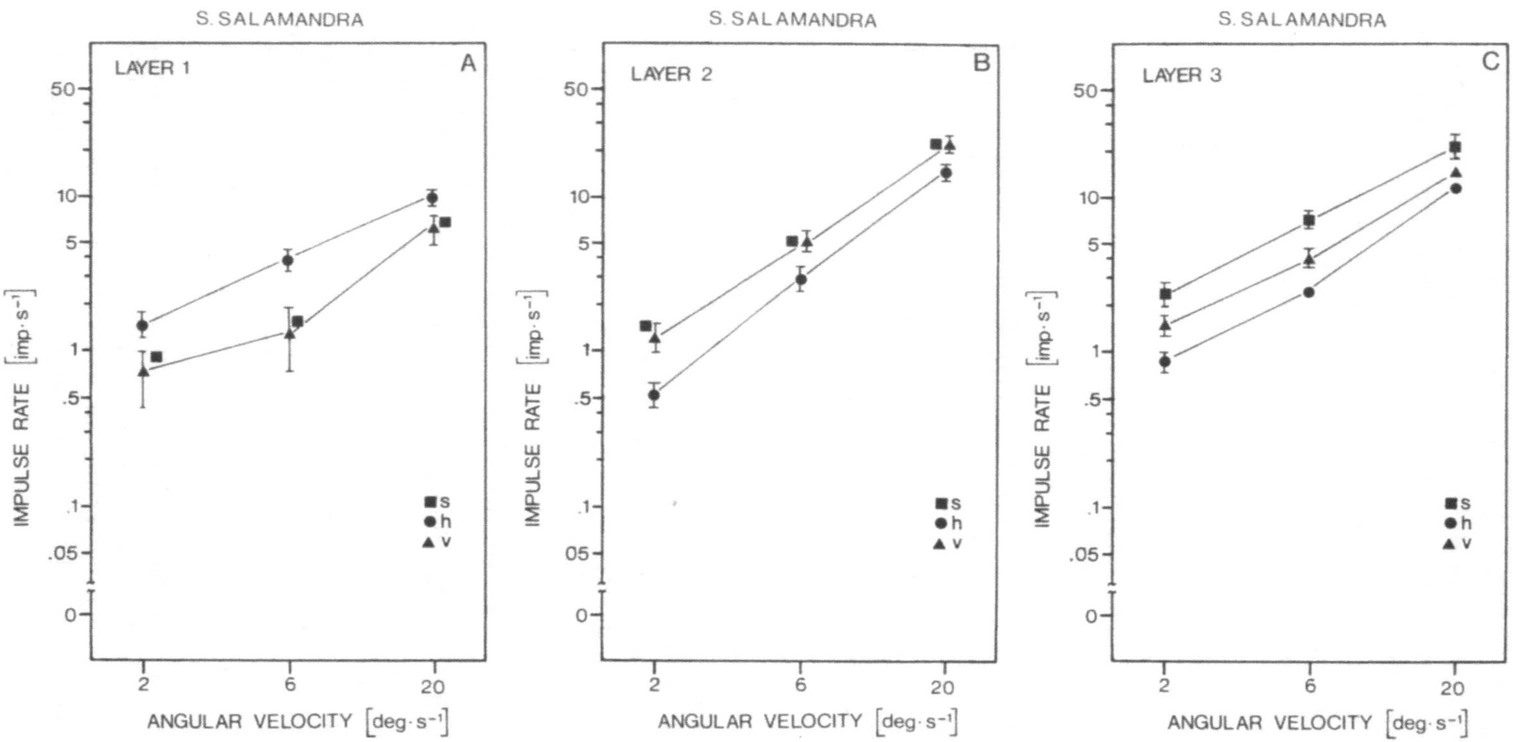

Fig. 5.6 A–C. Response characteristics of retina ganglion cells in *S. salamandra*. Stimuli were 2 × 8° rectangles oriented horizontally (*filled circles*) and vertically (*triangles*) and 8 × 8° squares (*squares*) moved at velocities of 2°, 6°, and 20°/s. **A** Layer-1 units; **B** layer-2 units; **C** layer-3 units. *Bars* indicate standard errors of the mean. (From Himstedt and Roth unpublished)

When the velocity of the stimuli is raised, the impulse rate also increases. This happens for all stimuli in about the same way, resulting in an approximately parallel increase of the curves for the three stimuli. Using logarithmic scales one obtains a linear function with an exponent of 0.82 for horizontal stimuli. Grüsser-Cornehls and Himstedt (1973) measured the velocity function of one layer-1 neuron by means of a square measuring 7 × 7° and found an exponent of 0.7. This value is rather close to that found in the experiments of Himstedt and Roth.

Figure 5.6B shows the responses of layer-2 neurons. Here again the discharge rate increases with increasing velocity of the stimuli and in a rather parallel way. In contrast to layer-1 neurons, the slope of the curves is steeper (with an exponent of 1.24). Additionally, layer-2 neurons show a reversed stimulus preference: the square and the vertical rectangle elicit stronger responses than the horizontal rectangle. This corroborates the assumption that the surround of these neurons possesses a lower inhibitory effect compared to layer-1 neurons and, thus, responds better to stimuli with larger vertical extension.

Figure 5.6C shows the responses of layer-3 neurons, again with parallel velocity functions, but with an exponent of 0.97. This value lies between those of layer-1 and layer-2 type. Important differences can again be found with respect to the stimulus preference. At all velocities there is a clear dominance of the square over the vertical and the horizontal rectangle. The differences are very clear between the responses to the square and to the vertical rectangle, a situation that is not found in layer-1 and layer-2 neurons. This shows that the diameter of the ERF of layer-3 neurons is larger than that of layer-2 neurons so that the clear preference for the square is a result of a larger excitatory area within the RF.

A comparison between these results and those of Grüsser-Cornehls and Himstedt (1973) reveals some differences. Himstedt and Roth were unable to distinguish two subgroups of layer-2 neurons. Furthermore, the response characteristics of layer-3 neurons were somewhat different. The after-discharges of layer-3 after the stimulus had stopped within the ERF did not last for 4 to 6 s. Unlike the former experiments, Himstedt and Roth obtained only a few discharges. The exponent of the velocity function was greater (0.97 in contrast to 0.85), and the extremely low threshold values for minimal stimulus size and minimal contrast could not be found. The latter stimulus parameters were not tested quantitatively during the experiments of Himstedt and Roth, but even with stimuli moved by hand the different properties could be clearly recognized.

These differences between the results of Himstedt and Roth and those of Grüsser-Cornehls and Himstedt (1973) may have several reasons. Apart from differences in the maintenance of the subjects and the conduction of the experiments, the geographic differences between the respective subjects might

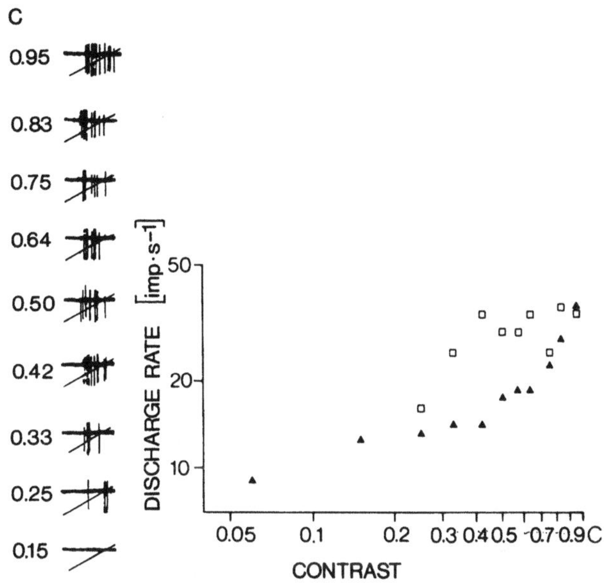

Fig. 5.7. Relationship between different stimulus-background contrasts and impulse frequency in a layer-2 neuron (*open squares*) and in a layer-3 neuron (*triangles*) in *S. salamandra*. The stimulus was a 3.1° spot of varying shades of gray, moving at 18.4°/s (layer-2 neuron) and 9.6°/s (layer-3 neuron) through the receptive field of the neurons. *Left*: Original recordings of a layer-2 neuron at different stimulus-background contrasts. Calculation of contrast: $C = (I_s - I_b)/(I_s + I_b)$; I_s = stimulus luminance, I_b = background luminance. (After Grüsser-Cornehls and Himstedt 1973)

be an important factor. Grüsser-Cornehls and Himstedt recorded the layer-3 neurons exclusively from members of the subspecies *Salamandra salamandra terrestris* from Western Europe, whereas in the present experiments only members of the subspecies *Salamandra salamandra salamandra* from Yugoslavia were used. Behavioral studies showed clear differences in prey-catching behavior between these two subspecies as well as between geographically separated populations of the eastern subspecies (from Austria and Yugoslavia). Such differences may be ascribed to genetic differences as well as to different early visual experience with prey animals (see Chap. 2).

Grüsser-Cornehls and Himstedt (1973) also measured the role of stimulus-background contrast. The visual contrast was calculated as $C = (Is - Ib)/(Is + Ib)$, where Is is the stimulus luminance and Ib is the background luminance. As shown in Fig. 5.7, the threshold of the layer-2 neurons occurred at a contrast of about 0.25, and for layer-3 neurons at a contrast of less than 0.05 when stimulus spots of 2.8° were used. Thus, layer-3 neurons are much more sensitive to very low visual contrast compared to layer-2 neurons. This may be of great importance for feeding under conditions of dim light. In layer-2 neurons the neuronal activity increased with increasing contrast between the moving stimulus and the background whereas layer-3 neurons quickly reached saturation.

Comparison with anurans. The response characteristics of the retinal neurons of *Salamandra salamandra* show fundamental similarities to the retinal neurons of frogs and toads (e.g., Lettvin et al. 1959, Grüsser and Grüsser-Cornehls 1968, Ewert and Hock 1972). In the frog *Rana esculenta*, Grüsser and Grüsser-Cornehls (1968, 1976) found the following main response types of retina ganglion cells: Class-1 cells had an ERF diameter of 2–4°, class-2 cells 2.5–5°, class-3 cells 6–8°, and class-4 cells about 14°. Neurons of class-1 continued to discharge when a stimulus moving within the ERF stopped. The activity disappeared after turning the light off, but reappeared after turning the light on again ("non-erasibility" of activity, Lettvin et al. 1959). In class-2 neurons the same response occurred without reactivation after light on. Class-1 and class-2 neurons do not respond to changes in general illumination, class-3 neurons show an on-off response, class-4 neurons a prolonged off response, often with rhythmic discharges. Class-1 cells are found at a depth of 10–30 μm, class-2 cells at 30–90 μm, class-3 cells at 100–300 μm and class-4 cells at 300–400 μm. These depths correspond roughly to laminae B, D, F, and G of the anuran optic tectum. In the toad *Bufo bufo*, Ewert and Hock (1972) found results similar to those found in *Rana* except that class-1 cells could be recorded only rarely.

Thus, layer-1 neurons in salamanders are rather similar to class-1 and -2 neurons of anurans, layer-2 neurons correspond to class-3, and layer-3 neurons to class-4 neurons in anurans, when relative RF sizes, stimulus preferences, and response patterns to light on off are considered. However, in general, retina ganglion cells of salamanders seem to have larger RF sizes compared to frogs and toads. This is correlated with the size and number of photoreceptors and retina ganglion cells: anurans usually have at least ten times more retina ganglion cells than urodeles, and their cells are much smaller. However, the basic response properties seem to be similar. Grüsser-Cornehls and Himstedt, in their 1976 review paper on the salamander visual system emphasize that the properties of layer-1 retina ganglion cells found by the authors in *Salamandra* deviate from those found in *Triturus* (Cronly-Dillon and Galand 1966), as well as from those found in *Bufo* or *Rana*. In their opinion, *Triturus* resembles *Bufo*, whereas *Salamandra* resembles the goldfish (Cronly-Dillon 1964, Jacobson and Gaze 1964) in having no retinal afferents to the tectum that correspond to class-1 and class-2 of the anuran retina.

5.3 Color Coding in Retina Ganglion Cells

Himstedt and Fischerleitner (1975) and Himstedt et al. (1981) studied the responses of retina ganglion cells of *S. salamandra* and *T. alpestris* to stimuli differing in color (Figs. 5.8 and 5.9).

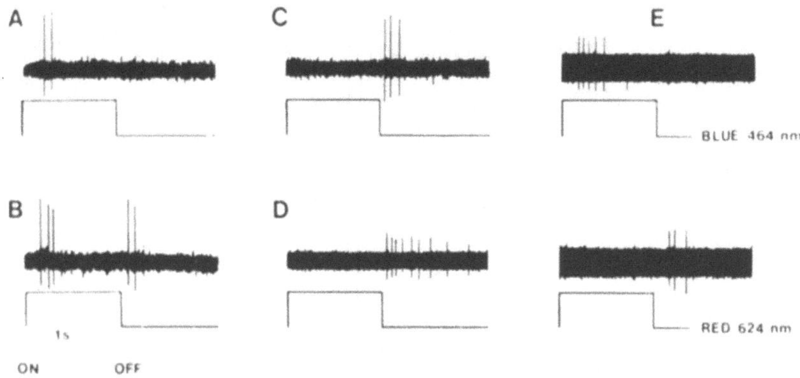

Fig. 5.8 A–E. Responses of different ganglion cell types in the retina of *S. salamandra*. **A** "On" cells; **B** "on-off" cells; **C, D** "off" cells; **E** opponent-color cells. (From Himstedt et al. 1981)

Extracellular recordings from retina ganglion cells of *Salamandra* made within the retina revealed the following response types:

1. on-cells that gave a relatively short response only at the onset of general illumination (Fig. 5.8 A);
2. on-off cells that were either chromatic or achromatic (Fig. 5.8 B);
3. off-cells with two subtypes: one with and the other without a slightly sustained reaction after light off (Fig. 5.8 C, D).

The chromatic on-off cells showed color-specific responses: stimulation with blue light elicited an on-response, stimulation with yellow or red light produced an off-response (Fig. 5.8 E).

Thalamus. Recordings from retina ganglion cell axons in the thalamus of *Salamandra* showed that from a total of 32 units, 25 showed an on-response, 5 an on-off response and 2 an off-response. Two types of on-neurons were found: nine had a sensitivity maximum in the yellow spectral region (around 560 nm) which does not correspond to the absorption spectrum of a single visual pigment, but rather to the sensitivity of the light-adapted eye (Fig. 5.9 A, above). During dark adaptation these yellow-sensitive neurons showed a Purkinje shift: the sensitivity to light strongly increased and the spectral sensitivity maximum shifted toward 500 nm which corresponds to the absorption spectrum of rhodopsin P 502 (= red rods). In six other on-neurons (Fig. 5.9 A, below) the maximum was in the blue region at about 430 nm which corresponds to the visual pigment P 432 (= green rods). A second peak occurred at about 560 nm. At longer wavelengths the sensitivity of these units corresponded to the photopic dominator of the frog. The on-off and off-fibers recorded in the thalamus showed the same spectral sensitivities as the on-units, i.e., they were either blue-sensitive (green rod curve) or

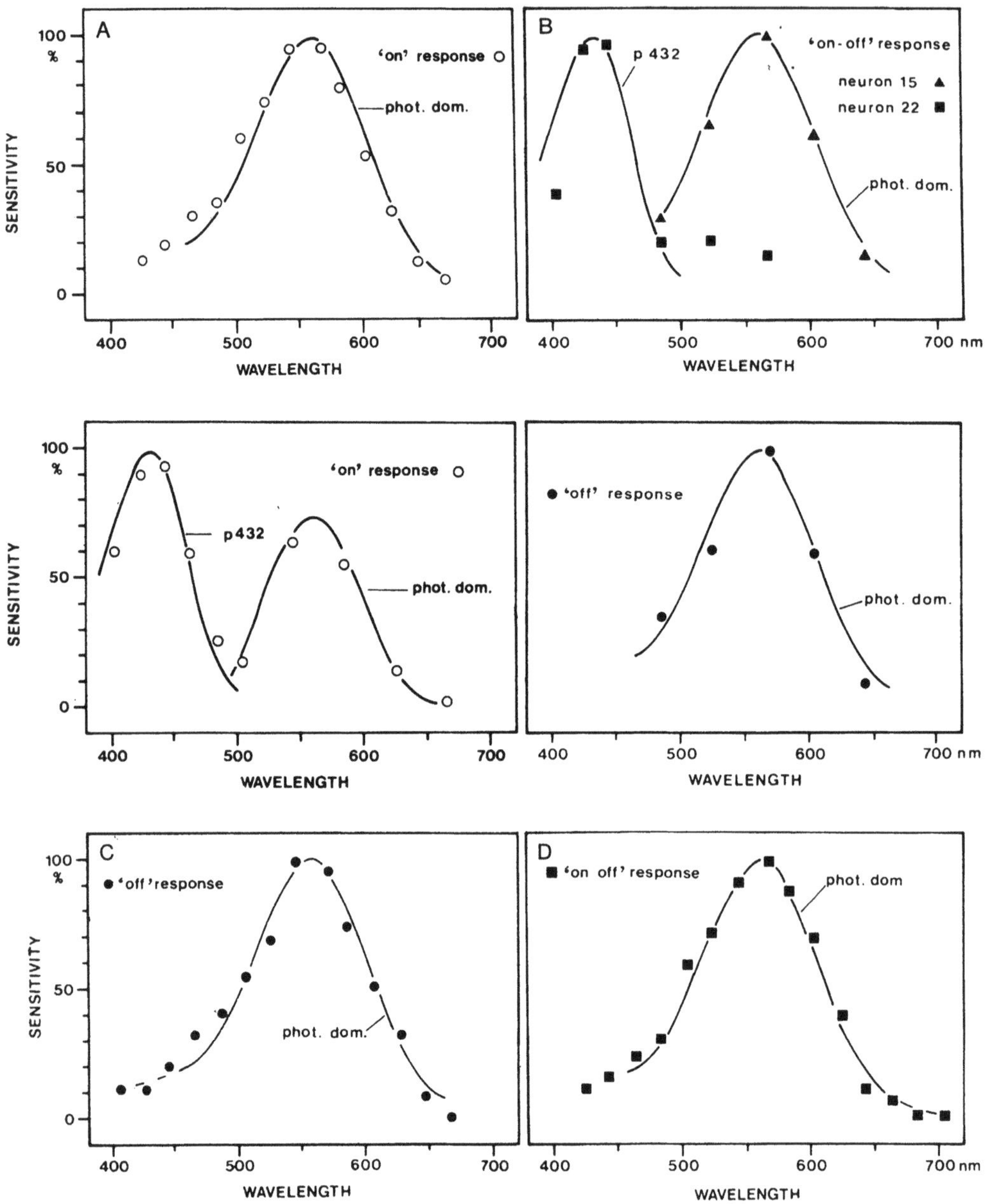

Fig. 5.9 A–F. Spectral sensitivity of neurons in *S. salamandra* (**A–E**) and in *Triturus alpestris* (**F**). A "On" fibers projecting to the thalamus. *Above*: Mean values of nine yellow-sensitive units. *Below*: Mean values of six blue-sensitive units. **B** Retinal afferents in the thalamus. *Above*: Two different "on-off" neurons. *Below*: Mean values of two "off" neurons. **C** "Off" fibers projecting to the tectum opticum; mean values of ten units. **D** "On-off" units projecting to the tectum opticum; mean values of 13 units. **E** Example of an opponent-color ganglion cell terminating in the tectum opticum. **F** "On-off" fibers projecting to the tectum opticum. Mean values of seven units recorded in aquatic newts in springtime (*squares*) and of ten units recorded in terrestrial newts in fall (*triangles*).

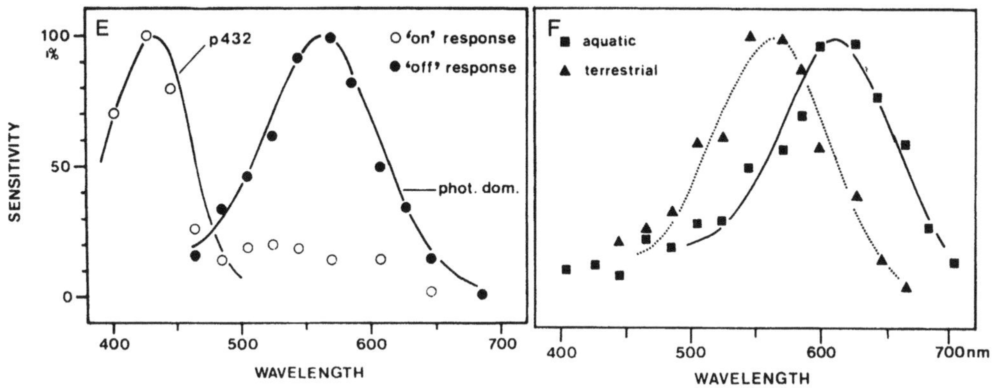

Fig. 5.9 E, F.

yellow-sensitive (photopic dominator) (Fig. 5.9 B, above). The two off-neurons followed the curve of the photopic dominator (Fig. 5.9 B, below). No color-opponent units were found in the thalamus of *Salamandra*.

Tectum. Among retinal afferents in the tectum of *Salamandra*, only on-off and off-units (n = 24) were found by the authors. The off-units were maximally sensitive to yellow colors, i. e., they fitted the curve for the frog photopic dominator (Fig. 5.9 C). Most on-off units recorded showed the same spectral sensitivity with a maximum at 560 nm (Fig. 5.9 D). During dark adaptation all these yellow-sensitive on-off and off-units again showed a Purkinje shift, i. e., in the scotopic state they became green-sensitive according to the curve of the pigment P 502. Some of the on-off cells differed from these cells in showing color opponency: the on-response was much more sensitive to short wavelengths than the off-response which in turn was more sensitive to long wavelengths. Here, the on-response is driven by the green rods (P 432), the off-response fits the photopic dominator (Fig. 5.9 E). During dark adaptation, the off-response became more sensitive to green light and shifted toward the sensitivity of rhodopsin, while the blue-sensitive on-response disappeared.

Similar to the situation in *Salamandra*, in the tectum opticum of *Triturus alpestris* only on-off and off-fibers were found, some of which were color-opponent. The sensitivity maxima, however, were found at different wavelengths depending on the annual life cycle of the newts (cf. Chap. 3). In springtime, when the newts are in the aquatic stage, the achromatic on-off neurons are maximally sensitive at about 620 nm, whereas during the terrestrial stage in the fall the maximum lies at about 560 nm. The same seasonal shifts took place in the off-neurons. The values for aquatic newts do not fit the curve of the photopic dominator of the frog, but correlate rather well with the curve of the tench (Granit 1942) (Fig. 5.9 F).

There were more color-coding neurons in *Triturus* than in *Salamandra*. The seasonal shift in spectral sensitivity is the reason that in springtime *Triturus* has blue-red units and in fall blue-yellow units.

5.4 Response Characteristics of Tectal Neurons

Grüsser-Cornehls and Himstedt (1973) gave a first, qualitative description of responses of tectal cells in salamanders. In the tectum opticum of *Salamandra* they found the following types:

1. Neurons with large RFs (> 30°) that responded to moving objects larger than 1° in diameter. Sometimes the RFs of these units covered the whole visual field of one eye. This type is similar to T-4 cells in the frog (see below).
2. Neurons responding preferentially to stimuli moving along the z-axis toward the eye, but only weakly to all other movement directions. This type corresponds to T-3 cells in frogs.
3. Neurons with RFs between 10° and 20° with rapid adaptation after repeated stimulation by a stimulus moving along the same pathway through the RF. This type corresponds to T-5 cells in frogs.
4. Neurons activated binocularly, in which the RF was identical in size with the binocular field of both eyes. This type corresponds to T-1 cells in frogs.
5. Multisensory neurons found primarily in the deeper tectal layers. Moving or stationary visual stimuli did not consistently activate these units, but vibratory and tactile stimuli did. The neurons did not respond to acoustic stimuli.

The first detailed single-cell recordings from the optic tectum of salamanders were carried out by Himstedt and Roth in *S. salamandra* (Roth and Himstedt 1978, Himstedt and Roth 1980), followed by recordings from the tectum of *Hydromantes italicus* by Roth (1982a). Additional studies on response properties of tectal cells in *Salamandra* were conducted by Finkenstädt and Ewert (1983a) and by Himstedt et al. (in press). Very recently, recordings were made from the tectum of *Bolitoglossa subpalmata and Batrachoseps attenuatus* as well as additional recordings from the tectum of *Hydromantes italicus* by Wiggers and Roth (in prep.).

Himstedt and Roth (cf. also Roth 1982b) as well as Roth (1982a) used the same stimulus program as in their study on responses of retina ganglion cells presented above (cf. Fig. 5.10), i.e., black squares measuring 8 × 8° (S) and rectangles measuring 2 × 8° oriented with their long axis parallel to the direction of movement ("horizontal" rectangle, H) or perpendicular ("vertical" rectangle, V) to it. The stimuli were moved horizontally at velocities of

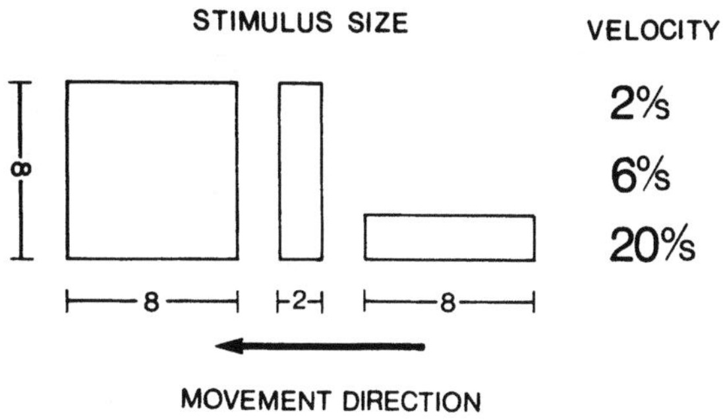

Fig. 5.10. Stimulus program for recording experiments in *S. salamandra* (Himstedt and Roth 1980), *Hydromantes italicus* (Roth 1982a) and *Bufo bufo* (Roth and Jordan 1982)

1, 3 and 10 cm/s (2°, 6°, and 20°/s). The recordings were made with glass micropipettes. In the two series of experiments with *Hydromantes* about half of the recorded neurons were marked iontophoretically by means of a NaCl-alcian-blue dye.

5.4.1 *Salamandra salamandra*

In *Salamandra salamandra* Himstedt and Roth (1980) studied a total of 78 tectal cells with the full stimulus program. The following response types were found (Fig. 5.11 A–G):

1. Neurons (n = 14) which preferred the square (S) to the horizontal rectangle (H) and this stimulus to the vertical rectangle (V) at all velocities (type S > H > V). In this group the responses to S and H tended to converge at high velocity. The mean RF size was 42° (Fig. 5.11 A).

2. Neurons (n = 15) in which again S was the preferred stimulus, but now V was better than H (type S > V > H). The RF size was 48° (Fig. 5.11 B).

3. Neurons (n = 12) in which H was always preferred to S and that stimulus to V. At high velocity the curves for S and V converged (type H > S > V). The mean RF size was 32° (Fig. 5.10 C).

4. Neurons (n = 7) in which V was dominant over S and H. However, the differences were significant only at low velocity, and at high velocity all curves converged (type V > S > H). The mean RF size was 34° (Fig. 5.11 D).

5. Neurons (n = 5) in which S was always the dominant stimulus, but a preference inversion took place between V and H such that at low and medium velocity V was preferred to H, but that at high velocity H was strongly preferred to V and was almost as effective as S (type S > VXH). The mean RF size was 34° (Fig. 5.11 E).

6. Neurons (n = 7) in which a preference inversion took place between H and S such that H was preferred at low and S at medium and high velocity. V was always the least effective stimulus (type HXS > V). The mean RF size was 27° (Fig. 5.11 F).

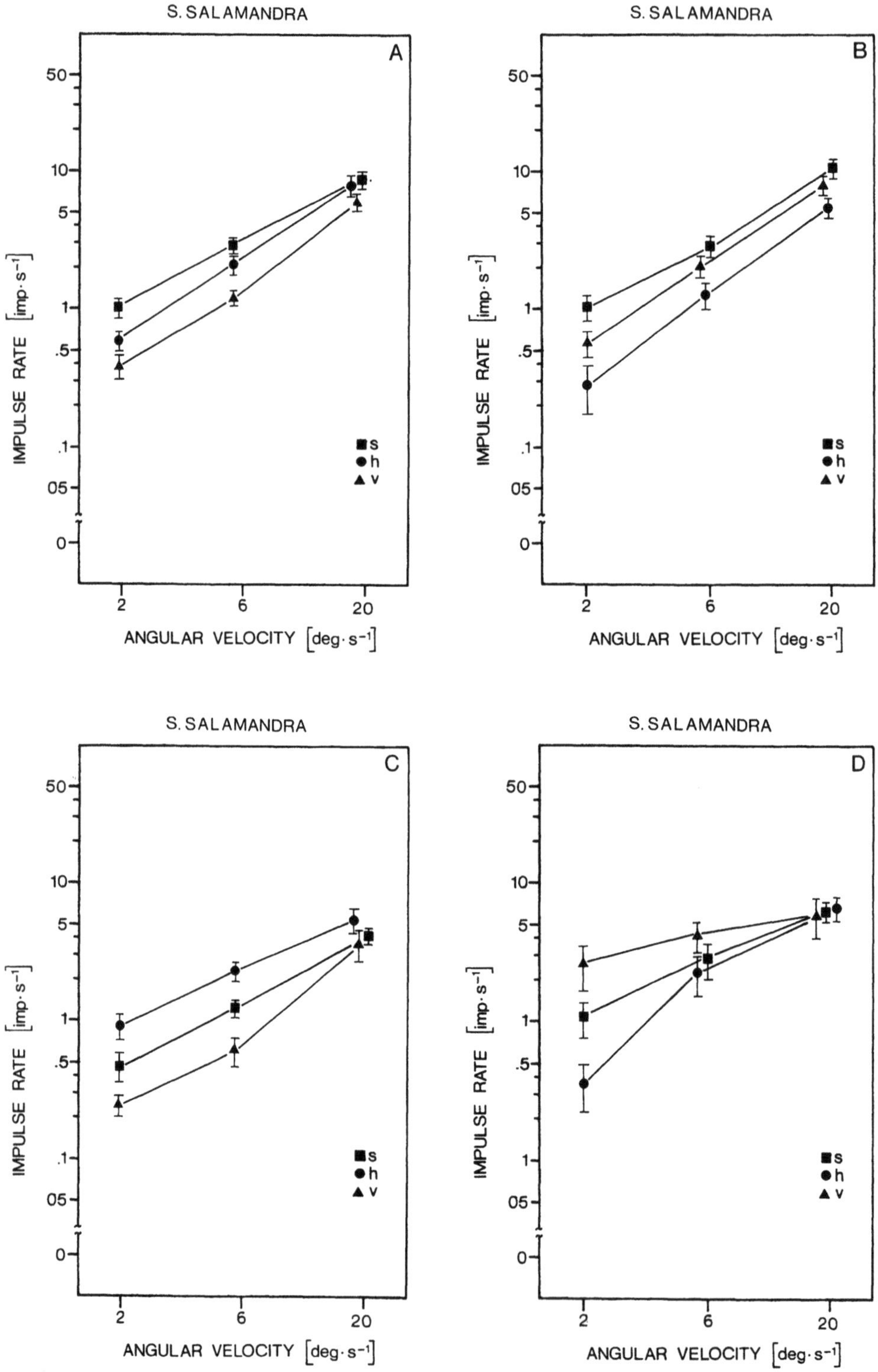
S. SALAMANDRA
A
50
10
5
1
.5
.1
05
0
IMPULSE RATE [imp·s⁻¹]
s
h
v
2
6
20
ANGULAR VELOCITY [deg·s⁻¹]
S. SALAMANDRA
B
50
10
5
1
.5
.1
05
0
IMPULSE RATE [imp·s⁻¹]
s
h
v
2
6
20
ANGULAR VELOCITY [deg·s⁻¹]
S. SALAMANDRA
C
50
10
5
1
.5
.1
05
0
IMPULSE RATE [imp·s⁻¹]
s
h
v
2
6
20
ANGULAR VELOCITY [deg·s⁻¹]
S. SALAMANDRA
D
50
10
5
1
.5
1
05
0
IMPULSE RATE [imp·s⁻¹]
s
h
v
2
6
20
ANGULAR VELOCITY [deg·s⁻¹]

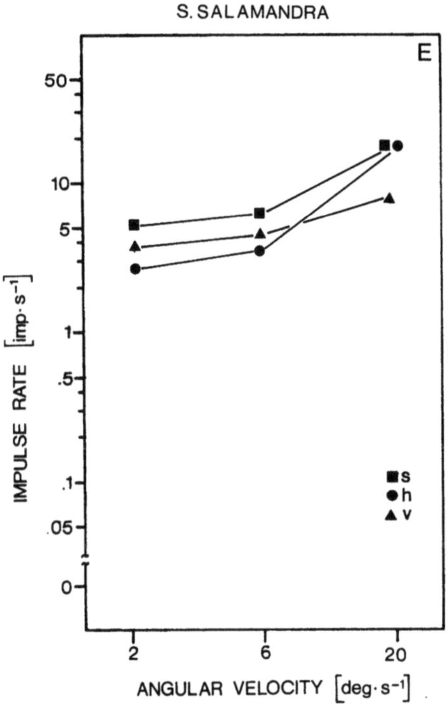

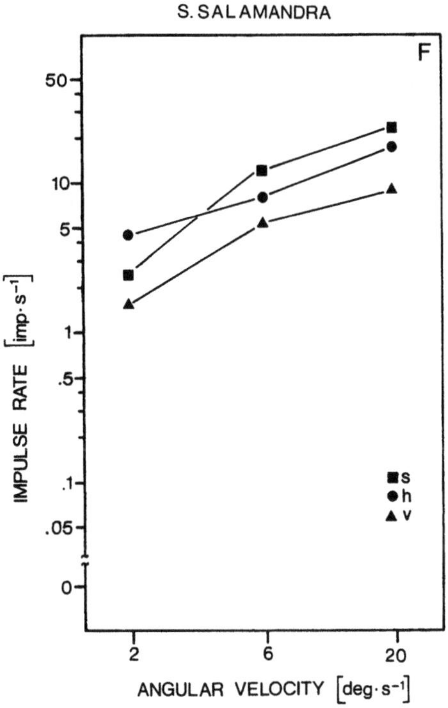

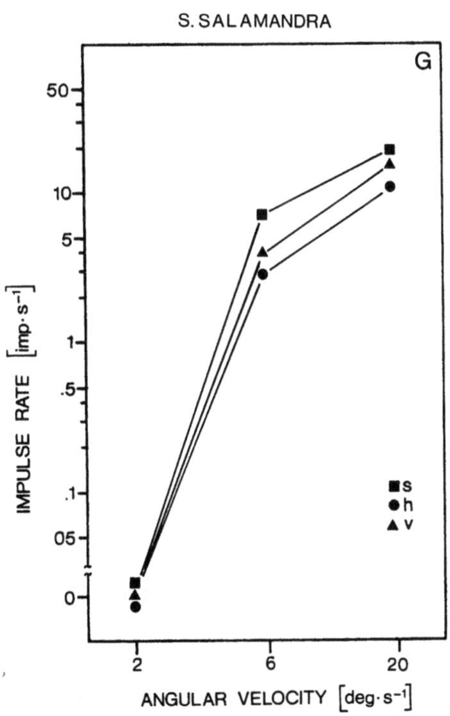

Fig. 5.11 A–G. Response types in the tectum of *S. salamandra*. **A** Neurons (n = 14) with stimulus preference S > H > V. **B** Neurons (n = 15) with preference S > V > H. **C** Neurons (n = 12) with preference H > S > V. **D** Neurons (n = 7) with preference V > S > H. **E** Neurons (n = 5) with preference inversion S > VXH; one characteristic neuron is shown. **F** Neurons (n = 7) with preference inversion HXS > V; one characteristic neuron is shown. **G** "High-velocity" neurons (n = 6); one characteristic neuron is shown.
Squares: Square stimulus; *filled circles*: horizontal rectangle; *triangles*: vertical rectangle. *Bars* indicate standard error of the mean. (After Roth 1982b)

7. Neurons (n = 5) which showed no responses to any of the three stimuli at low velocity, but responded well to them at medium and high velocity, where the response type was S > V > H. The mean RF size was 29° (Fig. 5.11 G).

Although direction specificity of these neurons was not studied systematically in these experiments, a few neurons were found which responded only to the square when moved by hand over the head of the subjects in a caudorostral direction.

A total of 65 cells could be classified in this manner with respect to their responses to the three stimuli at the three velocities. The remaining neurons showed no significant differences in their responses to the stimuli or rather bizarre behavior such as double preference inversions.

Finkenstädt and Ewert (1983a) studied tectal neurons of *Salamandra* under somewhat different recording and stimulus presentation conditions. These authors used laquer-coated steel or tungsten microelectrodes. Their stimulus set consisted of a black square measuring 2 × 2° which was systematically enlarged in size by the factors 2, 4, 8, and 16 either along its horizontal or its vertical edge or along both edges which resulted in rectangles of increasing horizontal or vertical extension and in squares of geometrically increased size (cf. Fig. 5.12, above left). These stimuli were moved horizontally at a velocity of 7.6°/s.

The total number of recorded cells was 47. As shown in Fig. 5.11, seven types of responses were found which were classified according to the scheme of Ewert and von Wietersheim (1974) and Schürg-Pfeiffer and Ewert (1981):

1. T-5.1 neurons (n = 14) in which S was preferred to H and to V at edge lengths of 4° and 8°. At 16° and 32° H was slightly, but not significantly better than S. The optimum edge length was 8° for S and 16° for H. V elicited only a few responses at any edge length. The RF size was about 34°.

2. T-5.3 neurons (n = 14) in which V was preferred to S at edge lengths of 4–8°, but toward greater edge lenght S was now strongly preferred to V with a maximum at 16°. In these neurons the responses to H were very low and did not increase in frequency with increasing edge length of H. The RF size was about 34°.

3. T-5.4 neurons (n = 4). These neurons responded only to S. The response frequency increased with increasing size of S up to an edge length of 16°. The RF size was 20–30°. These neurons showed strong adaptation and did not respond to a small 2 × 2° square.

4. T-2.3 neurons (n = 10). Here, the responses to S and V were more or less the same up to an edge length of 8°; at 16°, S strongly dominated over V, whereas at 32° the curves of S and V converged. The response frequency to H did not substantially increase with increasing edge length and remained very low. The RF-size was about 100°.

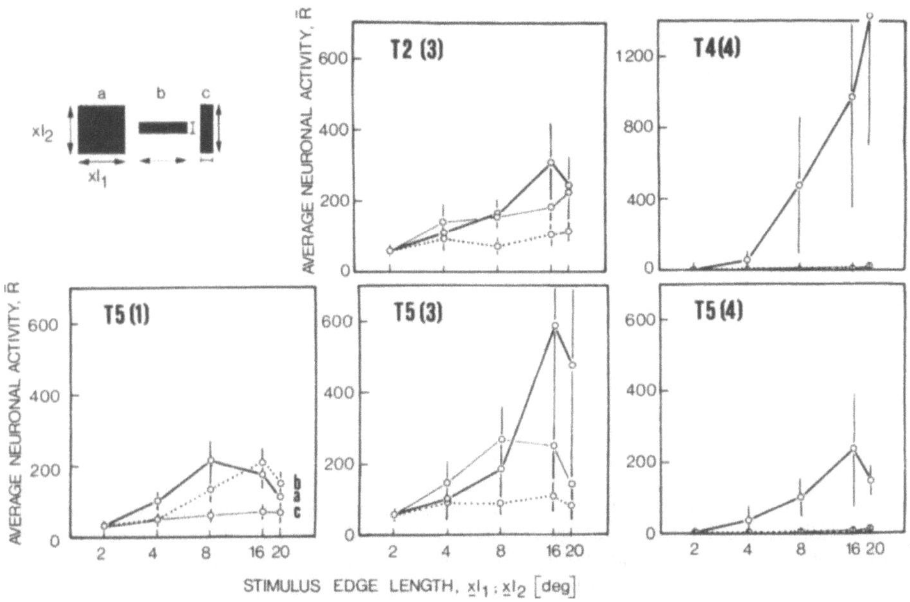

Fig. 5.12. Response types in the tectum of *S. salamandra*. As shown *above left*, rectangular stimuli were used which were increased in edge length either in direction of movement (*b*), or perpendicular to it (*c*), or in both directions (*a*). The stimuli were moved at 7.6°/s through the receptive field. For further explanation see text. (From Finkenstädt and Ewert 1983a)

5. T-4.4 neurons (n = 5). The excitatory receptive fields of these neurons were about 180° and covered most of the visual field of the contralateral eye. The response properties were similar to type T-5.4, but differed from them by even larger RF sizes and by optimal stimulus size which was equal to or greater than 32°. Some of these neurons responded to tactile and vibration stimuli, too. They were situated mostly in deep tectal layers.

6. T-7 neurons (n = 3) which showed similar stimulus preferences as those described in frogs (Grüsser and Grüsser-Cornehls 1970, 1976, Schürg-Pfeiffer and Ewert 1981), i.e. they preferred small squares to the other stimuli at edge lengths of 3° and 4°. The RF size was very small with about 4°.

7. T-3 neurons (n = 4). These neurons, like those described in the frog (Grüsser and Grüsser-Cornehls 1970, 1976) responded best to a stimulus approaching the eye along the z-axis.

The authors reported that they were unable to find T-5.2 cells which in toads prefer H to S and to V at all edge lengths (cf. Ewert and von Wietersheim 1974) and which were described in *Salamandra* by Himstedt and Roth (type H > S > V, cf. above).

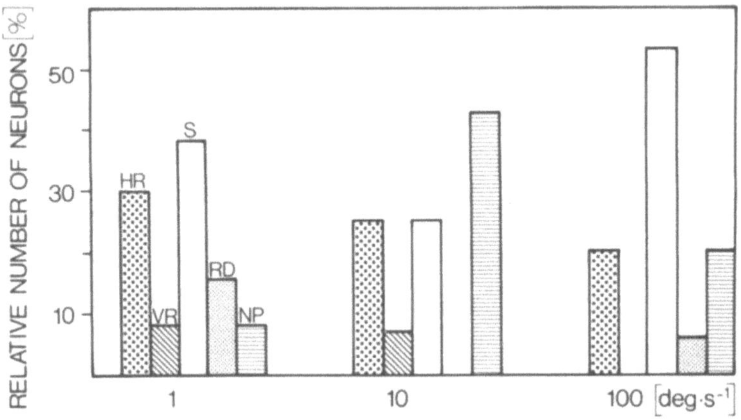

Fig. 5.13. Response types in the tectum of *S. salamandra*. Neurons were stimulated with 3 × 15 mm rectangles oriented horizontally (*HR, coarsely dotted columns*) or vertically (*VR, diagonally hatched columns*), with a 16 × 16 mm square (*S, void columns*) and with a 60 × 60 mm random dot stimulus (*RD, finely dotted columns*). The stimuli were moved through the receptive field at velocities of 1°, 10°, and 100°/s. The graph shows the relative number of neurons with different stimulus preferences satisfying 20% criterion (see text). *NP* No stimulus preference satisfying criterion. (From Himstedt et al. in press)

A third study of the responses of tectal cells in *S. salamandra* was carried out by Himstedt et al. (in press). Here, similar to the experiments of Himstedt and Roth (1980), three stimuli, a 15 × 15° square, a 3 × 15° horizontal rectangle and a 15 × 3° vertical rectangle were used and presented at three different velocities which, however, covered a much wider range with 1°, 10°, and 100°/s. Because the authors also studied thalamic and pretectal areas assumed to be involved in stimulus-background discrimination, a 60 × 60° random-dot stimulus ("Julesz-pattern") was also used, moved at the same velocity as the other stimuli.

The authors used a classification scheme similar to that used by Himstedt and Roth (1980), Roth (1982a, b), and Roth and Jordan (1982), but introduced a criterion of 20% minimum difference in impulse frequency elicited by the best and second best stimulus. The recording method was the same as in Himstedt and Roth (1980) and Roth (1982a) with the exception that a different perimeter device was used in which the stimuli were not mechanically moved, but projected onto a screen and moved by means of electronically controlled movable mirrors as explained above. The neurons were marked iontophoretically.

As shown in Fig. 5.13, in the tectum opticum Himstedt et al. found the following groups of visual neurons (n = 31):

At low velocity (1°/s), five neurons preferred S to V and to H (S > V > H), four neurons preferred H to S and to V (H > S > V), and one neuron preferred V to H and to S (V > H > S >). Two neurons responded best to the Julesz-pattern.

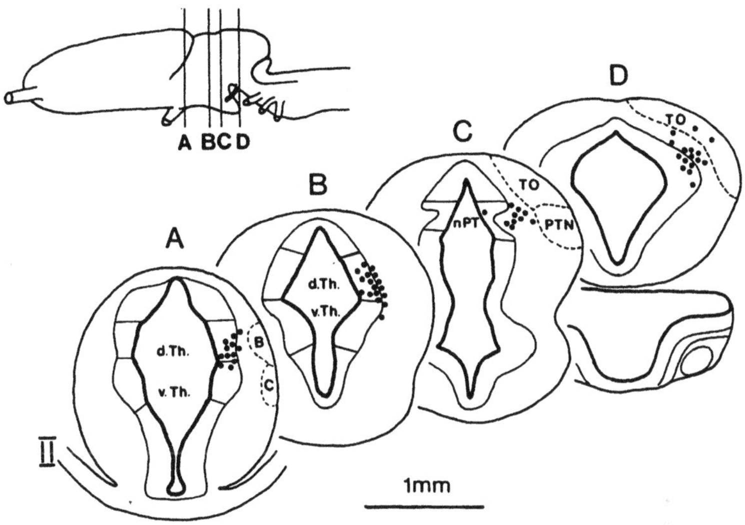

Fig. 5.14A–D. Location of representative recording sites recorded by Himstedt et al. in the visual centers of *S. salamandra*. **A** rostral thalamus; **B** caudal thalamus; **C** praetectum; **D** tectum opticum. Abbreviations: *B* neuropil Bellonci; *C* corpus geniculatum thalamicum; *d.Th.* dorsal thalamus; *nPT* pretectal nucleus; *PTN* posterior thalamic (pretectal) neuropil; *TO* tectum opticum; *v.Th.* ventral thalamus; *II* optic nerve. (From Himstedt et al. in press)

At medium velocity (10°/s), seven neurons belonged to the type S > H > V >, seven neurons to type H > S > V, and two neurons to type V > H > S. At this velocity the Julesz-pattern elicited little or no response.

At high velocity (100°/s), eight neurons were found that showed preference type S > H > V, three neurons showed a preference H > S with no responses to V and one neuron responded best to the Julesz-pattern.

In both groups neurons with preference inversion were found. In five neurons of the first group only H elicited responses at 1°/s, whereas at 10° and 100°/s S was better than H. V elicited responses only at 10°/s. These neurons resemble type HXS > V described above. In another neuron of this group an inversion took place with respect to V and S between 1° and 10°/s, which corresponds to type H > VXS. In the latter group the neurons mostly showed complex response patterns, again with preference inversion among V and H as well as among S and H.

As shown in Fig. 5.14D, most of the recorded and marked neurons were situated in the upper layers of the gray matter and in the deeper layers of the white matter. Some of them were situated rather superficially in the white matter. The recording depths were mostly between 50 and 370 μm.

Most neurons showed an RF size between 20° and 35° (with eight neurons < 20° and one neuron > 50°). Himstedt et al. did not find any correlation between response properties, recording site, and RF size.

The results from the three series of experiments on response properties of tectal cells in *Salamandra* reported here are somewhat difficult to compare, at least those from Himstedt and Roth and from Himstedt et al. on the one hand, and from Finkenstädt and Ewert on the other, since in the former studies stimulus velocity was changed, but only three types of stimuli were used, whereas Finkenstädt and Ewert tested 13 different stimuli but only at one velocity. However, if we restrict our attention to the stimuli with an edge length of 8° and 16°, which in the experiments of Finkenstädt and Ewert usually elicited maximal responses and corresponded to the edge lengths tested by Himstedt and Roth (8°) and Himstedt et al. (16°) and if we consider the stimulus velocity of 7.6°/s used here to be not too different from the medium velocities used in the studies of Himstedt and Roth (6°/s) and Himstedt et al. (10°/s), then the data become comparable:

Type T-5.1 of Finkenstädt and Ewert corresponds to type S > H > V of Himstedt and Roth at 8° stimulus edge length (EL), whereas type T-5.3 corresponds to V > S > H at 8° EL or to S > V > H at 16° EL. T-2.3 corresponds to S > V > H at 16° EL. As already mentioned, Finkenstädt and Ewert were unable to find a type T 5.2 corresponding to type H > S > V, although this type was found rather frequently both by Himstedt and Roth at 8° EL and by Himstedt et al. at 16° EL. In the study of Finkenstädt and Ewert, at 16° EL type T-5.1 shows a preference similar to type H > S > V which, therefore, would correspond to type T-5.2. However, in type T-5.1 the differences between H and S are not significant as would be characteristic for type T-5.2.

If one takes all data together, one can easily see that the types found and described by the different authors in *Salamandra* are more or less the same, although they were not always found at the same edge length. The dominating types seem to be S > H > V or T-5.1 and S > V > H or T-5.3. Therefore, the 8 × 8° square is by far the most effective stimulus. Type H > S > V or T-5.2 is almost as frequent as the two other types at least in the studies of Himstedt and Roth and of Himstedt et al. Interestingly, there is a certain number of neurons that prefer V to S and H, a type that was not found in other amphibians (see below). The number of "inversion neurons" is rather high in *Salamandra* in comparison to other amphibians.

5.4.2 *Hydromantes italicus*

Recordings from the optic tectum of *Hydromantes italicus* were done by Roth (1982a) under the same conditions as in the experiments of Himstedt and Roth on *S. salamandra*, i.e., using an 8 × 8° square, a 2 × 8° horizontal and a 8 × 2° vertical rectangle, moved horizontally at 2°, 6°, and 20°/s (cf.

Fig. 5.10). A total of 110 neurons were studied with the stimulus program. The following response types were found:

1. Neurons (n = 24) with a preference S > V > H. The differences between the responses to the three stimuli were all significant except for V and H at high velocity where the two curves converged. The mean RF size was 22° (Fig. 5.15A).

2. Neurons (n = 16) with a preference S > H > V. Here all differences were again significant except between H and V at high velocity. The mean RF size was 31° (Fig. 5.15B).

3. Neurons (n = 18) showing a preference S > VXH, i.e. they preferred S at all velocities, but responded better to V at low and better to H at high velocity, such that a preference inversion between V and H occurred. At medium velocity V and H were equally effective. All differences are significant. The mean RF size was 21° (Fig. 5.15C).

4. Neurons (n = 12) that did not respond to the stimuli at low velocity, but responded with rather high frequency discharge at medium and high velocity. Now, the preference was S > V > H. The increase of impulse frequency from 6° to 20° was rather strong, such that neurons of this type can be considered as "high-velocity neurons". The mean RF size was 19°. With respect to their stimulus preference and their RF size they may be regarded as a subgroup of type 1 (S > V > H) (Fig. 5.15D).

5. Neurons (n = 6) in which at low velocity only S elicited any response. At medium velocity the neurons responded rather well to all three stimuli with a preference S > V > H. Toward high velocity a preference inversion between the responses to V and H took place, such that the preference was now S > H > V. While toward high velocity the impulse frequency of the responses to H further increased, the responses to V sharply decreased, sometimes down to 0. The mean RF size was 19° (Fig. 5.15E).

6. Neurons (n = 5) which at low velocity responded only to S and at medium and high velocity additionally either to H or V, such that either H (three neurons) or V (two neurons) never elicited any response (Fig. 5.15F).

7. Neurons (n = 6) which, contrary to all other neurons recorded, showed a moderate to strong decrease of impulse frequency toward higher stimulus velocity. Thus, these neurons could be regarded as "low velocity neurons". The preference was mostly S > H > V. Often, there was no response to H and/or V at high velocity. The RF size ranged from 8° to 30° (Fig. 5.15G).

The remaining eleven neurons either showed no significant differences between the stimuli presented or they showed rather complex response patterns.

In an ongoing second series of experiments with *Hydromantes italicus*, dealing with neural mechanisms of depth perception, Wiggers and Roth (in

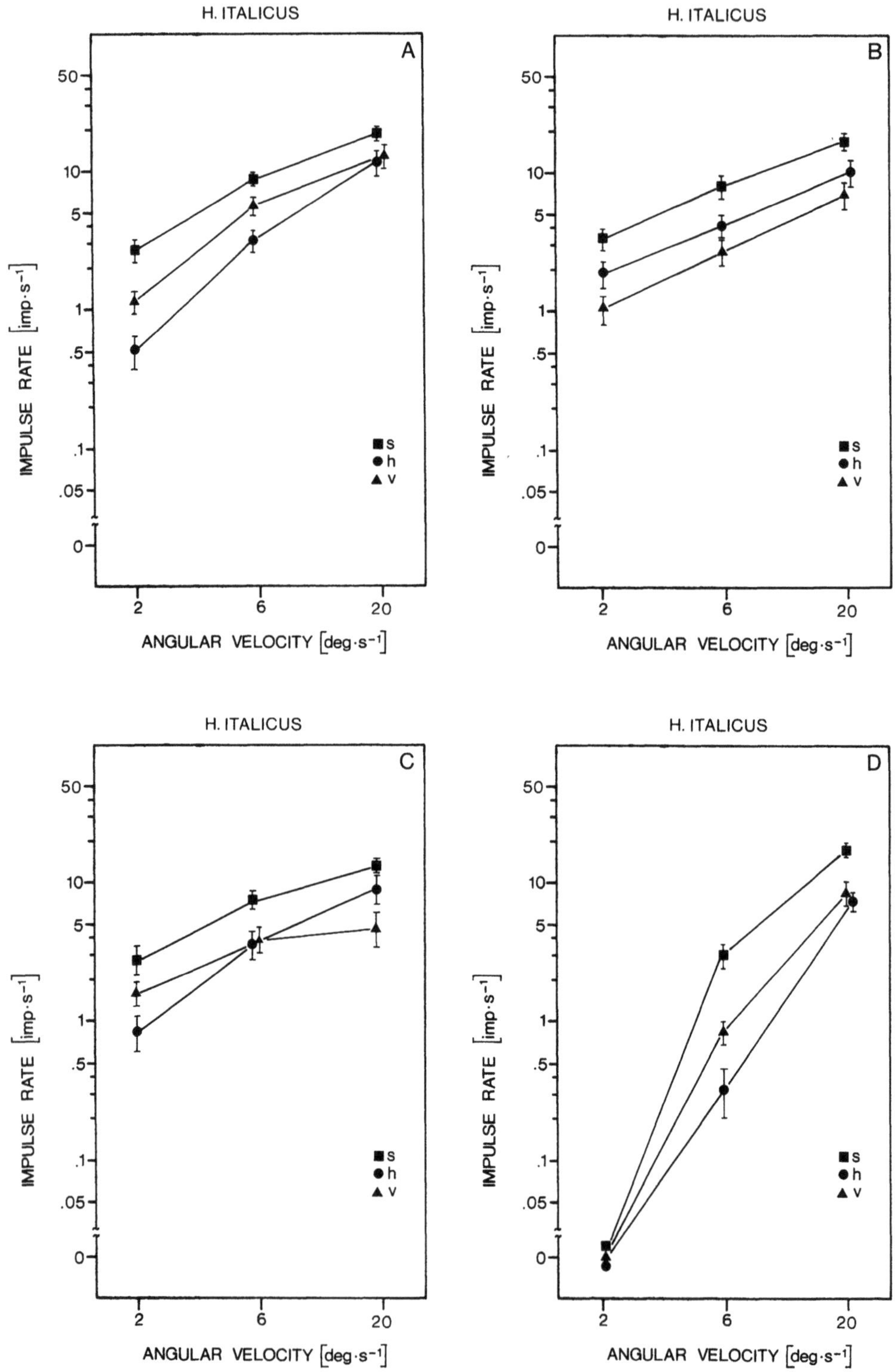
H. ITALICUS
A
50
10
5
1
.5
.1
.05
0
IMPULSE RATE [imp·s⁻¹]
s
h
v
2
6
20
ANGULAR VELOCITY [deg·s⁻¹]
H. ITALICUS
B
50
10
5
1
.5
.1
.05
0
IMPULSE RATE [imp·s⁻¹]
s
h
v
2
6
20
ANGULAR VELOCITY [deg·s⁻¹]
H. ITALICUS
C
50
10
5
1
.5
.1
.05
0
IMPULSE RATE [imp·s⁻¹]
s
h
v
2
6
20
ANGULAR VELOCITY [deg·s⁻¹]
H. ITALICUS
D
50
10
5
1
.5
.1
.05
0
IMPULSE RATE [imp·s⁻¹]
s
h
v
2
6
20
ANGULAR VELOCITY [deg·s⁻¹]

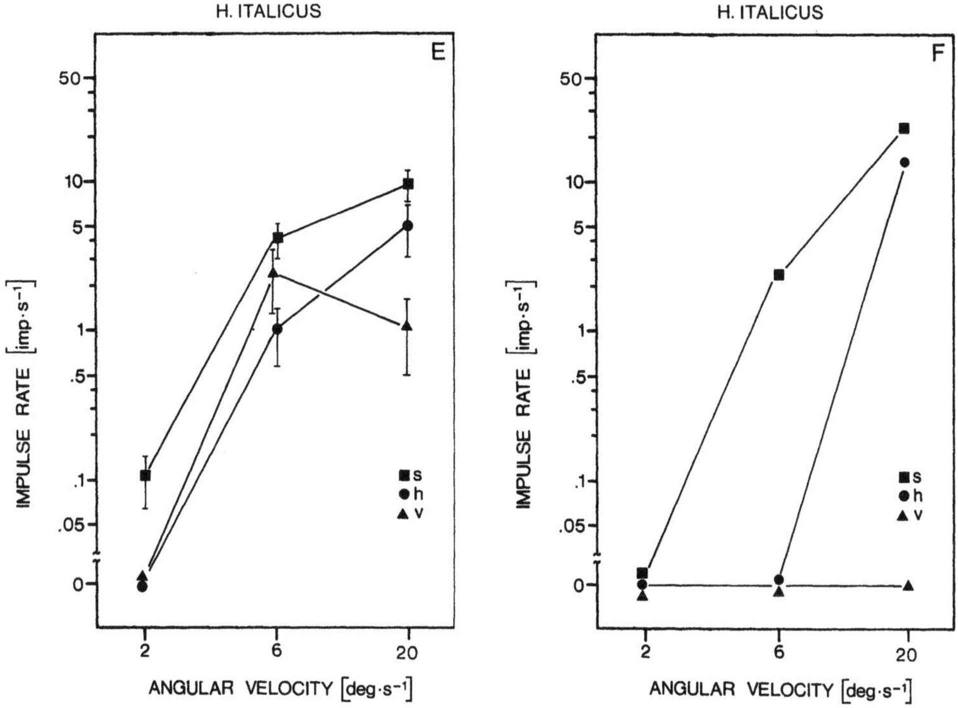

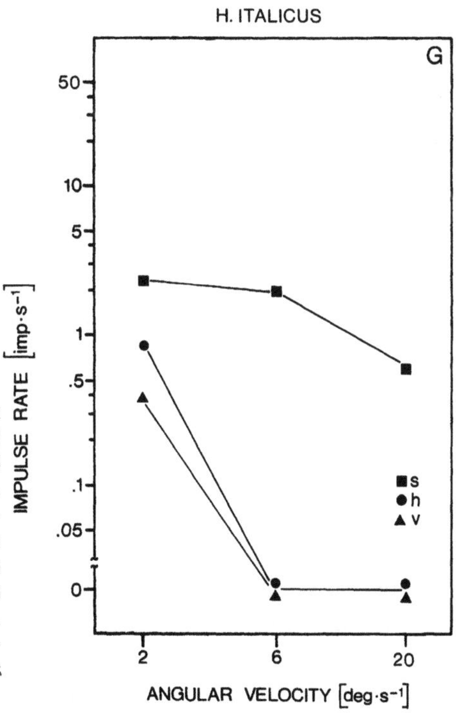

Fig. 5.15 A–G. Response types in the tectum of *Hydromantes italicus*. **A** Neurons (n = 24) with stimulus preference S > V > H. **B** Neurons (n = 16) with preference S > H > V. **C** Neurons (n = 18) with preference inversion S > VXH. **D** "High-velocity" neurons (n = 12). **E** "High-velocity" inversion neurons (n = 6). **F** Neurons (n = 5) responding only to S at medium and to S and H at high velocity. One characteristic neuron is shown. **G** "Low-velocity" neurons (n = 6). One characteristic neuron is shown (After Roth 1982a)

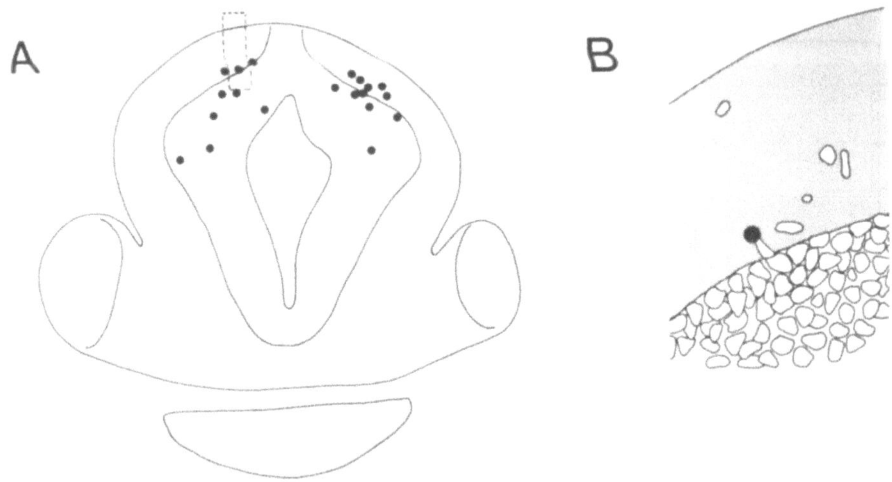

Fig. 5.16. A Distribution of recorded cells in the tectum opticum of *Hydromantes italicus*. Cells that are distributed throughout the entire tectum are projected onto a cross-section through the mid-tectum. **B** Labeled cell that has migrated slightly outside the gray matter. Other migrated cells are also present. (Wiggers and Roth unpublished)

prep.) used square stimuli of 2°, 4°, 8°, and 16° edge length and moved at velocities of 2°, 6°, and 20°/s. 151 neurons have been recorded. After recording, the neurons were marked iontophoretically. The preliminary results of these experiments are the following:

Half of all recorded neurons preferred the 16° square to the other stimuli, 38% preferred the 8° square, 9% the 4° square, and 3% the smallest (2°) square. The differences in stimulus size preference were clearest at high velocity. Toward low velocity, the largest square often decreased in effectiveness, and became the second or third effective stimulus, whereas the 4° square strongly increased in effectiveness for more than double its value. Thus, in almost 80% of all cases the differences between the effectiveness of the four stimuli were greatest at high velocity and smallest at low velocity.

The majority of the neurons recorded (74%) showed ERF sizes between 2° and 50°, with a maximum (38%) between 15° and 20°. Only about 11% showed ERF sizes equal to or larger than 100°. As in the recordings of Himstedt et al. from the tectum of *Salamandra*, in both studies on *Hydromantes* tectal response types, the recorded cells were mostly situated at or near the border between the cellular and the fiber layer of the optic tectum; some were "migrated" cells situated in the fiber layer (Fig. 5.16).

Wiggers and Roth measured the response latencies of tectal neurons in *Hydromantes italicus*, *Bolitoglossa subpalmata*, and *Batrachoseps attenuatus*. At binocular stimulation with a photo flash, in *Hydromantes italicus* (n = 33) the mean latency was 97 ms (minimum 70 ms, maximum 152), in *Bolitoglossa subpalmata* (n = 20) the mean latency was 131 ms (minimum 70 ms, maximum 317 ms), and in *Batrachoseps attenuatus* (n = 39) 121 ms (minimum 90 ms, maximum 161 ms). The mean of all tectal cells of the three species together measured at binocular stimulation (n = 95) was 114 ms, at monocular *contralateral* stimulation (n = 28) the mean was 112 ms (minimum 65 ms, maximum 175), at monocular *ipsilateral* stimulation (n = 70) 136 ms (minimum 70 ms, maximum 810). It is interesting to see that with monocular contralateral and ipsilateral stimulation, the minimal response latency varied only slightly (due to the existence of many direct ipsilateral retinotectal fibers), while some monocular ipsilateral responses had extremely long latencies indicating polysynaptic tectal stimulation.

The results obtained from the tectum of *Hydromantes italicus* show similarities as well as remarkable differences to those from *Salamandra*. Most interestingly, almost all neurons preferred the square to the other stimuli at all tested velocities. In comparison to *Salamandra*, the types H > S > V and V > S > H are missing. Furthermore, type S > V > H is much more numerous than type S > H > V. Another difference is that in *Hydromantes* the recorded tectal neurons showed a considerably stronger preference for the high stimulus velocity as well as a generally higher level of impulse frequency. Some neurons of *Hydromantes* had a maximum impulse frequency beyond 50 impulses/s which was never found in *Salamandra*.

Similar to the study of Finkenstädt and Ewert on *Salamandra*, the experiments of Wiggers and Roth on *Hydromantes* revealed the astonishing result that most tectal neurons recorded prefer stimuli with edge lengths of 8° and 16° which, with regard to their absolute size, are far too large to be optimal in eliciting feeding behavior. The significance of this fact will be discussed in detail in Chapter 6. It is unclear so far why in most tectal neurons of *Hydromantes* the largest (16°) stimulus decreased in effectiveness with decreasing velocity while the second smallest (4°) stimulus strongly increased.

Comparison with Response Characteristics of Anuran Tectal Cells. On the basis of recordings from tectal cells in *Rana esculenta*, *Rana pipiens*, *Hyla septentrionalis*, and *Bufo bufo*, Grüsser and Grüsser-Cornehls (1970, 1976) described the following main response types (Fig. 5.17):

1. Class T-1 neurons. The ERF of these cells is oval-shaped and 15–30° in size. They are found only in the binocular visual field, but some of them respond only to monocular stimulation. These neurons are preferentially activated by small preylike stimuli.

2. Class T-2 neurons. These neurons have large ERFs (> 90°) and respond only to monocular stimulation. In addition, they display directional selectivity by increasing their neuronal activity when the stimulus moves in a temporonasal direction within the visual field. Frequently, these neurons show spontaneous activity. The neurons respond best to small, preylike

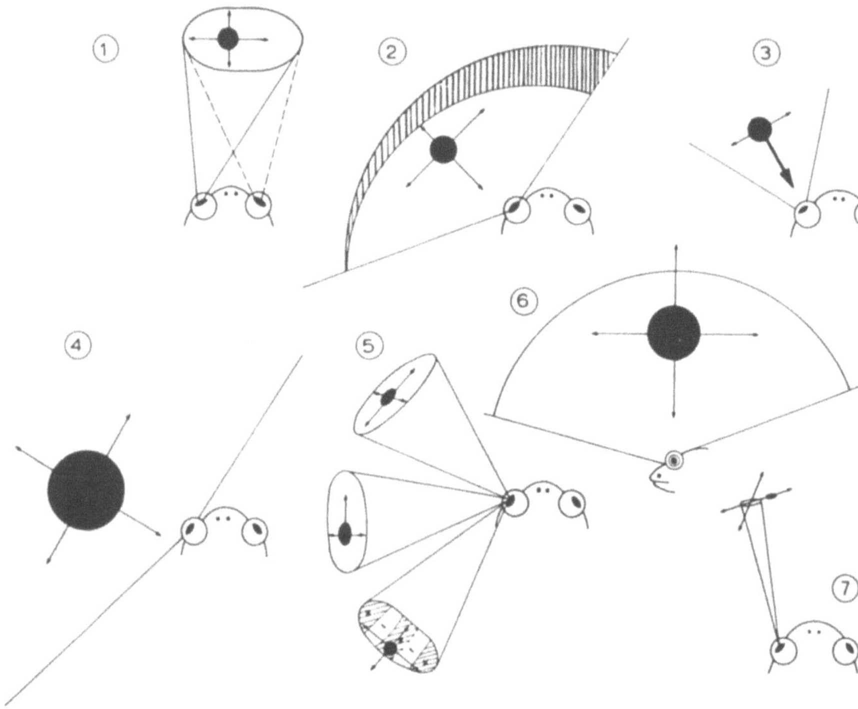

Fig. 5.17. Different classes of tectal neurons (T1–T7) in *Rana temporaria*. For further explanation, see text. (From Grüsser and Grüsser-Cornehls 1976)

objects 2–6° in size, but strong responses occur also when stimulated with larger objects.

3. Class T-3 neurons. These neurons respond well to objects larger than 3° in size moving toward the eye along the z-axis, whereas stimulus movements in the opposite direction slightly inhibit the spontaneous activity; movements in a tangent plane elicit no or only very weak responses. If simultaneous movement of two objects in the ERF occurs, one toward and one away from the eye, then the activity level elicited by the object approaching the eye is dominant.

4. Class T-4 neurons. The RF of these neurons covers the entire visual field or a major part of it. The neurons respond both to small (< 5°) and larger objects (up to 30°). Some of them are multisensory (tactile and auditory). They are located primarily in the deeper layers of the anuran tectum.

5. Class T-5 neurons. These cells correspond to the "newness" neurons described by Lettvin et al. (1961). They have an ERF of 8–30° and respond best to moving objects within a size range of 3° and 15°, but a stimulus as small as 0.1° is still able to evoke a response. Optimal stimulus size, however, is dependent on velocity and configuration. T-5 neurons show great variability with regard to their response properties. Some of them are directionally

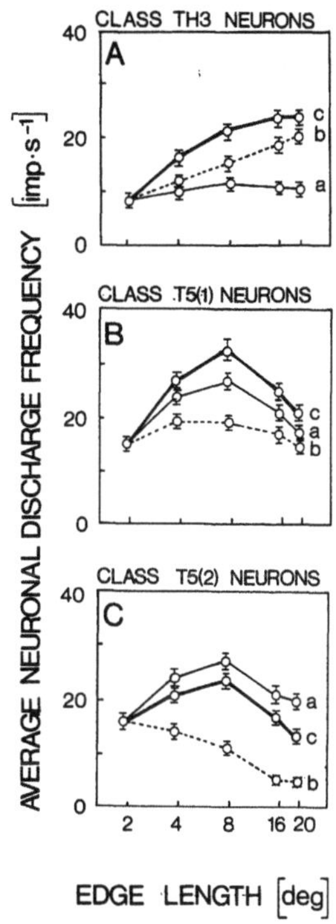

Fig. 5.18 A–C. Response characteristics of thalamo-pretectal (TH3, **A**) and tectal neurons (T-5.1, T-5.2, **B** and **C**) in *Bufo bufo*. Generation of stimuli as described in Fig. 5.12. *a* Horizontal rectangles; *b* vertical rectangles; *c* squares. (After Ewert and von Wietersheim 1974 and Ewert 1984)

sensitive, others show a "striated" receptive field in which excitatory zones are interrupted by "silent" zones.

6. Class T-6 neurons. The ERF of this type of neuron is located above the animal. It measures at least 120° in its rostrocaudal extension and 90° in its horizontal extension. It responds best to stimuli larger than 8°.

7. Class T-7 neurons. These neurons have very small ERFs (2–3°) and respond best to very small moving objects (1–2°). The angular velocity threshold is very low (0.01°/s).

Ewert and von Wietersheim (1974) and Roth and Jordan (1982) in the toad *Bufo bufo*, and Schürg-Pfeiffer and Ewert (1981) in the frog *Rana temporaria* have further investigated the response characteristics of T-5 cells using stimuli of different configuration. The experimental conditions in the studies of Ewert and von Wietersheim and Schürg-Pfeiffer and Ewert were more or less the same as in the study of Finkenstädt and Ewert in *Salamandra*, and those of the study of Roth and Jordan were the same as in the studies of Himstedt and Roth in *Salamandra* and Roth in *Hydromantes*.

In these studies three subtypes of type T-5 were distinguished: T-5.1 (S > H > V), in which at an edge length of 8° the square is preferred to the horizontal rectangle and to the vertical rectangle; T-5.2 (H > S > V), in

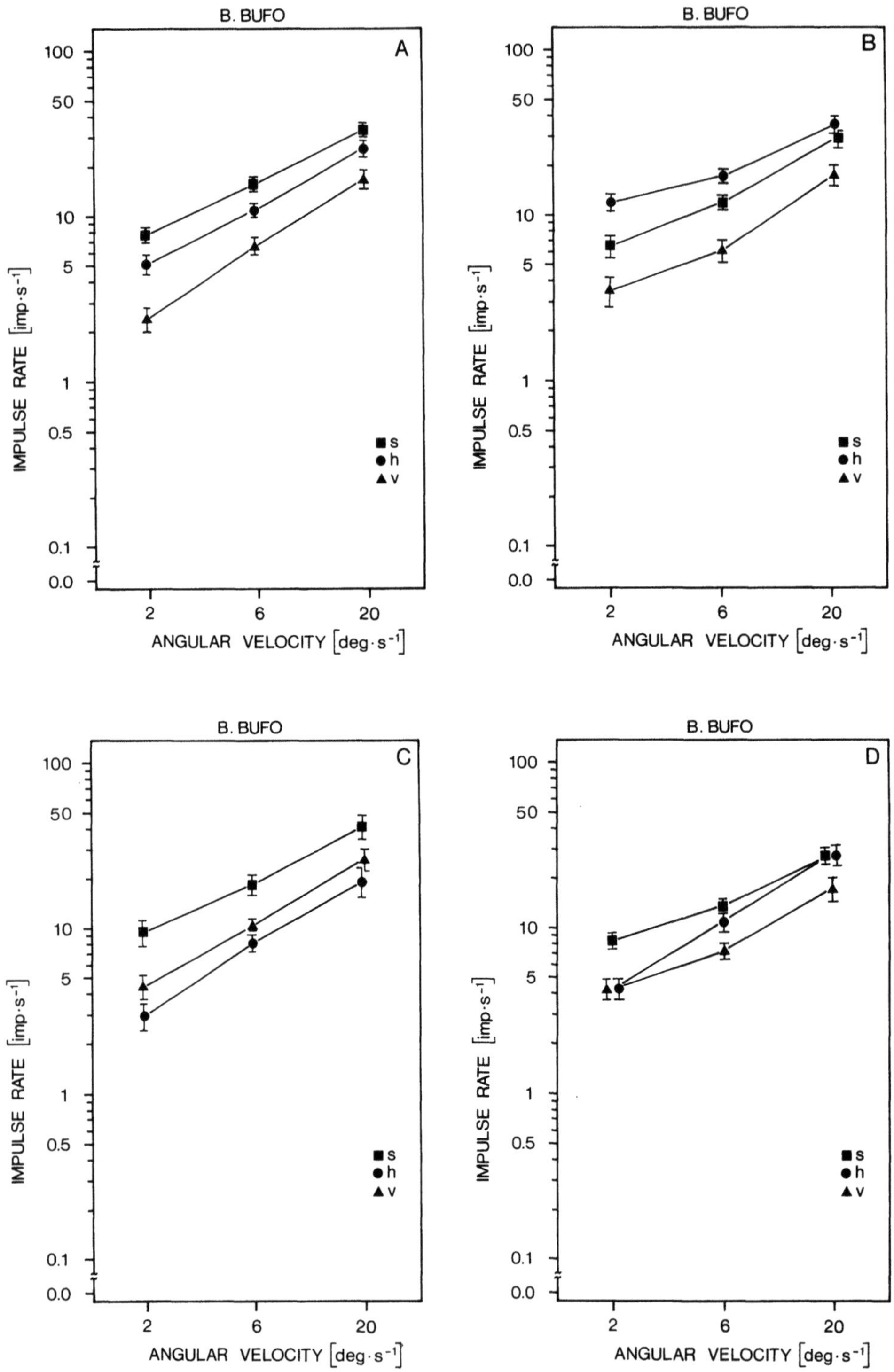
B. BUFO
A
B
C
D
IMPULSE RATE [imp·s⁻¹]
ANGULAR VELOCITY [deg·s⁻¹]
100
50
10
5
1
0.5
0.1
0.0
2
6
20
s
h
v

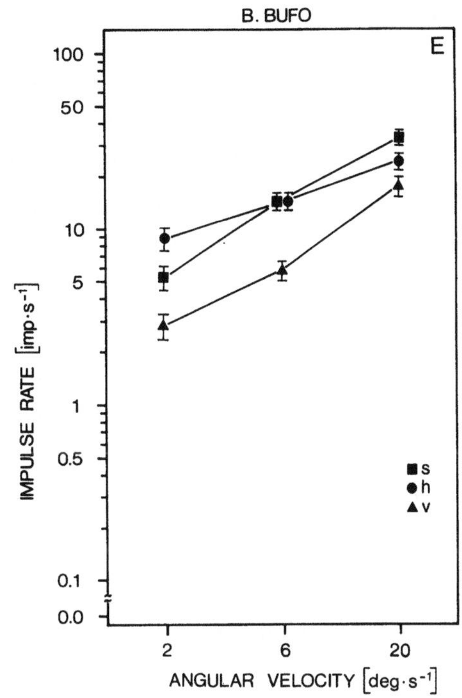

Fig. 5.19 A–E. Response characteristics of tectal neurons in *Bufo bufo*. Experimental conditions as indicated in Fig. 5.10. **A** Neurons (n = 34) with preference S > H > V (T-5.1). **B** Neurons (n = 25) with preference H > S > V (T-5.2). **C** Neurons (n = 13) with preference S > V > H (T-5.3). **D** Neurons (n = 22) with change in preference regarding H. **E** Neurons (n = 16) with preference inversion HXS > V. (After Roth and Jordan 1982)

which the horizontal rectangle is better than the square and the vertical rectangle; and T-5.3 (S > V > H), in which – after the square – the vertical rectangle is better than the horizontal one (Figs. 5.18 B, C and 5.19 A–C). The frogs *Rana esculenta* and *R. temporaria* differ from the toad by the absence of the "worm-preference" type T-5.2 (Schürg-Pfeiffer and Ewert 1981, Grüsser-Cornehls 1984). In the toad, Roth and Jordan additionally described two types that are velocity-dependent in their stimulus preferences and show a "preference inversion". In the first type S > VXH the square is preferred at all measured velocities, but V is better than H at low velocity and H better than V at high velocity. In the second type HXS > V, H is superior to S at low velocity and S to H at high velocity (Fig. 5.19 D–E).

A comparison between tectal response types in urodeles and anurans shows that there are great similarities, despite the large anatomical differences. Using the same experimental conditions, rather similar types can be found in *Salamandra* and *Bufo*, on the one hand, and in *Hydromantes* and *Rana*, on the other. Here, the absence of the "worm-preference" type T-5.2 (H > S > V) in the frog tectum corresponds to the absence of the same type in the tectum of *Hydromantes*. The significance of this fact will be discussed in the next chapter.

5.5 Recordings from Diencephalic Visual Areas

To date, only Manteuffel (1982, 1984 b) and Himstedt et al. (in press) have studied the response characteristics of diencephalic visual neurons in urodeles. Himstedt et al. studied the following three diencephalic visual centers:

1. rostral thalamus, at the level of the neuropil Belloncí and the corpus geniculatum thalamicum, thus the region of the "nucleus Bellonci" of Herrick;
2. caudal dorsal thalamus, rostral to the pretectal neuropil and to the commissura posterior;
3. praetectum, below and immediately behind the commissura posterior and below the rostal tectum opticum.

The experimental conditions were the same as in the recordings of Himstedt et al. from tectal neurons as reported above. The results were the following:

5.5.1 Rostral Thalamus

In the rostral thalamus 26 neurons were recorded. Their RF size was mostly between 36° and 50°; five neurons showed RF-sizes beyond 50°. These RF-sizes are considerably larger than those of tectal cells.

At a stimulus velocity of 1°/s, four neurons showed the preference type H > V/S, i.e. H dominated over V and S which were of equal efficiency. Four neurons showed type V > S/H, and one neuron showed type S > H (with no response to V). There were no responses to the Julesz-pattern. Of the recorded cells 25% responded more or less equally to all four stimuli, i.e. they did not satisfy the preference criterion. At 10°/s seven neurons showed type H > S/V, six neurons showed type S > H > V, and three neurons responded best to the Julesz-pattern. Now 36% of the neurons did not satisfy the preference criterion. At 100°/s two neurons responded to H only. Three were of type S > H > V. One neuron responded best to the Julesz-pattern (Fig. 5.20).

With regard to the effect of stimulus velocity, the neurons showed the following behavior: half of the neurons that could be recorded with the full stimulus program showed a response maximum at 10°/s and then a more or less strong decrease occurred toward high velocity, sometimes down to 0 for one to all stimuli. Some of the neurons showed preference inversions between S and V in both possible ways. The other half of the neurons further increased their impulse frequency toward high velocity at least with respect to one stimulus. Here, too, preference inversions occurred, but now between H and S.

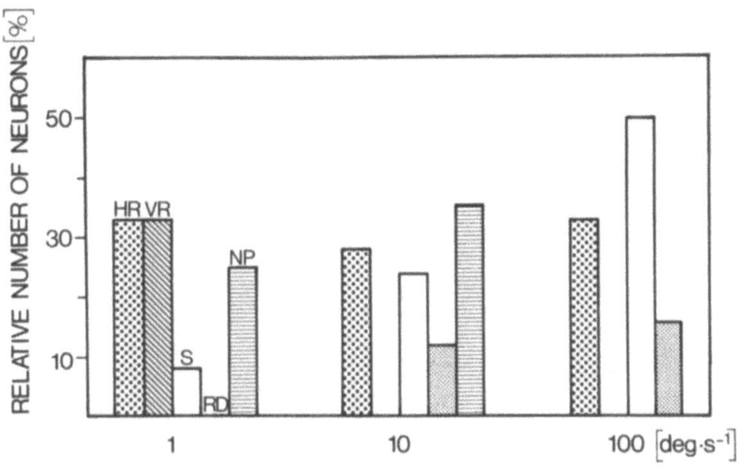

Fig. 5.20. Relative number of neurons in the rostral thalamus of *S. salamandra* with or without significant preference for one of the used stimuli. For further explanation, see Fig. 5.13. (From Himstedt et al. in press)

The marking experiments showed that most recorded cells were situated in the periventricular cellular layer, some of them rather deep; only two "migrated" neuron were stained. The neurons were found at depths between 317 and 590 μm, situated around the sulcus medius of the diencephalon and immediately above it, and between 400 and 800 μm rostral to the commissura posterior near the neuropil Belloni and the corpus geniculatum thalamicum, which corresponds to the position of the nucleus Bellonci of Herrick (Fig. 5.14A).

5.5.2 Caudal Dorsal Thalamus

In the caudal dorsal thalamus, 32 neurons were recorded. Their RF sizes mostly ranged between 36° and 50°; in eight neurons the RF size was beyond 50°. Therefore, the RF sizes are comparable to those measured in the rostral thalamus. There was generally no response to the Julesz-pattern.

At 1°/s six neurons showed a preference H > V > S, and three a preference S > V/H. About 18% of the recorded neurons showed no significant preference for one stimulus. At 10°/s eleven cells showed a preference H > V > S and the same number a preference S > H > V. One neuron showed a preference V > H/S. About 28% of the neurons showed no significant preference. At 100°/s four cells showed the preference H > S > V and the other four a preference S > H > V (Fig. 5.21).

The recorded cells were found again around the sulcus medius thalami and immediately above it, but now posterior to the nucleus Bellonci

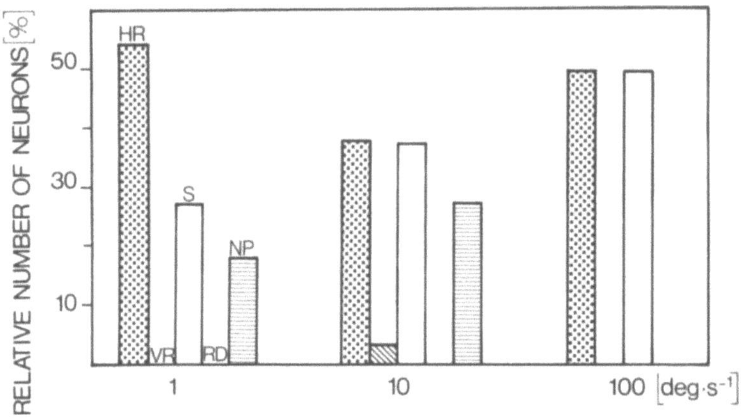

Fig. 5.21. Relative number of neurons in the caudal thalamus of *S. salamandra* showing significant stimulus preferences. For explanation, see Fig. 5.13. (From Himstedt et al. in press)

270–30 µm rostral to the commissura posterior and more dorsal, at a depth between 230 and 587 µm. Most of the neurons were situated near the border between the cellular and the fiber layer; about half of them were "migrated" cells. Only one cell was found deep in the periventricular gray matter (Fig. 5.14 B).

5.5.3 Praetectum

Thirteen neurons were studied in the praetectum, of which only three could be tested with the complete stimulus program. The RF sizes were mostly beyond 50°.

At 1°/s one neuron showed a preference for the Julesz-pattern and one a preference for S. The third one showed no significant stimulus preference. At 10°/s four neurons preferred J to S > V > H, one neuron showed the preference S > V > H > J. More than half of the neurons showed no significant preference. At 100°/s two neurons showed a significant preference for S, but the other neurons did not significantly prefer any stimulus (Fig. 5.22).

The recorded cells were situated 250–600 µm caudal to the commissura posterior at a depth of 284–645 µm, thus mostly in the zone of transition between the praetectum and the basal optic nucleus. With one exception, the marking points were all situated outside the periventricular gray matter, partly adjacent to "migrated" cells and partly away from any cell bodies. These recordings most probably were made from unipolar cells in the praetectum that have their axon hillocks outside the gray matter (Fig. 5.14 C). Similar results were obtained by Manteuffel (see below).

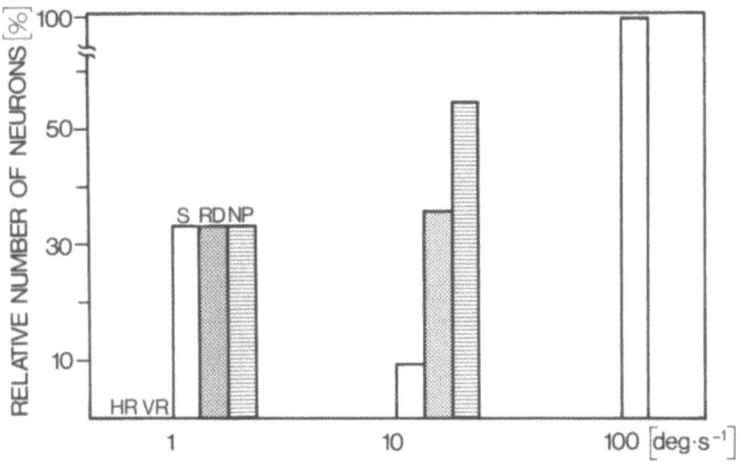

Fig. 5.22. Relative number of neurons in the praetectum of *S. salamandra* with significant stimulus preferences. For explanation, see Fig. 5.13. (From Himstedt et al. in press)

In summary, the data show that there are remarkable differences between diencephalic cells and tectal cells, on the one hand, as well as among the diencephalic neurons themselves, which could indicate different functions of the different recording sites. The RF sizes of all thalamic and pretectal cells were considerably larger than those of the tectal cells. They had RF sizes between 36° and 50° with a substantial number beyond 50°. The largest receptive fields were found in the praetectum.

With regard to stimulus preference, remarkable differences exist. In the rostral thalamus, the number of neurons preferring H was rather high, at least at low and high velocity. Furthermore, at low velocity there was a surprisingly high percentage of neurons preferring V.

In the caudal dorsal thalamus, again many neurons preferred H. At medium and high velocity the number of neurons preferring H or S was equal. The percentage of preferences for V was very low.

In the praetectum, in contrast, the neurons preferred S or the Julesz-pattern. This corresponds well with the much larger RF sizes. No neuron preferring H or V was found. This corroborates the idea that the caudal praetectum is not or not primarily involved in prey recognition or guidance of feeding.

According to lesion experiments, the praetectum has proven to be essentially involved in controlling gaze-stabilizing reflexes in various amphibians (Lázár 1973, Montgomery et al. 1982, Manteuffel et al. 1983). The system has a close relationship to the eye-muscle motor neurons of cranial nerves III and probably to those of nerves IV and VI.

Detailed studies on responses of pretectal cells to optokinetic stimulation in *S. salamandra* were carried out by Manteuffel (1979, 1982, 1984b). In the

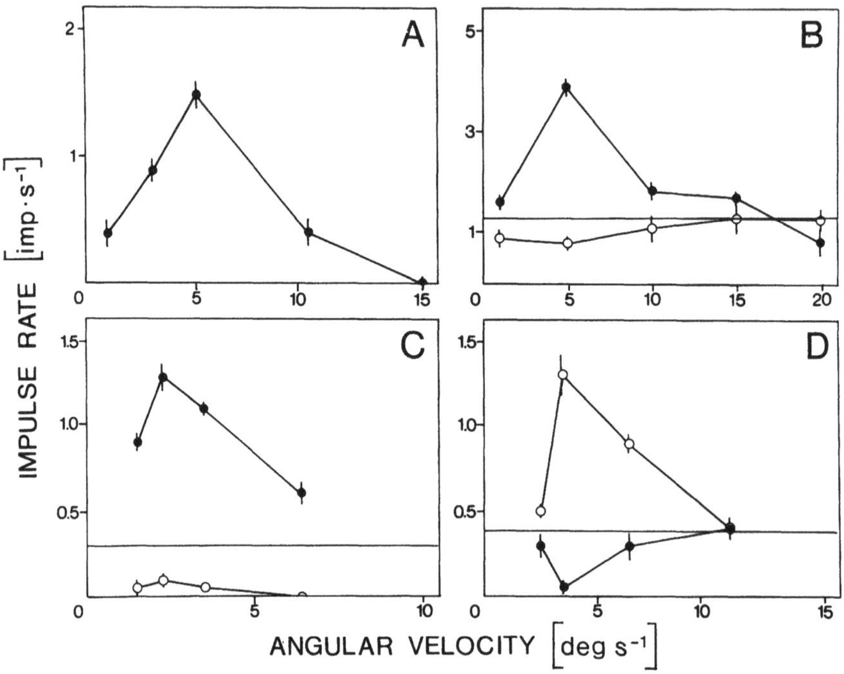

Fig. 5.23 A–D. Response characteristics of pretectal neurons to optokinetic stimulation. *Open circles*: Responses of neurons to nasotemporal (N-T) stimulus movement; *filled circles*: responses of neurons to temporonasal (T-N) stimulus movement; resting discharge activity of neurons is indicated by *horizontal line*. **A** Non-spontaneous type. **B** Spontaneously active cell with weak suppression in N-T direction. **C** Low velocity unit with strong suppression in the N-T direction. **D** Unit with preference for N-T direction with mirrored suppression in the T-N direction. (From Manteuffel 1984a)

latter study, the stimulus consisted in a rotating gray-white striped pattern. The stripes had a width of 3.3°. The stimulus was moved linearly in one direction at different velocities, or sinusoidally with an amplitude of 30° and frequencies of 0.01–1 Hz. The recorded neurons were marked iontophoretically by means of alcian blue, some neurons were morphologically identified by injection of HRP.

Fourty-one direction-sensitive neurons were studied. Most of them were situated around the comissura posterior at a depth of less than 400 μm, seven cells were situated more deeply within the zone of the basal optic nucleus (BON), and three were found in the zone of the nucleus of the oculomotor nerve.

About two-thirds of the cells showed a sensitivity for temporonasal movement of the stimulus within the visual field (Fig. 5.23 A–D). They usually possessed large receptive fields with a mean horizontal diameter of 82° and a mean vertical diameter of 34°, and their center was always situated within the contralateral visual hemifield. Spontaneous activity was either low (Fig. 5.23 B–D) or absent (Fig. 5.23 A). The majority of these cells preferred

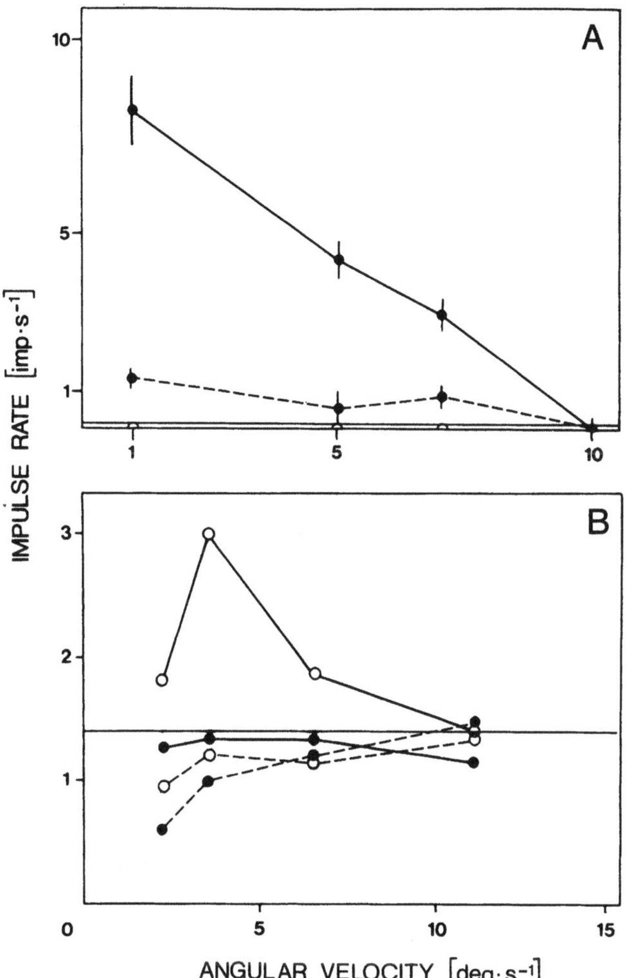

Fig. 5.24 A, B. Response characteristics of pretectal neurons to optokinetic stimulation. **A** Pretectal unit showing clearly better responses to binocular stimulation in T-N direction (*filled circles, solid line*) than to contralateral monocular stimulation (*filled circles, broken line*) in the same direction. Spontaneous activity was low (0.2 Hz) and was fully depressed at nasotemporal stimulus movement (*open circles*) regardless of binocular or monocular stimulation. **B** Strictly binocular pretectal cell. A significant response occurred only at binocular stimulation in N-T direction (*open circles, solid lines*). In T-N direction (*filled circles, solid line*) the resting discharge activity (1.33 Hz) was nearly maintained. Contralateral monocular stimulation (*broken lines*) elicited weak depression. With ipsilateral stimulation, spontaneous activity remained unaffected. (From Manteuffel 1984a)

stimulus velocities lower than 10°/s: 30% were optimally activated by velocities below or equal to 1°/s, and 50% showed best activation at stimulus velocities between 1° and 5°/s.

In those cases where spontaneous activity occurred, it was either unaltered (Fig. 5.23 B) or suppressed (Fig. 5.23 C, D) when the stimulus moved

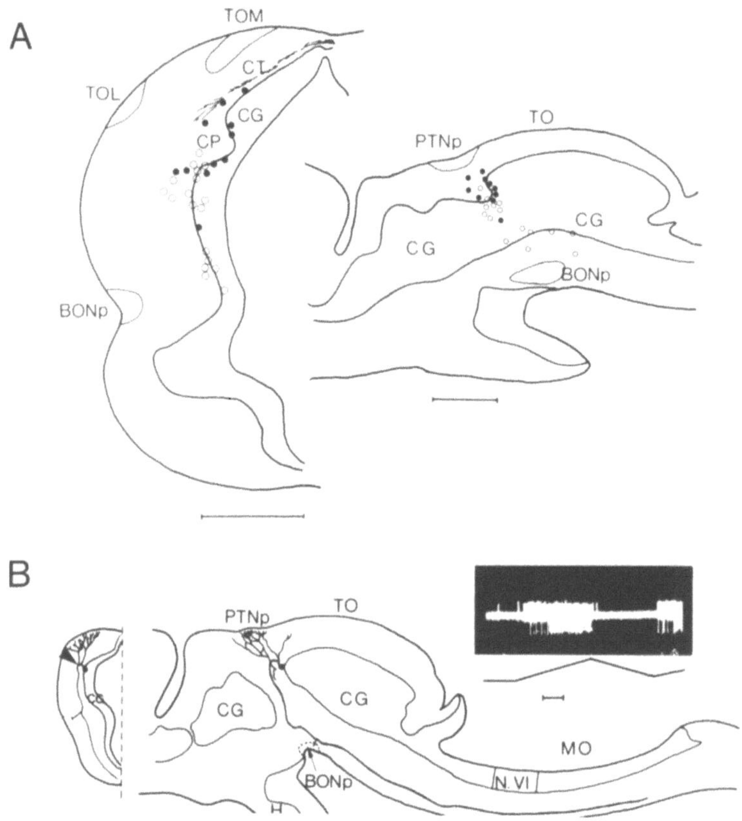

Fig. 5.25 A, B. Anatomical distribution and characteristics of pretectal neurons in *S. salamandra*. **A** Recording sites in the pretectal region. *Open circles*: Contralateral, monocularly driven units. *Filled circles*: Binocularly influenced units. **B** Reconstruction of a single HRP-stained pretectal cell. The reconstruction was made from sagittal sections and is projected onto a parasagittal plane on the *right*. On the *left* a simplified reconstruction in a coronal projection is shown. *Upper right, inset*: Recording showing sensitivity to temporonasal (T-N) stimulus movements in front of the contralateral eye at a velocity of 1°/s; *bar* = 5 s. Recording and HRP injection occurred at the origin of the long axon (*arrowhead*). The cell represents a long-distance efferent type which sends its axon caudal to the medulla oblongata without crossing. It gives off a short collateral to the basal optic neuropil (*BONp*).
Abbreviations: *CG* central gray matter; *CP* commissura posterior; *CT* commissura tecti; *BONp* basal optic neuropil; *H* hypothalamus; *MO* medulla oblongata; *N VI* nucleus of abducens nerve; *PTNp* pretectal neuropil; *TO*: tectum opticum; *TOL* lateral optic tract; *TOM* medial optic tract. Bar = 500 μm. (From Manteuffel 1984a)

in one direction, and strongly enhanced when moving in the other. Thus, these neurons are direction-*specific*.

Nearly 50% of the cells located in the praetectum were binocular cells: they did not respond to ipsilateral stimulation, but were strongly activated by binocular and less vigorously by contralateral (monocular) stimulation. The strictly monocular neurons responded equally to binocular and monocular stimulation (Fig. 5.24 A, B).

In the BON, four cells were found to be sensitive to the temporonasal movement direction of the stimulus and one to be sensitive for the opposite direction. No binocular neurons were found in the BON or the oculomotor nucleus.

In *S. salamandra* those cells which are sensitive to horizontal optokinetic stimuli form a band that starts rostrally at the commissura posterior and is directed caudally and ventrally to the region of the basal optic nucleus (Fig. 5.25 A). This band involves the nucleus praetectalis, the nucleus Darkschewitsch, and the nucleus fasciculi longitudinalis medialis. However, at deeper and more caudal sites close to the basal optic neuropil, cells that respond preferably to vertical stimulation appear to be more numerous. The optokinetically sensitive pretectal cells ramify within the pretectal neuropil and seem to receive inputs from several retinal fibers. Their main efferent projection, as revealed by extracellular filling of the recorded cells with HRP (Manteuffel et al. 1983), reach the contralateral praetectum, the ipsilateral basal optic neuropil, and the ipsilateral caudal medulla oblongata (Fig. 5.25 B). These results are in good accordance with data from the frog (Montgomery et al. 1981).

Optokinetic cells with binocular properties are predominant mainly in the pretectal nucleus. These cells may receive inhibitory input from the ipsilateral eye via direct or indirect ipsilateral fibers, as was shown by intracellular recordings from pretectal cells, when both optic nerves were stimulated (Manteuffel in press). In an earlier study (Manteuffel 1979), the author had shown that the rostral part of the praetectum plays a role in the fixation process. In this context, the binocular optokinetic units may interact with cells that have been shown to be best activated by a stimulus moving nasally in front of the animal.

Apart from the optokinetic response, the praetectum is also assumed to be involved in the constitution of specific tectal response properties in the context of prey recognition. This possible function will be discussed in detail in the last chapter.

5.6 Electric Stimulation Experiments

Finkenstädt and Ewert (1983 b) carried out experiments with *S. salamandra* in which various parts of the brain were stimulated electrically using chronically implanted steel electrodes. The results were the following:

Stimulation of the Tectum. In seven of ten cases, motor patterns of prey-catching behavior could be elicited. In the absence of any prey stimulus orienting, approaching, fixating, and snapping or, more often, biting was activated. During stimulation of different positions within the tectum, the turning responses elicited showed various orientations corresponding to the retinotopic map. In three cases tectal stimulation at higher levels of electric current, an avoidance reaction such as creeping backward and ducked body posture occurred. However, because of the strength of the current one cannot exclude that in these experiments the underlying tegmentum was stimulated.

Stimulation of the Dorsal Thalamus (Praetectum). In four cases, electric stimulation of the caudal dorsal thalamus (praetectum) elicited avoidance behavior. The same occurred after stimulation of the rostral thalamus.

Stimulation of the Telencephalon. Here, different sites in the medial and lateral pallium were stimulated. They all resulted in a variety of behavioral responses such as ducking, creeping backward, raising the rostral trunk, or opening and closing the jaws.

5.7 Brain Lesion Experiments

Finkenstädt and Ewert (1983b) carried out brain lesion experiments in *Salamandra salamandra* using anodal DC current, radiofrequency current, Kainic acid microinjections or microknife cuts. The authors obtained the following results:

Ablation of the optic tectum abolished any visually guided prey-catching and predator-avoidance behavior. After unilateral lesions in the caudodorsal thalamus (praetectum), the animals responded with feeding toward moving objects of any size and configuration when the object was shown within the visual field contralateral to the lesioned brain region. However, the "worm/antiworm" discrimination was strongly impaired. Predator-avoidance behavior could not be elicited, i.e., the animals responded more or less equally to elongated stimuli oriented parallel and perpendicular to the direction of movement. Thus, the salamanders showed the thalamo-pretectal disinhibition phenomenon ("TP-phenomenon") known from toads and frogs (Ewert 1968, Ingle 1973). The size of the area in the visual field where disinhibition occurred corresponded with the size of the lesions. Using the axon-sparing neurotoxin Kainic acid, the authors could show that the disinhibition phenomenon was due to a destruction of the somata of pretectal cells and not of fibers originating in other brain regions and passing through the praetectum.

Recordings from tectal T-5.1 cells after DC coagulation in the ipsilateral praetectum resulted in an increase of ERF size from about 30° to about 50° and a strong deficit of the recorded cell with regard to "worm/antiworm" discrimination.

After lesions in the rostrodorsal thalamus the same disinhibition phenomenon occurred, but affected only the binocular visual field. Lesions in the ventral thalamus did not result in any disinhibition effect with respect to feeding. However, the animals exhibited abnormal body postures such as extreme body bending. Spontaneous motor activity was increased.

Lesions in the medial pallium of the telencephalon again led to strong inhibition of feeding, often combined with "aggressive" behavior. Escape behavior could not be elicited. Spontaneous motor activity could be observed. After lesions in the lateral pallium the animals responded more readily with escape behavior to rather small objects that normally would not elicit escape reactions.

According to the authors, these results show that the pretectal and the rostrodorsal thalamic region as well as the medial pallium are essentially involved in the constitution of prey-enemy distinction. After destruction of these regions all kinds of objects are regarded by the animal as prey.

Manteuffel et al. (1983) reported an abolishment of the head optokinetic nystagmus and a dilatation of the ipsilateral pupil after destruction of the pretectal neuropil and the basal optic neuropil, but not after destruction of the tectum.

5.8 Responses of Tectal Neurons to Monochromatic Light

In the newt *Triturus alpestris*, Henning (1985) studied the responses of tectal cells to monochromatic light of different wavelengths. After successful extracellular recording, some of the cells were filled with HRP such that a morphological characterization was possible.

The following response types were found: (1) ON-neurons, (2) OFF-neurons, (3) color-unspecific ON-OFF neurons, and (4) color-opponent neurons. The ON-neurons either showed no differences in their responses to different wavelengths or preferred long wavelengths, i.e., they were red-sensitive. The HRP-filled cells of this group showed medium-wide dendritic arborizations with dendrites reaching the superficial parts of the stratum album. In the OFF-neurons which allowed a morphological reconstruction the dendritic tree was wide, and in some cases the dendrites did not reach the upper parts of the album. The different cells either showed a response minimum at 563 nm or responded only to stimuli of this wavelength.

Within the group of color-unspecific neurons, one subgroup showed no wavelength preference, the neurons of a second subgroup were yellow-sensitive, and those of a third subgroup were blue-sensitive. The dendritic trees of the first subgroup were rather small, those of the second one were medium wide, and those of the third one very wide, with perikarya situated immediately below the border between stratum album and stratum griseum. Within the group of color-opponent neurons, some were blue- or red-sensitive with an additional sensitivity maximum, whereas others were truly color-opponent with blue-sensitive ON-responses and red-sensitive OFF-responses. The dendritic trees of these neurons ranged from small to extremely wide.

The latencies of the ON-responses varied between 100 and 140 ms, whereas the OFF-responses had latencies around 200 ms. In ON-OFF neurons, the OFF-latency was always longer than the ON-latency. It is interesting that the points of recording and injection of HRP were always found in the stratum album, mostly within its lower part. Presumably, the recordings were made from the axon hillocks being situated in this part of the stratum album.

6 Conclusions and Speculations on the Neural Guidance of Visual Behavior in Salamanders

The preceding chapters have shown that a considerable amount of data have been accumulated on behavior, neurophysiology, and neuroanatomy related to vision in salamanders, especially with regard to visually guided feeding behavior. The main question will now be to what degree this knowledge can be put together to form a reasonable concept for neural guidance of visual behavior. Feeding behavior will again be in the center of these considerations.

6.1 Neural Guidance of Feeding Behavior

The visual system – as far as it is involved in the guidance of prey-catching – has to deal with the following tasks: (1) identification of an object as prey; (2) localization of the prey including estimation of the distance of the prey; (3) control of the appropriate turning and approach movements; (4) exact determination of the reach of snapping distance; (5) triggering of the snapping reaction and of the engulfment movements.

6.1.1 Prey Object Identification

The process of prey recognition has to precede and accompany the whole sequence of prey-catching. From the observation of feeding behavior we can conclude that a rather precise distinction of the visual system between prey and nonprey takes place before the first turning or approach movement: in most cases in which a stimulus elicits these movements, a complete sequence of feeding will follow provided the prey does not disappear suddenly or escape or become motionless such that the predator loses interest. Thus, the animal identifies an object as prey almost completely before it fixates on it and approaches it. Since turning and fixation can be elicited by a prey object within the entire visual field, we have to assume that the whole tectum has "recognition properties" in the sense that recognition units are distributed over the entire tectum. This was shown by those experiments in which part of the tectum was destroyed, and the animal was still able to orient toward prey objects within the visual field corresponding to the spared part of the tectum (Sperry 1945).

Furthermore, it is evident that prey recognition is accompanied by localization of the prey object within the visual field which determines the angle of turning movement, and by a first and maybe rough estimate of object distance. The latter is of importance because the decision "prey/nonprey" necessarily involves estimation of absolute size of the prey at least within certain limits of size. Otherwise it would not happen that a salamander may ignore a near, very small object and instead turns toward a larger and more distant object of the same angular size, but of an absolute size closer to the prey-size optimum. It is also necessary to distinguish a near prey object from a distant enemy of the same angular size.

We have to assume, too, that the "recognition units" are active not only prior to the first turning or approach movement but also during the whole feeding sequence. There is little evidence that the feeding sequence acts as a reflex chain in the sense that once this sequence has been triggered, it proceeds automatically to its end. The reason for this assumption is that the feeding sequence can stop at any point as soon as the prey object loses a certain important feature, e. g., when an insect becomes suddenly motionless (as crickets often do), and it can often be observed that the approach velocity of the salamander, as a good indicator of motivation, changes with respect to a change in certain features of the prey such as velocity or movement pattern.

6.1.2 Neural Mechanisms of Prey Recognition

With regard to the question of the neural mechanisms of prey recognition we are confronted with the following basic question:

Which are the basic neural units of prey recognition? Is prey recognition performed by specific single "detector-cells" or is it the result of the interaction of different kinds of cells within smaller or larger networks, or the result of the "mass activity" of the whole optic tectum?

Barlow (1953) and later on especially Lettvin, Maturana and co-workers (Lettvin et al. 1959, 1961, Maturana et al. 1960) who were the first to study the amphibian visual system systematically with regard to object recognition, concluded from their experiments that the retina, at the level of retina ganglion cells, performs the decisive steps of object recognition and that the visual centers in the brain are mostly involved in transformation of visual neural activity into motor patterns. The retina ganglion cells, in their opinion, were specific detector cells, dependent on their receptive field properties. Since, for example, class-2 (small-field) retina ganglion cells (convexity detectors) of the frog show a preference for small dark moving compact objects, these units were regarded by the authors as "bug detectors". In contrast, class-3 and -4 cells with larger receptive fields were considered to be involved in "enemy" and "dimming" detection.

Later, extensive studies on retina ganglion cells of amphibians, done mainly by Grüsser and Grüsser-Cornehls (cf. Grüsser and Grüsser-Cornehls 1970, 1976) and by Ewert and co-workers (Ewert and Hock 1972, Ewert et al. 1983) in anurans, and by Grüsser-Cornehls and Himstedt (1973) in urodeles showed, however, that the differences in response properties of the retina ganglion cells are by no means sufficient to explain the observed prey preferences in amphibian behavior. The arguments are the following: (1) As Grüsser and Grüsser-Cornehls have stated, every stimulus presented to the visual system which is not too small or too large to be of biological significance activates at least two types of retina ganglion cells, and the same stimulus under different conditions, for example at different velocities or distances, may activate different sets of retina ganglion cells. As Grüsser and Grüsser-Cornehls wrote with regard to the visual system of the frog: "... one can cautiously conclude that the probability of prey-catching behavior ... depends on a combined overall activation of the different retinal neuronal classes. An increased (probability) is correlated with an increased activity of class-1, -2 and perhaps -3 neurons, whereas an increased activity of class 0 and 4 neurons (perhaps also of class-3 neurons above a certain excitatory level) leads to a reduction in (probability)" (Grüsser and Grüsser-Cornehls 1976, p. 364). (2) When activated by prey stimuli, retina ganglion cells show rather stereotyped patterns that are by no means comparable to the complexity of response patterns of tectal and diencephalic cells as presented in Chapter 5, although there are some differences among different groups of amphibians such as between frogs and toads with regard to class-2 retina ganglion cells (Ewert 1976) or between newts and salamanders (Grüsser-Cornehls and Himstedt 1976). Whereas some response properties of tectal and diencephalic cells can be explained as a more or less simple summation of the activity of retina ganglion cells, those of other tectal cells have to be considered – as will be shown later – as a function of further processing of retinal activity in the tectum, and parts outside the tectum may also be involved in that processing. (3) There is a contrast between the similarity of response characteristics of retina ganglion cells of all amphibians studied so far and the great differences in prey preferences, for example between frogs and toads, or between *Salamandra* and bolitoglossine plethodontids, which, therefore, cannot be explained on the mere basis of retinal activities.

These arguments do not only show that prey recognition is unlikely to be the work of specific retinal "detectors", as was originally assumed, but also that the retina, working as a network consisting of different overlapping response types, is only a first stage within the recognition process. There is sufficient evidence that the visual centers in the brain, mainly the optic tectum, but also thalamic and pretectal centers and probably parts of the forebrain, are involved in this process:

1. Ablation of the optic tectum abolishes all feeding and predator avoidance behavior (Bechterev 1884), though not all visual functions such as the optokinetic nystagmus and barrier avoidance disappear (cf. Ingle 1973). Local lesions of the tectum produce "scotomata" in the visual field within which no prey identification and localization takes place (Sperry 1945). In contrast, lesions in the thalamus-praetectum region as well as in the medial and lateral pallium strongly facilitate feeding behavior and abolish prey-enemy discrimination (Ewert 1968, Finkenstädt and Ewert 1983b).
2. The sequence of feeding behavior can be released completely or partially by electric stimulation of the optic tectum (Ewert 1967, Finkenstädt and Ewert 1983b, Satou et al. 1985). The part of the visual field to which the animal turns or snaps, corresponds roughly to the stimulated part of the tectum.
3. Most tectal and many diencephalic neurons studied respond well to prey stimuli, but often in a different and mostly much more complex way compared to retina ganglion cells.

These facts indeed show that the visual centers of the brain, primarily the optic tectum, are the place of later, and more specific steps of the prey recognition process. They do not, however, by themselves rule out the possibility that prey recognition is done by specific tectal "detector" cells, especially in the light of the much larger number of different tectal response types in comparison to the retina. On the basis of the assumption that in amphibians visual prey preferences are highly inborn, one could develop a concept in which in the optic tectum of frogs, toads or salamanders for any major natural prey type there is a specific "detector" cell, for example, a "bug", "millipede", "fly", "worm", "slug", "libella", etc. detector which is activated when such a prey type enters the visual field. This detector, by its neural activity, would then act as as a "command" unit and initiate certain motor responses.

Such a concept, although it has gained some popularity among neuroethologists, is confronted with the same difficulties as the original "retinal" detector concept. No cell has been found in the amphibian brain that would respond exclusively to just one type of prey or stimulus, and, accordingly, all natural or artificial visual prey stimuli studied so far elicit responses in more than just one type of tectal cell. For example, in the tectum of most amphibian species, a squarelike stimulus of a size of $8 \times 8°$ moving at an appropriate velocity will activate more or less optimally at least half of all tectal cells described in Chapter 5. In certain amphibians, this may be true for elongate, "wormlike" stimuli, too: if we take the major response types found in the tectum of *Salamandra* or *Bufo*, then this "wormlike" stimulus will activate again almost all those types although not always to the same strength as the square stimulus. And even a bar oriented perpendicular to the direction of movement will activate at least all T-5 neurons, although to much lower degrees of strength. Thus, all the tectal response types described so far cannot act as true "prey detectors".

Furthermore, a stimulus of a certain shape, size, and orientation with respect to the direction of movement may differently activate different types of tectal or diencephalic cells depending on its actual speed, movement pattern or direction of movement. Since a fly is able to move slowly or fast, for example when crawling on the surface or when flying, according to the "detector" concept there should exist at least two different fly detectors, one for slow and one for fast movement, in the tectum of a frog or a bolitoglossine salamander, two animals that respond well both to slow and fast moving dipterans. Again, some types of prey, especially elongate ones such as caterpillars, can change their shape, length or orientation to the direction of movement, and specific detector cells for all those cases should exist. Additionally, the aspect of prey objects may vary with respect to their relative position to the predator, for example, when such a prey moves parallel to the predator or toward it. Thus, in order to explain the different prey preferences in amphibians, we would end with postulating an infinity of different "prey detector" cells.

What we find experimentally, however, is that all those tectal and diencephalic cells that are well activated by stimuli with an edge length or diameter of 6–16° and, therefore, are possibly involved in the process of prey recognition, exhibit only *gradual* differences within the strength of their responses to different prey stimuli, and, therefore, cannot function as single "prey detector" units of sufficient reliability. As will be shown later, this fact of only gradual differences of tectal cells is a rather strict consequence of the same situation at the level of retinal cells. Therefore, we have to assume that prey recognition is not based on the activity of single "detector cells" corresponding to just one type of prey, but rather on the simultaneous activity of at least several types of cells with partly overlapping response characteristics. Possible mechanisms of prey recognition based on such small networks will be discussed later.

The third possibility, namely that instead of single detector cells or small networks prey recognition is the work of the whole tectum or at least large parts of it, can be ruled out by the following facts:

1. When we record from tectal cells by means of microelectrodes we normally find rather restricted areas that can be activated by a prey stimulus

from a certain position within the visual field. Apparently, there is no major horizontal "spread" of activity within the tectum. This may be due to an inhibitory influence from outside the tectum (see below).

2. As already mentioned, small lesions in the tectum will produce circumscribed "scotomata" with regard to prey recognition. In contrast, if one destroys most of the tectum sparing few parts, prey recognition is still possible, if the object is situated in that part of the visual field which corresponds to the "spared" part of the tectum.

6.1.3 Classification of Tectal Cells Involved in Prey Recognition

In contrast to the situation found in the retina, where a rather limited number of more or less clearly distinguishable response types of retina ganglion cells occurs (with some species-specific differences), in the optic tectum – and to a similar degree in the diencephalon – we find a large variety of response types differing with respect to their preferences of stimulus size, orientation, shape, velocity, receptive field sizes, directionality, etc. However, in the present context of prey recognition we can reduce this variety by considering only the preferences for stimulus size, orientation, shape, and velocity. This procedure is legitimate insofar as one can understand differences in receptive field size, directionality (if there is any), etc. as the product of additional specific connectivity among tectal cells.

In all amphibians studied so far the large majority of tectal cells prefer round or squarelike stimuli the optimum of which ranges between 6° and 16° edge length or diameter. Two subgroups can be distinguished: one subgroup in which, after the square, elongate stimuli with an optimum size of 6–16° edge length and oriented parallel to the direction of movement are clearly preferred to the same stimulus configuration oriented perpendicular to the direction of movement, and another one which shows the opposite preference for elongate stimuli. The first subgroup corresponds to class T-5.1 (S > H > V) and the second one to class T-5.3 (S > V > H).

A different group of tectal response characteristics is represented by those cells that prefer elongate stimuli extended parallel to the direction of movement to squarelike or round stimuli and to elongate stimuli of the opposite orientation. This group corresponds to class T-5.2 (H > S > V). In contrast to types T-5.1 and T-5.3, class T-5.2 is not universally found in amphibian tecta. It is, for example, not clearly present in *Rana* and absent in *Hydromantes* (cf. Chap. 5).

A final group comprises cells that prefer elongate stimuli oriented perpendicular to the direction of movement to round and squarelike stimuli and to elongate stimuli of orientation parallel to the direction of movement. To date, this response type V > S > H was found only in *Salamandra*. It may well be, as the data of Finkenstädt and Ewert (1983a) suggest, that this type has to be regarded as a subtype of class T-5.3: in *Salamandra*, neurons of this class showed the preference V > S > H at an edge length of 8° and the preference S > V > H at 16°.

As already mentioned, several other tectal types described by different authors can be understood as variants of these types. For example T-2.3, described by Finkenstädt and Ewert (1983a), is more or less identical with respect to its stimulus preference to T-5.3 at an edge length of 16° and corresponds at this edge length to type S > V > H described by Himstedt et al. Type T-5.4 of Finkenstädt and Ewert, which responds only to squares with an optimum of 16° edge length and was described first by Roth and Himstedt (1978), may be a variant of class T-5.1. Also, the "preference inversion" neurons HXS > V and S > VXH can be understood as variants of the above types (see also below).

The majority of tectal cells showing these stimulus preferences may differ with respect to the following features:

1. Receptive field sizes, which may vary between 6° and beyond 180°. Since many of the medium and large-field neurons (RF > 30° and RF > 90°, respectively) had the same stimulus preferences as those described for the so-called small-field tectal T-5 cells, one may consider them as belonging to the same class. Ingle (in litt.) found that neurons of the "sameness cell" type of Lettvin et al. (1961) show preferences for 4° stimuli and no responses to stimuli larger than 10° despite an ERF size of 60–90°. As Ewert et al. (1983) stated, there is no correlation between optimal stimulus size and receptive field size. One can easily imagine that tectal neurons with large receptive fields but with an optimal stimulus size of 4–16° are the product of convergence of small- and medium field cells.

2. Movement directionality. Whereas many (mostly small- and medium-field) tectal neurons show no or almost no direction sensitivity, large-field neurons more often show directional sensitivity or even selectivity, as was described by Himstedt and Roth (1980). However, in the tectum directional sensitivity or selectivity is not found as often as in the thalamus and especially in the praetectum.

3. Velocity dependence. In all amphibian tecta in which velocity dependence of neurons was tested, some of the types showed no substantial change of preference with a change of velocity, whereas another part showed the phenomenon of preference inversion, either between S and H or H and V, and in both directions. Most neurons increased their response frequency with increasing stimulus velocity, at least up to a velocity of 40°/s. Only a few cells were found (cf. Roth 1982a, with respect to *Hydromantes*), that clearly preferred low velocities (e.g., 2°/s) to higher ones.

6.1.4 The Constitution of Tectal Response Properties

The easiest way to understand the above described basic preference types is the assumption that they are the result of more or less simple integration processes of the activity of retina ganglion cells. An der Heiden and Roth (1983 and submitted) have developed a mathematical model for the constitution of the different response types in the optic tectum. They were able to show that only two neural principles are necessary for the formation of all observed tectal prey preference types: (1) summation of the activity of the different retinal response types, especially class-2 and class-3; (2) lateral recurrent inhibition among tectal neurons.

Indeed, as shown in Fig. 6.1, some tectal response types can be understood as the result of a simple summation process of retinal activity: in this way, the first type, T-5.1 or S > H > V, is constituted by a summation of the activity of class-2 rgc (Fig. 6.1 A), and the second type, T-5.3 or S > V > H by a summation of the activity of class-3 rgc (Fig. 6.1 B). For the constitution of T-5.2 cells that prefer horizontal bars, however, one has to assume that between tectal cells that either sum up the activity of retina class-2 cells or class-3 cells, lateral recurrent inhibition occurs, in the sense that cells, when activated, suppress the activity of their neighboring cells (Fig. 6.1 C, E). This may occur either through direct inhibitory coupling or via inhibitory interneurons, as indicated in Fig. 6.1 F. Both mechanisms may occur simultaneously in the tectum.

If the level of inhibitory coupling between different "first-order" tectal cells is varied, we obtain an interesting continuum of response types: for example, if we have tectal cells summing up the activity of retina class-3 cells with *no* recurrent inhibition, then we obtain tectal type T-5.3 (Fig. 6.1 B). If there is a *medium* degree of recurrent inhibition, then T-5.1 arises (Fig. 6.1 D), and with *strong* recurrent inhibition T-5.2 arises (Fig. 6.1 E). This shows that there is no need to assume discrete T-5 response types in the optic tectum of amphibians. It may well be that all belong to one or two basic types, e. g. one constituted by summation of class-2 rgc and another one constituted by summation of class-3 r. c, which are modified to the observed response types by different degrees of lateral inhibition. With such an assumption, the existence of intermediate response types which are not infrequent, could be explained. However, since we do not observe a linear continuum of preference types, we have to postulate that there is some nonlinearity in the realization of lateral inhibition. The most important point, however, remains that it is unnecessary to assume that discrete response classes exist in the tectum.

This model has not yet been explicitly developed for tectal types other than those mentioned above, for example, for the "preference inversion" neurons, but their constitution could be shown qualitatively by the authors. For example, one can assume the inhibitory coupling to be established by

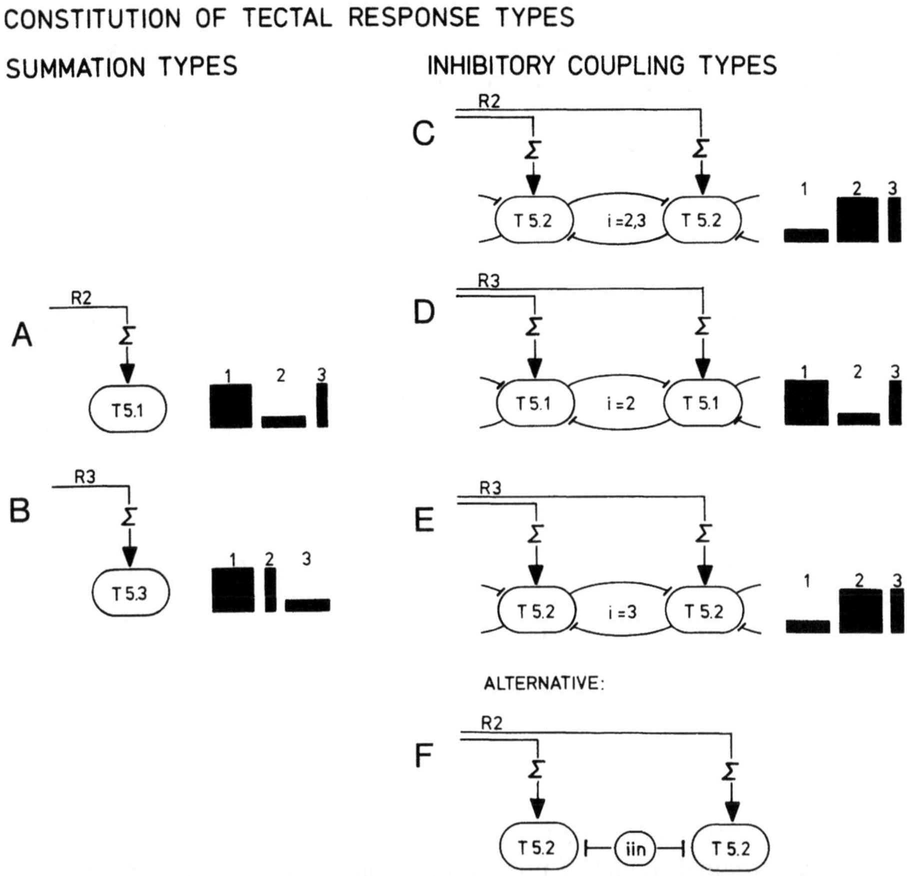

Fig. 6.1 A–F. Constitution of tectal response types according to the model developed by an der Heiden and Roth (1983). *Left*: Response types constituted by mere summation of activity of retina ganglion cells (rgc) are given. *Right*: Response types constituted by summation of rgc plus lateral recurrent inhibition between tectal cells are shown. *i* Strength of inhibitory coupling expressed in arbitrary units. Row of black stimuli indicates the respective stimulus preferences. In the *bottom line* an alternative for inhibitory coupling through inhibitory interneurons (*iin*) is presented. *R2* rgc class-2; *R3* rgc class-3. For further explanation, see text. (From Roth 1986)

interneurons that are by themselves controlled by other neurons. Then, a velocity dependence of these latter neurons would be sufficient to increase or decrease the amount of inhibitory coupling among T-5 cells via influencing the activity of interneurons (as shown in Fig. 6.1 F) and, thus, change their stimulus preferences. If, for example, the cell controlling the inhibitory activity of an interneuron has a preference for low stimulus velocity, then inhibitory coupling between T-5 cells would be sufficiently strong at low velocity and thus constitute type T-5.2. Toward higher stimulus velocity, the inhibitory coupling would become weaker and turn the response characteristics into type T-5.1. Thus, a preference-inversion type similar to HXS > V would be the result. With opposite velocity-dependent characteristics of the inhibition-controlling interneuron, other types of preference inversion would come up. Furthermore, it could be shown by an der Heiden and Roth (in prep.) that different types of overlap of retina ganglion cells and different time constants of excitation and inhibition in retinal elements can constitute other tectal response types known from recording experiments.

6.1.5 Tecto-Pretectal Interaction

The model also leads to a reconsideration of the so-called TP-phenomenon. This phenomenon was first observed by Ingle in *Rana pipiens* with an increase in ERF size from 15–20° to 30° in TP-lesioned frogs. As discussed in Chapter 5, in the toad a destruction of thalamic and pretectal (TP) regions leads to a loss of "worm/antiworm" discrimination ability in prey-catching behavior, i.e., the ability to distinguish between elongate stimuli oriented parallel or perpendicular to the direction of movement. Ewert and co-workers observed that after such lesions, in the toad's tectum T-5.2 characteristics vanished and transformed into a response property similar to the thalamic (or pretectal) response type TH-3, a type of neuron that responds best to large squares and to vertically oriented bars (cf. Fig. 5.18 A). Additionally, the receptive field sizes were enlarged from about 27° to more than 40°. The same "TP-phenomenon" was also observed by Finkenstädt and Ewert (1983b) in *Salamandra*, here with an influence on the T-5.1 response type. Interestingly, both frogs and toads return to "seminormal" behavior after some weeks (Ingle 1980, Ewert et al. 1983).

In their original model of prey recognition in the toad, Ewert and von Seelen (1974) assumed that the T-5.2 characteristics are exclusively due to a subtractive inhibitory interaction between tectal T-5.1 characteristics and pretectal TH-3 characteristics. If this were the case, then destruction of the thalamo-pretectal region including TH-3 cells and, therefore, an elimination of their inhibitory influence should turn T-5.2 characteristics back into T-5.1 ones. This, however, is not observed: the "disinhibited" T-5.2 cells now show preferences for very large objects (> 32°). Furthermore, the "disinhibition" effect occurs also in those animals that do not have clear T-5.2 characteristics, as is the case in *Rana pipiens* (Ingle 1973).

We can avoid the difficulties arising with the assumption of a *specific* pretectal inhibition of T-5.2 neurons by assuming that the inhibitory influence on which the "TP-phenomenon" is presumably based, is a *nonspecific* one, and that the constitution of type T-5.2 occurs through recurrent inhibition between T-5.1 cells within the tectum according to the an der Heiden-Roth model. There is some anatomical evidence that two types of inhibition exist within the tectum: an intrinsic one produced through tectal stellate cells, and an extrinsic one produced through fibers coming from pretectal neurons

(Székely and Lázár 1976). Székely and Lázár discussed both types of inhibitory processes as being involved in the restriction of lateral spread of local activation over the whole tectum. It may well be, however, that the extrinsic inhibitory influence consists in controlling and restricting the lateral spread of activity and with this the receptive field size of tectal cells, whereas the intratectal inhibitory stellate cells produce the specific inhibitory effect among T-5 cells.

According to this assumption, after destruction of the praetectum, the disinhibition effect is then caused by a general increase in receptive field size and optimal stimulus size and, therefore, loss or impairment of the response properties of T-5 neurons. Due to this increase the "worm/antiworm" discrimination would vanish, because the interaction of excitatory and inhibitory processes within the receptive field as a basis of prey discrimination activity has disappeared. The "disinhibition" effect would be an unspecific one and not the result of the destruction of specific inhibitory connections between T-5.2 and TH-3 cells. However, we cannot exclude the possibility that the inhibitory tectal interneurons are themselves influenced by specific thalamo-pretectal cells. What is important is to recognize that there seems to be a general inhibitory influence of the thalamo-pretectal region on the tectum independent of the T-5.2 characteristic that controls receptive field size and discrimination abilities of tectal cells. This assumption is strongly supported by the fact that recovery from TP-phenomenon occurs, as mentioned above.

The studies of Manteuffel and co-workers on the anatomy, physiology and the behavioral relevance of the praetectum and the accessory optic system (cf. Chap. 5 and below) have helped to further clarify the question of the interaction between tectum and praetectum with regard to prey recognition. As was shown by these authors as well as others in other vertebrates, the praetectum, together with the system of the basal optic nucleus/neuropil (BON) is involved in the recognition of the motion of large structured visual patterns and in the guidance of compensatory optomotor movements.

Of course, stimulus situations in which large parts of the environment move in a coherent manner without movement of the animal are very rare and in nature may occur only in plant-dwelling or -climbing animals, for example, when a salamander or frog sits in front of moving twigs or leafs. Normally, the optokinetic nystagmus and the neural mechanisms of the praetectum and BON seem to play a role mainly (1) in recognition and control of body/head movements relative to the visual environment in addition to vestibular and proprioceptive information, and (2) in object-background discrimination. When a salamander moves toward a prey that is itself moving, a movement of the salamander relative to the environment and the prey, and of the prey relative to the animal and the environment occurs. For the retina, then, two motions exist, one of a large structured area, dependent on its own body or head movements, and an independent movement of a smaller object in front of it (or of a larger object in the case of a conspecific or an enemy). Thus, a prey object is recognized by its small size and its movement as being incoherent with the motion of a large background structure. It is interesting in this context that according to recent experiments of Manteuffel et al. (1986), tongue-projecting salamanders like *Hydromantes* which are able to catch prey by fast lateral head movements and lateral tongue protrusion, show the best optomotor responses of all amphibians tested so far. For these animals and their specific feeding behavior, precise discrimination between prey movement and head movement relative to the background is of great importance.

On the basis of the inhibitory influence of the praetectum onto the tectum, we may assume that during pursuit of a prey object in front of a structured background when the praetectum is activated by movement of a large patterned area relative to the body and/or head of the salamander, the activity of the whole tectum is suppressed except for the area activated by the prey object. In this way, the tectum could "concentrate" on the prey object. The suppression of tectal activity could be achieved by establishing intratectal inhibition resulting in relatively small ERFs in those tectal cells that are involved in prey recognition. This would increase visual discrimination ability. After destruction of the praetectum, two phenomena could be expected to occur: (1) enlargement of the ERFs of T-5 (and maybe other) neurons, (2) increase in overall tectal activity. Both phenomena would necessarily lead to a hyperactivity with regard to feeding due to a spread of activity throughout the tectum and a loss of prey-stimulus discrimination.

This assumption is corroborated by the finding of Manteuffel (1984) that the nucleus praetectalis responds best to a rotatory movement of the head (which induces opposite retinal shifts of the image of the environment in both eyes, e.g. a nasotemporal one in one eye and a temporonasal in the other), while the nucleus Darkschewitsch responds best to a translatory (forward-backward) movement of the head and body (which in both eyes induces retinal shifts of the image of the environment in the same direction). As mentioned in Chapter 4, both nuclei project to the tectum, the n. praetectalis bilaterally, the n. Darkschewitsch ipsilaterally. In a recent paper, Manteuffel (1986) hypothesized that during feeding the two nuclei suppress (via global inhibitory interaction) the responsibility of tectal units to relative (rotatory and translatory) movements of the environment so that they are activated only by a prey object moving in front of the environment.

Thus, it is unnecessary to assume specific inhibitory influence of pretectal cells onto only one sort of tectal cell. Furthermore, we can explain why the pretectal inhibition phenomenon occurs in animals, too, that do not have the tectal T-5.2 response type. The present hypothesis is that the role of the praetectum with regard to the guidance of feeding behavior is not the constitution of a specific inhibitory effect, but to establish or improve (prey) object-background discrimination.

6.1.6 The "Recognition Module" Concept

The model I propose here for prey recognition is the following: a certain prey object activates a spatially restricted part of the optic tectum, depending on where in the visual field the object moves. As indicated in Fig. 6.2, this part of the tectum comprises a certain number of neurons involved in prey recognition, such as the T-5 neurons presented above. I will call this tectal space "recognition module". It has to be emphasized that these modules are not understood as anatomically stable "columns": there is no evidence that the amphibian tectum is anatomically organized in a columnar fashion; the idea is that a minimum set of T-5 neurons differing in response characteristics are activated simultaneously by a prey stimulus, and the borders of this set can shift anywhere according to the position of the stimulus within the visual field. The properties of such a module are determined by the following features:

1. the qualitative composition of the module with respect to the different neural preference types, i.e., which of the T-5 cell types are present,
2. the quantitative composition of the module, i.e., the numerical percentage of the different preference types within the module; and
3. the relative contribution of the different preference types to the overall activity of the module, depending on the activity level of the types as well as certain connectivity properties and the synaptic transmission efficiency.

Such a module can act as a universal "prey analyzer": any prey object will activate all components of this module, due to the fact that the response properties of the components are overlapping, although different objects will do this to a different degree . For example, as shown in Fig. 6.2D, a square-like object measuring 8 × 8° would activate T-5.1 and T-5.3 cells to a high, and T-5.2 cells to a medium degree, and the overall activity level would be rather high. A horizontal rectangle would activate T-5.2 to a high, T-5.1 cells to a medium, and T-5.3 cells to a low degree, as shown in Fig. 6.2E, and the overall activity of the module would still be high, but lower than at stimulation with the square. Even a vertical rectangle stimulus would activate all types present in the module, but the level of activation would be low except for type T-5.3 where medium activation will occur (Fig. 6.2F). For the sake of simplicity, in Fig. 6.2 "inversion" neurons as well as other types involved in prey recognition are omitted.

Thus, the components of this module act as semispecific analyzers with respect to single characters of the prey object as explained above. They "decompose" the objects into those characters, for example, orientation and

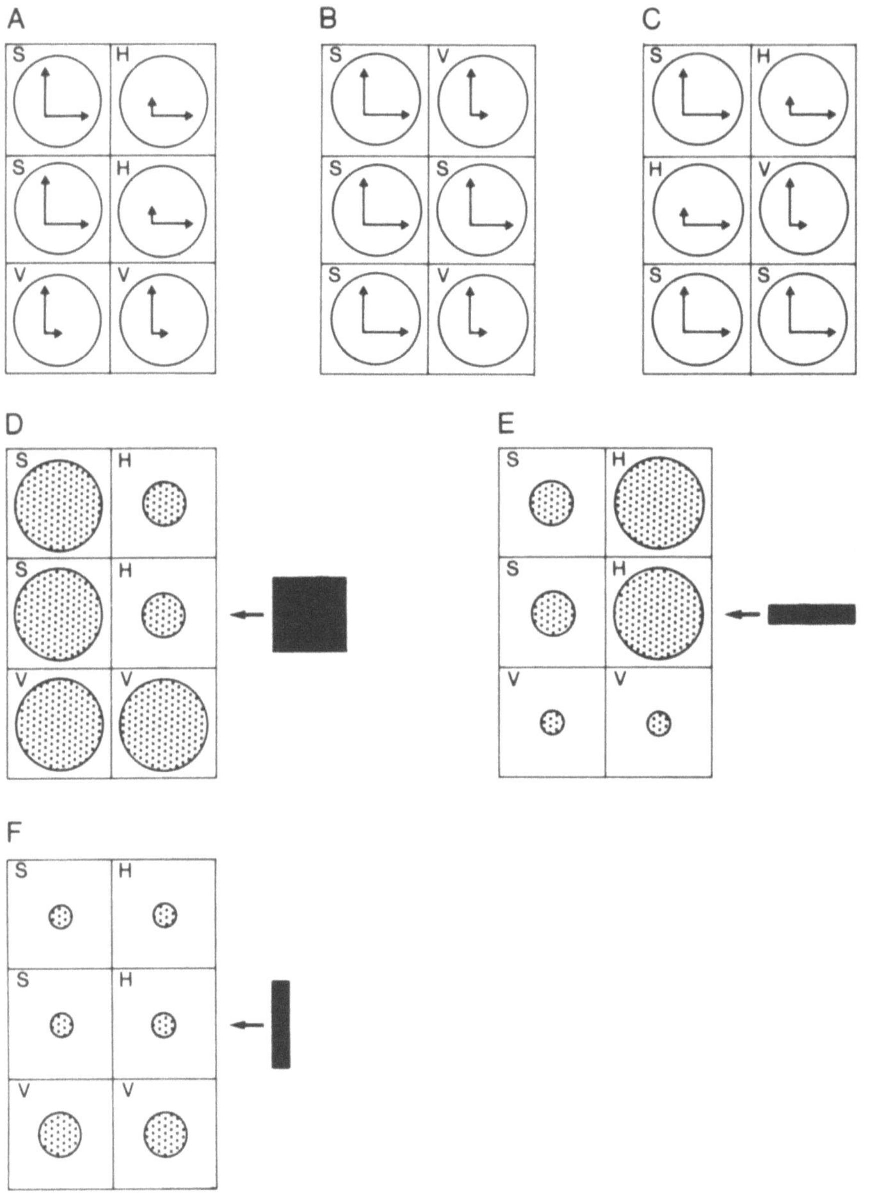

Fig. 6.2 A–F. Diagrammatic representation of the "recognition module" concept. **A–C** Recognition modules differing in qualitative and quantitative composition by tectal response types. Only three of the tectal response types mentioned in the text, namely T-5.1 (*S*), T-5.2 (*H*) and T-5.3 (*V*) are considered. These types differ in their responses to horizontal and vertical edges, as indicated by *arrows*. **A** Module with equal number of *S*, *H*, and *V*; **B** module with only *S* and *V*; **C** module in which *S* is most abundant, followed by *H*. **D–F** The states of activity of module A during stimulation by a square (**D**), a horizontal rectangle (**E**), and a vertical rectangle (**F**). The size of the *circles* indicates the strength of the responses of the respective response types to the stimulus

horizontal and vertical extension of edges, and their simultaneous activity, then, represents the prey object at least with regard to its major features. Accordingly, different prey objects will activate the different components of the module differently depending on the properties of the prey, and, therefore, elicit a distinct overall activity pattern within the module. Accordingly, the output of the module will be different. There is no need for a large number of discrete prey-detector cells in the tectum.

As will be discussed further below, the assumption of such a "recognition module" does not necessarily imply that its activity is completely summarized already within the tectum, i. e., that the module has just one output line. Indeed, as recent experiments have shown, several, if not all known T-5 neurons seem to project separately down to bulbar and perhaps spinal sites (Satou et al. 1985, Satou and Ewert 1985).

In summary, we can explain the formation of the different tectal response types involved in prey recognition by a combination of two processes: summation of activity of different retinal response types and lateral recurrent inhibition among tectal cells, probably partly mediated by pretectal influence. This enables us to understand the formation of different behavioral prey preferences: different amphibians such as *Bufo, Rana, Salamandra*, and *Hydromantes* – besides their differences in response properties of retina ganglion cells – show characteristic differences in the composition of their tectal "prey recognition modules" in which the presence and the degree of lateral inhibition play a decisive role. In amphibians that prefer wormlike prey as *Bufo* and *Salamandra*, we have to assume lateral inhibition to be realized to a higher degree leading to the formation of specific response types such as T-5.2, whereas in amphibians that prefer more compact prey lateral inhibition among specific tectal cells seems to be weak, as in *Rana*, or absent, as in *Hydromantes*, resulting in an incomplete formation or absence of T-5.2 characteristics. The T-5.2 characteristics, thus, have to be understood as analyzers for stimulus orientation parallel to the direction of movement, not as a "worm detector". The presence or absence of the T-5.2 type will, then, change the overall activity pattern of the recognition module and determine the behavioral prey preferences.

6.1.7 The Problem of Stimulus Size Recognition

One unsolved problem in the understanding of the amphibian visual system is the mechanism that underlies prey-size estimation. Within a certain range of stimulus distance, amphibians do not confound objects that are of the same angular size but differ in absolute size and distance; for example, they do not confound a large and distant object (an obstacle or enemy) with a small and nearby object (a prey item).

Ingle (1968) and Ingle and Cook (1977) showed that in frogs size constancy exists for prey size but not for barrier-gap size. Ewert and Gebauer (1973) showed that binocular toads prefer a square stimulus measuring 10 × 10 mm, to stimuli of other sizes independent of stimulus distance within a range of 13 to 230 mm. In salamanders, many experiments in which stimulus preference was tested showed that the preference for a certain stimulus size, in *Bolitoglossa* and *Hydromantes*, for example, 2–5 mm edge length (Roth 1976, Wiggers and Roth in prep.), was independent of the distance at which the animals first saw the stimulus and thus from the angular size of the prey, i.e., the size of the retinal image of the prey object.

The angular size of a prey object varies considerably during feeding. For example, when a salamander approaches a prey from a distance of 100 mm until 10 mm snapping distance, then the retinal image of a stimulus increases tenfold by size, i.e., a square stimulus measuring 5 × 5 mm subtends a visual angle of about 3° at a distance of 100 mm at which the first turning movement may occur, and of 30° at snapping distance of 10 mm. Thus, mechanisms have to be postulated that compensate for such variation in angular size.

The problem is now that tectal neurons believed to be involved in prey recognition, such as T-5 neurons, do not show response characteristics that could be correlated to the phenomenon of size constancy. These neurons seem to respond to the angular size rather than to absolute size of the stimuli presented (although not in a very precise way, as discussed below). Thus, when stimuli of different absolute size are presented at different distances such that their retinal image subtends the same visual angle, then these neurons show the same activity. The neurons seem to "ignore" the differences in absolute stimulus size as well as in stimulus distance. The apparent absence of size constancy in T-5 cells, for example, in *Bufo* and *Rana*, has been a point of major criticism with regard to the assumption that these cells act als "command neurons" in the guidance of prey-catching behavior (cf. Grüsser and Grüsser-Cornehls 1976).

Different tectal response types show somewhat different size preferences. In *Salamandra* (Finkenstädt and Ewert 1983a), *Bufo* (Ewert and von Wietersheim 1974) and *Rana* (Schürg-Pfeiffer and Ewert 1981) T-5.1 cells respond best to 8° square stimuli, T-5.3 cells best to 8–16° squares. T-5.2 cells respond best to 8° squares in *Bufo* (Ewert and von Wietersheim 1974) and to 4° squares in *Rana* (Schürg-Pfeiffer and Ewert 1981). However, despite the existence of such response maxima, all these neurons show a rather broad range of high-frequency responses to variation in stimulus size within a range of 2° and 16°. In *Bolitoglossa*, *Batrachoseps* and *Hydromantes* (Wiggers and Roth, in prep.) the majority of tectal T-5 cells respond best to 16° squares, but a considerable number of neurons respond best to 8° squares and others are maximally activated by 4° squares. Most of these cells had a rather broad tolerance with regard to differences in prey size, i.e., the differences in response to stimulus sizes between 4 and 16° were not significant. Thus, these neurons are, at least under the condition of paralysis during recording, by no means reliable instruments for measuring absolute or relative prey size.

A further difficulty arises when we compare behavioral prey-size preferences with size preferences of tectal cells. We find that the absolute size of the stimuli, which in recording experiments optimally activate T-5 neurons, is mostly much larger than those which optimally elicit feeding responses in behaving animals. For example, at a distance of 25–30 cm, an 8 × 8° square which in *Bufo* and *Salamandra* is optimal for most T 5 neurons, has an absolute size of 4 × 4 cm. This is a stimulus size which in behavioral experiments very rarely elicits responses and, furthermore, in nature represents a prey that could never be swallowed by a salamander or a toad. Similarly, in *Bolitoglossa* and *Hydromantes* the preferred angular sizes of most neurons are 8 × 8° and 16 × 16° at a stimulus distance of 15 cm. This corresponds to an absolute stimulus size of 2 × 2 cm and 4 × 4 cm, respectively. These bolitoglossine salamanders, however, show a behavioral preference for square stimuli of 0.2–0.5 cm in size.

Generally, in all neurophysiological experiments with amphibians, tectal cells showed a preference for stimulus sizes that are considerably larger than those being preferred in their feeding behavior. If size preferences of tectal cells corresponded in a direct way with behavioral preferences, then at a stimulus distance of 15 cm, *Hydromantes* or *Bolitoglossa* tectal neurons should respond best to angular stimulus sizes of 1–4° edge length (i.e., absolute stimulus sizes ranging from 2–10 mm edge length), and the same would be true for *Bufo* and *Salamandra* at a stimulus distance of 30 cm (i.e. absolute stimulus sizes ranging from 5–20 mm edge length). However, in most species the preferred angular stimulus sizes were 8 and 16°.

There are two alternative explanations for the apparent lack of size constancy within the responses of tectal neurons and for the differences in size preference of tectal T-5 neurons on the one hand, and behavior, on the other:

1. T-5 cells involved in prey recognition normally possess size constancy, but the underlying mechanisms are suppressed or disturbed by the immobilizing drug (e.g. curare, or succinyl-choline).
2. Tectal T-5 cells do not show size constancy even under normal conditions. Size constancy is based rather on the activity of tectal mechanisms other than T-5 cells or of nontectal centers.

With regard to the first explanation it is often assumed that size constancy is based on a change in receptive field size relative to the object distance. Since objects increase in angular size when approaching the eye, it should be expected that the RF size of T-5 "recognition neurons" showed an analogous increase during stimulus approach, thus "maintaining" the same spatial relationship to angular stimulus size. For example, such an increase could be due to an influence of the accomodation mechanism on these neurons, i.e., during approach to a prey object, the increase in tension of the accomodation muscle could lead to an increase in RF size of T-5 neurons. Such an effect of prey-distance dependent RF size would be suppressed by the use of immobilizing drugs affecting accomodation.

As has been discussed in Chapter 3, in *Salamandra* the immobilizing drug succinyl-choline affects the accomodation mechanism. It first leads to a convulsion of the protractor lentis muscle, resulting in an extreme short-sightedness of the eye, but afterwards leads to a complete relaxation of the muscle and, therefore, to far-sightedness. Thus, the relaxed state corresponds to a situation when the salamander fixates on an object within a distance of 10–30 cm (according to Werner and Himstedt 1985, in the relaxed eye of *Salamandra* the far point lies at a distance of 30 cm). In all neurophysiological experiments discussed here the stimulus distance varied between 15 and 30 cm; this means that even in the state of immobilization the accomodation mechanism seems to be in a "normal" state, i.e., being far-sighted with regard to the stimulus distance used in the recording experiments, and the RF size of the neurons should be "normal", too.

However, we cannot exclude the possibility that the "estimation" of absolute size by tectal T-5 neurons is somehow distorted resulting in the observed differences between neural and behavioral size preferences. If the stimulus sizes preferred by tectal neurons are too large compared to behavior this could mean that the neurons "see" the stimuli nearer than they really are. One can easily calculate that the optimal neural stimulus size observed in *Hydromantes* and *Bolitoglossa* would fit the behavioral ones at a distance of about 37 mm instead of 150 mm (with a stimulus of 5 mm = 8° edge length). Thus, the eye of the immobilized salamander should "see" the stimulus four times nearer than it really is (namely, 150 mm), i.e., at about or close to snapping distance. However, the fact that in the immobilized state the accomodation is in a natural state with regard to stimulus distance contradicts the assumption of such a heavy distortion of distance estimation.

Furthermore, recordings from amphibians which were immobilized not by application of drugs but by destruction of the spinal cord did not show dramatic changes in ERF size or neuronal size preferences of tectal cells that would fit the behavioral ones (Roth and Himstedt unpublished results). The same was true for recording experiments carried out when the effect of the immobilizing drug vanished and the subjects were ready to "wake up" (Wiggers and Roth in prep.). Unfortunately, data from recording experiments with freely moving amphibians concerning the problem of size constancy are lacking.

Assuming the second alternative, namely, that prey-size recognition is based on tectal cells other than T-5 cells, one could take into consideration the response properties of the T-3 cells described in *Rana esculenta* (Grüsser and Grüsser-Cornehls 1976), in *Bufo bufo* (Ewert and von Wietersheim 1974) and in *S. salamandra* (Grüsser-Cornehls and Himstedt 1973, Finkenstädt and Ewert 1983a). These T-3 cells, as described in Chapter 5, respond mostly or exclusively to stimuli that approach the retina along the z-axis. The mechanisms underlying these response properties are unclear. Since these cells were recorded from the tectum of immobilized amphibians their response properties cannot be coupled to eye accomodation, because under the assumption that the protractor lentis muscle is indeed relaxed by the immobilizing agent during the recording experiments, the state of accomodation cannot change during the approach or removal of a stimulus. On the other hand, it is unlikely that the T-3 effect is merely due to the increase in stimulated retinal area during the approach of the stimulus, because this would be inconsistent with the inhibitory effect onto the spontaneous activity during the removal of the stimulus along the z-axis as reported by Grüsser and Grüsser-Cornehls (cf. Chap. 5). The significance of the response characteristics of these T-3 cells for the present problem remains to be studied in greater detail, especially in unrestrained animals.

A substantial argument against the influence of the state of accomodation on RF size and size preference of tectal neurons is that, as has recently been shown by Douglas et al. (1986), the accomodation state of the eye is controlled and monitored by the oculomotor nerve. The motor nucleus of this nerve is situated in the midbrain tegmentum and has no direct connection to the tectum. Furthermore, as mentioned in Chapter 2, under binocular conditions amphibians do not need accomodation for correct object distance estimation; they rely rather on computation of the binocular input. Amphibians will largely disregard accomodation cues when these contradict binocular ones (Collett 1977, Jordan et al. 1980). As will be shown below, binocular disparity cues do not require any distance-dependent change in RF size of T-5 cells.

In summary, it seems rather unlikely that tectal neurons, especially of class T-5, show size constancy under normal conditions connected to a change of RF size or similar phenomena, and that this ability is suppressed by the specific experimental conditions. However, this question has to be studied in unrestrained animals.

The alternative hypothesis presented here concerning the problem of absolute size recognition is based on the distinction between size estimation and stimulus-distance estimation which are normally viewed together. I assume that it is unnecessary for amphibians to precisely estimate absolute size of visual objects. As shown in Chapter 2, all salamanders studied so far have a rather large range of accepted prey size. Under different ecological conditions salamanders are often forced to switch from rather large prey to small or very small prey, and depending on prey density they may deliberately change prey size preferences. Similarly, all tectal T-5 cells of salamanders (as well as of anurans) show a wide "tolerance" with regard to changes in angular as well as in absolute size. In Bolitoglossini studied in our laboratory, this tolerance ranges between 2° and 16° being equal to a range of 2–16 mm in absolute size.

The idea is now that the T-5 system gives only a rough estimate of angular size in the sense that an object lies within or outside a certain size range. What is necessary, however, at least for bolitoglossine salamanders, is a precise estimation of prey distance. The depth perception mechanism (which will be discussed below) would add to the rough information about angular size coming from T-5 cells the precise information about object distance. These two types of information together will result in a reliable decision between prey and nonprey (obstacle or enemy). For example, an object that has an angular size of 2° is optimal prey for *Bolitoglossa subpalmata* or *Hydromantes italicus* (namely a size of 5 mm) when 150 mm away, but rather suboptimal when 10 mm away (namely, 0.3 mm, i. e. a small collembolan which is, however, still well perceived). On the contrary, an object that has an angular size of 16° represents an optimal prey when 10 mm away (namely, 3 mm in size), but represents an obstacle or an enemy (depending on whether and how it moves) at a distance of 150 mm (namely, 40 mm in size). Thus, with the combination of both types of information, a sufficient prey-nonprey decision is possible for the animals. In the case of binocular vision, this combination most probably occurs in the tectum itself, whereas under conditions of monocular depth perception, depending on lens accomodation, the combination seems to occur at tegmental or bulbar levels.

An interesting aspect for a somewhat different concept for the constitution of absolute size estimation comes from recent findings by Ingle in the frog *Rana pipiens*. Ingle (1983) was able to show that destruction of the uncrossed tecto-bulbar tract, just caudal to the tectum, eliminates snapping but not turning toward the prey, whereas destruction of the crossed tecto-bulbar tract abolishes turning but not snapping (i. e., the frog snapped in an undirected manner). Ingle was able to show by means of the HRP method that the uncrossed tract originates from tectal neurons that have contact with class-3 rgc and the crossed tract from class-2 rgc. Accordingly, after destruction of the tectal fiber layer containing fibers from class-2 rgc the orienting/turning responses disappear. Ingle deduced from these findings that turning toward prey, including prey recognition, involves cells activated mainly by class-2 rgc, whereas snapping is elicited by cells activated mainly by class-3 r. g. c. activity.

Now we have to consider the fact that class-2 rgc are best activated by objects of small angular size (2–6°) and class-3 rgc by medium angular size (8–12°). Then we can assume that a prey of small absolute size activates class-2 rgc when distant, and class-3 rgc when near, e. g., at snapping distance. Interestingly, as was shown by an der Heiden and Roth in their model presented above, we can assume that two subclasses of T-5 cells are present in the tectum, one driven by class-2 rgc and one driven by class-3 rgc, but showing the same response properties. Therefore, it is possible that neurons of the first subclass are active during fixation of distant prey and neurons of the other subclass are active during near fixation. In the frog tectum, Ingle (in litt.) found cells that arborize both at the level of class-2 as well as of class-3 afferents. Those dual-level neurons could act as "zoom lenses", without changes in RF size (i.e., at high intratectal inhibition they would be activated best by objects of small angular size such as distant prey, whereas at low or medium tectal inhibition they would be best activated by objects of medium angular size like nearby prey). This again would make it unnecessary to assume a depth perception based on the activity of T-5 cells, because it would give a first and rough estimation concerning prey-nonprey discrimination, because the "orienting channel" would not respond to obstacles and enemies. However, since the "snapping channel" would act as a prey detector at close distance as well as an enemy detector at larger distance, this system again needs further information about object distance for a reliable prey-enemy distinction.

6.1.8 Distance Estimation Mechanisms

Several authors have shown that two-eyed amphibians primarily use binocular cues for depth perception, whereas one-eyed animals seem to use accommodation (Collett 1977, Jordan et al. 1980, Douglas et al. 1986). For example, when accomodation cues were abolished by drugs or by the use of goggles, precision of snapping was only slightly or not impaired in binocular animals, whereas monocular amphibians were unable to snap correctly at prey under those conditions.

In the toad *Bufo marinus*, Collett (1977) showed that binocular animals use retinal disparity cues for distance estimation. Since toads, as all amphibians, lack both eye convergence and fixating eye movements, disparity was defined by Collett as the width of the two angles formed between the two lines connecting a target to the nodal point of each eye and the line in between connecting the target to the midpoint between the eyes. When the disparity angles were increased or decreased by the use of base-in and base-out prisms, the toads showed overshooting and undershooting, respectively. This means that with an increase in disparity angle the target seems to be more distant than it actually is, whereas with a decrease in disparity angle the target seems to be closer than it is. Since the prisms do not alter the accomodation state of the lens, the binocular animals largely override the "correct" information coming from the accomodation system.

With respect to the question of neural mechanisms of binocular depth perception, we have to consider the spatial representation of the target on the retina and the tectal surface. In the retina, the situation is the following: (1) When a salamander approaches a target straight ahead, or when a target moves straight toward the animal, then the two target images move symmetrically and with equal velocity over the retina from a more central to a more temporal position. (2) When a salamander moves its head relative to a target or a target moves through the visual field frontal to the animal, then the two images move in different, i.e., nonsymmentrical directions (a temporonasal one in one eye and a nasotemporal one on the other) and with different velocities over the retinas of each eye.

From this, it follows that the definition of disparity by Collett is somehow inappropriate, because in situation (1) when the target is approached straight ahead, its image is always located at symmetrical points of the retina and what changes is the *lateral eccentricity* of the two images with respect to the center of the retina; in situation (2) where the target is located lateral to the midline of the binocular visual field, true disparity in the sense of asymmetry of the two images occurs. As a consequence, in the following I will restrict the meaning of the term disparity only to the situation where the two retinal images of a target occupy asymmetric positions. The need for a distinction between lateral eccentricity and retinal disparity will become clear when we look at the situation in the tectum.

As explained in Chapter 4, there is a retinotopic map extended over the surface of the tectum such that the horizontal (nasotemporal) axis of the visual hemifield is projected onto the contralateral tectal hemispheres along a caudorostral line. The vertical, ventrodorsal meridian of the visual hemifield is projected onto the tectal hemispheres along a lateromedial line. Targets in front of the animal are, thus, represented in the rostral part of the tectum, and those in the lateral and caudal parts of the visual field are represented in the medial and caudal part of the tectum; targets above the visual horizon are represented in the medial parts and those below the horizon in the lateral parts of the tectum.

In addition to the contralateral projection of each eye to the tectum, there is an ipsilateral projection which in most amphibians is an indirect projection in the sense that the contralateral map is projected to the other tectal hemisphere via tecto-tectal and/or tecto-isthmo-tectal projections, and direct ipsilateral projections are sparse. As shown in Chapter 4, only bolitoglossine plethodontids show a larger degree of direct ipsilateral retinal projections to the tectum. However, both indirect and direct ipsilateral projections constitute a retinotopic map which is more or less in register with the contralateral one. Accordingly, a target located within the binocular visual field is represented twofold in each tectal hemisphere, with one representation coming from the contralateral eye and another one from the ipsilateral eye.

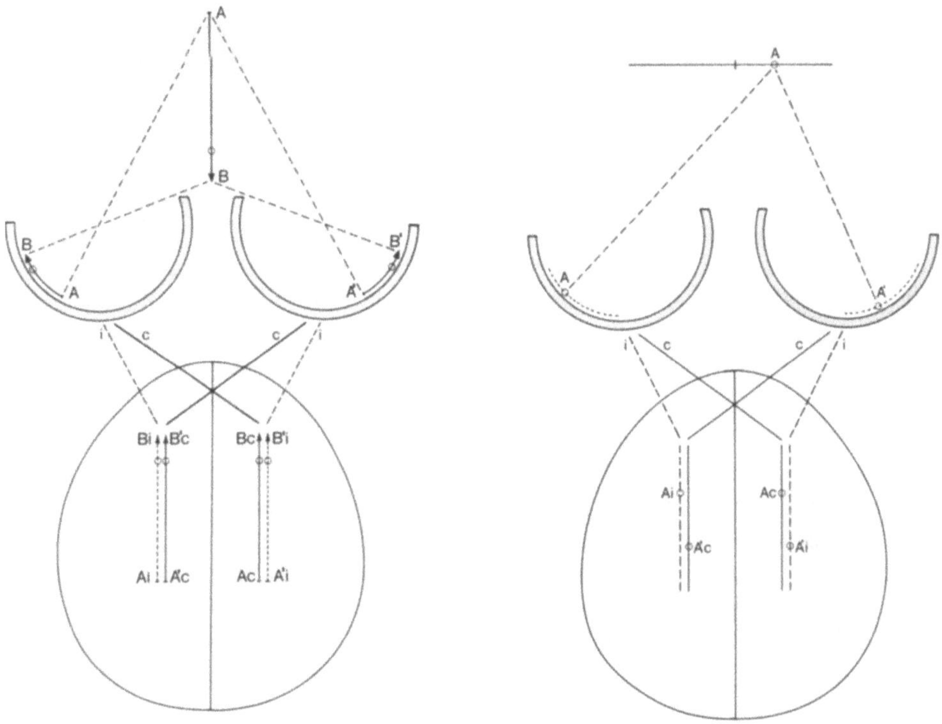

Fig. 6.3. The neural basis of depth perception in salamanders. *Left*: The situation in which a target approaches the salamander from *A* to *B* along the midline of the binocular visual field. Accordingly, the retinal images will move from a more central (*A*, *A'*) to a more temporal position (*B*, *B'*) and, thus, show an increase in *lateral eccentricity*. The retinal images are projected both contralaterally (*solid lines*) and ipsilaterally (*dashed lines*) to both tectal hemispheres. During approach, the tectal representations of the image move along the caudorostral axes of the ipsilateral (*Ai* – *Bi*, *A'i* – *B'i*) and contralateral (*Ac* – *Bc*, *A'c* – *B'c*) maps. For straight-ahead targets all four tectal representations are at the same caudorostral level, as indicated by open circles. Thus, they show no disparity, and distance is encoded only by the position along the caudorostral axis.
Right: The situation in which a target is located laterally within the binocular visual field. In this case the ipsilateral (*Ai*, *A'i*) and contralateral (*Ac*, *A'c*) projections from both eyes are not located at the same caudorostral level. The degree of disparity between *Ai* and *A'c* and *Ac* and *A'i* is proportional to the laterality of the target within the binocular visual field. Target distance is, thus, encoded by both disparity and eccentricity

Accordingly, as shown in Fig. 6.3, left, a target that is approached *straight ahead* subsequently activates the tectum along the caudorostral midline of each tectal half, i.e. along lines Ac – Bc and A'c – B'c for the contralateral retinotopic maps, and Ai – Bi and A'i – B'i for the ipsilateral ones. The points activated by the contralateral and by the ipsilateral pathways always coincide, or show zero disparity in each tectal half, and all these points of activity lie at the same rostrocaudal level when the two tectal hemispheres are compared. They are located more caudally when the target is distant, and more rostrally when it comes nearer or when the salamander approaches it. The different positions of the tectal images of the target along the caudo-rostral line correspond to different degrees of lateral eccentricity in the retina and, thus, code target distance along the rostrocaudal axis.

In all cases when a target is situated to the right or left of the midline of the visual field, within each tectal hemisphere the contralateral and the ipsilateral projection sites from each eye in each tectal half do not coincide, but show a disparity that is dependent on the degree of the lateral position of the target within the binocular visual field. In Fig. 6.3, right, such a disparity exists, for example, between Ai (ipsilateral projection from the left eye) and A′c (contralateral projection from the right eye) in the *left* tectal half, and Ac (contralateral projection from the left eye) and A′i (ipsilateral projection from the right eye) in the *right* tectal half. A disparitiy exists, too, between the projections in the left and right tectal halves, i. e. between A′c and Ac as well as between Ai and A′i. Of course, the degree of disparity is the same in the left and right tectal halves, and Ai and Ac, as well as A′c and A′i occupy the same positions along the caudo-rostral axes of the tectal halves.

The disparity between the ipsilateral projection from one eye and the contralateral projection from the other is larger the more lateral the target is situated in the visual field. Tectal projections of prey objects with positions above or below the horizon will lie on lines parallel to the rostrocaudal midline of the tectal halves, but otherwise everything said before holds for these situations, too.

With this sort of representation of the target, a combination of the degree of lateral eccentricity (expressed in rostrocaudal coordinates of the tectal map) and the degree of disparity (expressed in rostrocaudal distance between the contralateral and the ipsilateral projection of the target in each tectal half) unequivocally gives the exact location of the target within the visual space. The degree of lateral eccentricity is immediately given by the retinotopic map of the tectum. What is needed is a mechanism to measure the disparity of the contralateral and ipsilateral images of the target in each tectal half.

It is well known from studies on the visual system of mammals as well as of birds (Pettigrew 1972, van Sluyters and Stewart 1974, Pettigrew and Konishi 1976) that binocular neurons exist which respond optimally to a certain disparity of retinal images. Recently, Finch and Collett (1983) and Gaillard (1985) have demonstrated that binocular neurons exist in the frog tectum which at a given arbitrary distance between the eyes and a screen show different disparities between their contralateral and ipsilateral receptive fields. These, accordingly, are completely superimposed at different distances. In the study of Finch and Collett (1983), the neurons studied had very small receptive fields and small disparities and, thus seem to be tuned to distances beyond 25 cm. In the study of Galliard (1985) receptive fields ranged from "total-field" RFs to 4–6° in size, and showed disparities that corresponded to target distances from 6.3 to 300 cm. When activated, these cells could give the animal information about the absolute distance of an object.

However, for all targets which are located straight ahead of an animal there is no disparity between the ipsilateral and contralateral projections. Thus, for computation of the distance of straight-ahead targets, only the caudorostral coordinates of the target projection within the tectal map are needed. Projections in the caudal tectum mean that the target is far away, whereas those in the rostral tectum mean that the target is close. Disparity cues are, thus, important only for targets located laterally in the visual field. Since Gaillard studied neurons corresponding only to the central part of the binocular visual field, the meaning of his findings for depth perception remains unclear.

As already stated, one has to look for a mechanism inside or outside the tectum which is able to measure the degree of disparity between the ipsilateral and contralateral projections from each eye. The nucleus isthmi has often been assumed to play an important role in amphibian depth perception. This nucleus receives afferents from the ipsilateral tectum and projects mainly to the contralateral tectum via the commissura postoptica. It has been shown that in amphibians the isthmotectal map is in register with the retinotectal map; in the frog, isthmic afferents intermingle partly with the most superficial and partly with the deepest layers of the retinal afferents. Recently, Udin (1985) has shown in *Xenopus* that the precision of this mapping is partly dependent on visual experience. The degree of isthmotectal retinotopy seems to be rather good. At least the contralateral isthmotectal projections contain small-field fibers that are found in the superficial layer of retinal afferents with ERF sizes of 4–6° (Gruberg and Lettvin 1980). Ingle (in litt.) found that many of these neurons possess horizontally elongate ERFs of a size of 3 × 6° that respond best to spots moved along the long ERF axis. The size preferences of these units are about the same as of class-2 retinal fibers, so that the isthmotectal system is able to preserve the "fineness" of the retinal input. According to Gruberg and Lettvin, deeper isthmotectal units have responses that resemble those of class-3 plus class-4 retinal afferents.

However, there are experimental data that make it less likely that the nucleus isthmi is the only basis of stereoscopic vision. First of all, the isthmotectal system seems to constitute a considerable delay in the processing of retinotectal information. Gruberg and Lettvin (1980) found a delay of about 40 ms in the responses coming from the contralateral tectum via the contralateral nucleus isthmi compared to direct contralateral retinal afferents. This means that the activity from the ipsilateral eye comes 40 ms later than that from the contralateral eye. For binocular depth perception involved in localization of fast moving prey this delay may be too long. Second, in behavioral studies with *Bufo marinus*, Collett and Udin (1982) showed that bilateral destruction of the nucleus isthmi does not considerably impair stereoscopic vision, even when monocular cues are excluded. The authors found, however, that when confronted with pairs of objects, isthmus-lesioned toads often snap at a "ghost" image in the middle of the two prey stimuli which normal toads never do.

Thus, the role of the nucleus isthmi in stereoscopic vision remains unclear. Considering the "fine grain" isthmotectal projection on the one hand and the considerable delay of indirect ipsilateral projections via the nucleus isthmi, the nucleus isthmi may rather be a system for improvement of visual acuity by the overlap of two retinal maps. The isthmotectal map could also be involved in the control of intratectal inhibition. However, we cannot exclude that the nucleus isthmi is indeed involved to some degree in depth perception, although it does not seem to be a necessary prerequisite.

In view of the considerable delay of indirect ipsilateral retinal information to the tectum constituted by the isthmo-tectal projection system, it is not surprising that those amphibians that are able to exert very fast and precise feeding reactions, such as bolitoglossine salamanders, have developed (or strongly enforced) a system of direct and immediate comparison of the information from the two eyes via direct ipsilateral and contralateral retinotectal projections. Interestingly, in these salamanders the ipsilateral retinofugal projections to the tectum terminate exactly in the same layers as the isthmotectal do, namely in the uppermost and the deepest layer. In each half of the tectum of Bolitoglossini we, thus, find a rather complex arrangement of retinal afferents: three laminae constituted by the three different afferent fiber systems coming from the *contralateral* eye, two laminae constituted by direct *ipsilateral* afferents and overlapping with the superficial and the deep lamina of contralateral afferents, and finally two laminae constituted by fibers coming from the contralateral nucleus isthmi and overlapping with the two laminae of ipsilateral retinal fibers (not considered here is the fact that in addition a projection from the ipsilateral nucleus isthmi to the tectum exists which, however, seems to be rather weak). Thus, each tectal hemisphere of bolitoglossine salamanders receives a threefold projection from the retina, a direct contralateral, a direct ipsilateral and a indirect ipsilateral via the nucleus isthmi. However, the question of how disparity is measured in the tectum remains to be answered.

In summary, we can assume that in binocular depth perception rostrocaudal as well as medio-lateral disparity of the two retinal images of an object, combined with information about their absolute location within the retinotectal map is used for computation of prey localization as well as prey distance.

As already mentioned, *monocular* depth perception seems to be mostly based on accomodation cues. However, the exact way in which these cues are used within the visual system of amphibians remains unclear. Very recently,

Douglas et al. (1986) demonstrated that for monocular distance estimation, the accomodative state of the lens is monitored by the motor outflow of the accomodation system, i.e., the contraction effort exerted by the protractor lentis muscle in order to move the lens forward, instead of by sensory feedback. It is probable that accomodation is controlled by the oculomotor nerve supplying the ciliary ganglion. Since the motor nucleus of the oculomotor nerve is located in the tegmentum and sends no direct afferents to the tectum, this may be a reason for the above discussed fact that the tectal neurons are not influenced by the accomodative state, e.g., with regard to their receptive field sizes. It is, thus, more likely that depth perception cues based on lens accomodation join tectal information about location of prey objects in the visual space at tegmental or even more caudal centers.

6.1.9 Neural Guidance of the Feeding Sequence and the Problem of Sensori-Motor Integration

Prey recognition is without doubt the first step in the sequence of the feeding reaction. However, there has been a certain controversy about the principles of neural guidance of the whole feeding sequence. The question is whether prey recognition has to be considered as a separate process that takes place prior to or at the beginning of the feeding sequence and accompanies, as a separate process, the following reactions such as turning toward the prey, approach, fixation, and snapping, as is claimed by Ewert (cf. Ewert 1974), or whether prey recognition is represented by the *whole* feeding sequence as a unit, as Grüsser and Grüsser-Cornehls contended (e.g. Grüsser and Grüsser-Cornehls 1976). The latter concept is based on the findings of the authors that different classes of tectal cells are apparently involved in the guidance of different parts of the feeding sequence. For example, "small field" T-5 neurons, as shown above, may be involved in first prey recognition, T-2 neurons, which are optimally activated by temporonasal movement of small objects in the visual field, may be responsible for turning toward prey, T-3 neurons, which respond best to objects moving toward the eye, are involved in approach, T-1 neurons, which are activated only binocularly, may be responsible for binocular fixation, etc. *All* these types would need to have recognition abilities. Thus, the whole sequence of activity of T-5 → T-2 → T-3 → T-1 → T-3 neurons would lead to a behavioral sequence of recognition/turning → approach → binocular fixation → snapping.

Ewert has argued against this concept by emphasizing that the "recognition elements" (in the toad the T-5.2 neurons) as separate units, have to be active during the whole feeding sequence. He, therefore, constructed the sequence of neural activities in the tectum in the following way: T-4 (broad-field neuron) + T-5.2 (= "stimulus moving anywhere in the visual field"

+ "object recognized as prey") → ORIENT, PREY!; T-2.2 + T-5.2 + T-1.1 (= "stimulus in the frontal field" + "object recognized as prey located near the fixation area" + "stimulus far afield") → APPROACH!; T-5.2 + T-1.2 (= "prey near the fixation area" + "stimulus close to the toad") → FIXATE!; T-5.2 + T-1.3 + T-3 (= "prey inside fixation area" + "both retinas simultaneously stimulated". + "stimulus within snapping distance") → FIXATE and SNAP! (cf. Ewert 1984).

In this concept, the activity of T-5.2 neurons as the exclusive or main recognition element accompanies the activities of other specialized neurons. Ewert is correct in stating that the act of recognition of an object as prey has to precede all other acts, but this has not been denied by Grüsser and Grüsser-Cornehls: their sequence of neural activity also begins with T-5.2 neurons. Ewert, however, seems to disregard the experimental evidence that T-5 neurons other than T-5.2 are certainly involved in prey recognition and that neurons of classes other than T-5.2 can also have "recognition properties". Roth (1982a) in *Hydromantes* and Roth and Jordan (1982) in *Bufo* often recorded from large-field neurons such as T-4, or direction-specific neurons such as T-2, which, similar to T-5.2 neurons, showed clear discrimination between horizontally and vertically oriented bars. Therefore, it is not necessary to assume that the T-5.2 cells are always active during the feeding sequence, although this may well be the case. The argument of Ewert is based on recordings from freely moving toads (cf. Ewert 1984), in which T-5.2 cells were always active prior to an orienting and snapping movement "predicting" these events to some degree. However, this criterion is somewhat circular, because here all those neurons showing a "worm preference" are classified as T-5.2, although, as has already been pointed out, other cells than T-5.2 cells can also show a "worm preference". In this context, Ewert seems to unnecessarily revive the "detector neuron" concept contradicted by his own findings (see below). There is no doubt, however, that recordings from unrestrained subjects will give much more precise insight into the guidance of the feeding behavior.

A major difficulty of both concepts is that they leave the question of exact prey distance estimation unsolved. For example, a bolitoglossine salamander or frog must know exactly when the snapping or tongue projection distance is reached. This cannot be achieved by mere activity of certain tectal cells types such as T-1 or T-3, because even those cells show great "tolerance" with regard to variation in angular stimulus size.

Another important unsolved question is how and where in the brain the activity of tectal neurons involved in prey recognition is summarized and transformed into impulses activating the appropriate motor nuclei. As shown in the preceding section, different behavioral programs such as for locomotion and orientation during approach, fixation, and estimation of distance of the prey are additionally needed for the control of the feeding sequence and have to be combined with the activity of the networks involved in prey recognition. Furthermore, brain centers must exist in which information from other sensory modalities (e.g., olfactory, somatosensory, vestibular) are combined with visually guided activity. There are four possible levels of transformation of tectal visual activity into motor action patterns: 1) the tectum itself; 2) the rostral tegmental and/or the isthmic region in the ventral midbrain; 3) the reticular formation in the neighborhood of the motor nuclei in the medulla oblongata related to feeding; 4) the motor nuclei responsible for the motor actions constituting the sequence of feeding, i.e., orienting, approach, and snapping.

The original idea had been that the tectum itself is the place of complete object recognition and transformation of this information into action patterns guiding the activity of the motor nuclei involved in feeding. In a first model for visual guidance of feeding behavior in amphibians, Ewert had assumed that certain cells such as T-5.2 are at the top of a hierarchical organization of prey recognition networks in the sense that the activity of other tectal cells, such as T-5.1 and T-5.3 converges onto this type of cells (Ewert and von Wietersheim 1974, Ewert and von Seelen 1974). These second order neurons, thus, were believed to function as the main tectal output serving as a "command neuron". They should, therefore, exclusively constitute the tectal output with regard to prey recognition, and Ewert had always strongly emphasized the similarities between response properties of T-5.2 neurons in the toad's tectum and the toad's behavioral prey preferences (cf. Ewert 1976).

Anatomical and recent electrophysiological evidence (cf. Ingle 1983, Ewert 1984, Satou et al. 1985, Satou and Ewert 1985) contradicts the assumption that the tectum is the place of *complete* transformation of visual activity involved in prey recognition into final motor commands. Satou et al. carried out intracellular recordings from cells in the hypoglossal nucleus (which in toads – contrary to salamanders – controls both protraction and retraction of the tongue) and simultaneous electric stimulation within the "snapping area" in different parts of the tectum. They found that two polysynaptic pathways from the tectum to the motor nuclei exist, an inhibitory and an excitatory one. The excitatory responses of hypoglossal neurons became evident only after the tectum was repetitively stimulated with stronger stimuli, in contrast to the inhibitory responses which already appeared after single or double stimulations with weaker stimuli. The existence of polysynaptic pathways and, thus, of interneurons, was shown by "spatial facilitation" after simultaneous electric stimulation of two different points within the tectum as well as by "temporal facilitation" after repetitive stimulation. The authors assume a convergence of the excitatory and inhibitory pathways on interneurons situated in close proximity to the motor nuclei related to feeding.

Satou and Ewert (1985) carried out antidromic activation of tectal neurons by electric stimulation applied to the caudal medulla oblongata, in order to identify the tectal elements projecting to that area via the descending tracts. They found that various classes of tectal cells such as T-1, T-2, T-3, T-4, T-5.1, T-5.2, T-5.3, and T-5.4 could be activated from the contralateral medulla oblongata, and a limited number of T-5.1 cells could be activated ipsilaterally. The authors found T-5.2 cells to be activated most frequently, but the reason for this dominance was uncertain. The authors interpreted their results according to the concept of a "multiple-action system" in the sense that various tectal output elements project to the vicinity of feeding-related motor nuclei "where they form a substrate for the sensory-to-motor interaction ... and contribute to the initiation or coordination of the visually guided behavior, such as prey-catching".

As already discussed, these findings do not exclude that there is a high degree of interaction of different neurons involved in prey recognition or guidance of motor behavior, and that there is partial convergence, or a mixture of convergent and parallel processing in such a way that certain neurons converge with axon collaterals on second order tectum neurons (such as pyramidal and ganglionic cells in the anuran tectum) and at the same time project themselves outside the tectum. But it seems more likely that not all information necessary for prey/nonprey discrimination, estimation of size, velocity, movement direction and distance is already established within the tectum. Other nuclei, especially thalamic, pretectal, and tegmental ones regarding enemy and obstacle recognition, the state of eye accomodation, object-background discrimination as well as the distinction between object movement and self-induced movement, seem to join tectal information at "later", tegmental, isthmic, or bulbar levels. It is, therefore, crucial to further determine the connectivity and function of the efferent tectal tracts and their projection sites.

As shown in Chapter 4, the tectum has two main descending output channels, the crossed and the uncrossed tecto-bulbar tracts that are believed to be involved in the control of head and body orientation toward the prey and away from a threat, and in the control of approach and the release of snapping, respectively. Ingle (1983) could show that in *Rana pipiens* the first (crossed) output channel involved in head and body orientation receives retinal input mainly from small-field class-2 rgc, whereas the second (uncrossed) one involved in approach and snapping receives input mainly from class-3 rgc that have larger receptive fields. This is in good accordance with the anatomical findings of Lázár et al. (1983) who showed that the two descending tectal tracts originate from different types of large ganglionic cells receiving different retinal input. Since essentially the same situation is found in salamanders, we can assume a similar functional system in salamanders. However, more recent experiments lead to the assumption (Ingle, in litt.) that a double pathway involved in orientation exists, a direct uncrossed tecto-bulbar one from the contralateral tectum and an indirect,cdouble-crossing tecto-tegmento-bulbar one from the ipsilateral tectum. Grobstein et al. (1983) found that frogs (*Rana pipiens*) with a complete transverse hemisection of the neuraxis at a level between the optic tectum and cerebellum exhibited well-directed prey-catching movements for stimuli at any location to the side contralateral to the lesion. However, the subjects failed with stimuli at any location ipsilateral to the lesion. What remained in this latter case were sorts of forward-directed movements with or without snaps. However, it had been shown by Ingle (1983) and Grobstein and associates (Kostyk and Grobstein 1980, 1982) that unilateral tectal lobe ablation abolished turning responses to stimuli within the monocular (lateral) visual field of the eye projecting to the lesioned tectum, whereas it left turning toward stimuli within the whole visual field projecting to the intact tectal half unimpaired. Grobstein et al. (1983) derived from these findings that an ipsilateral (uncrossed) *and* a contralateral (crossed) tecto-bulbar pathway are involved in turns to the left or right side. The ipsilateral pathway is believed to originate from the rostral part of each tectum which corresponds to the ipsilateral binocular field. The crossed pathway which descends medial to the uncrossed one, is believed to originate from the more caudal part of the tectum which corresponds partly to the binocular and partly to the monocular contralateral visual field. Both pathways are assumed by Grobstein et al. to project on each side of the brain to a "pattern generating circuit", situated in the left and right medulla oblongata and controlling left and right turns, respectively. In addition, the authors assumed an unpaired center of "symmetric motor components" in the isthmic mesencephalon, to which both crossed and uncrossed tecto-bulbar pathways send collaterals and which, in turn, projects to the motor centers responsible for forward movement and release of snapping.

Ingle, in contrast, assumed that the monosynaptic *ipsilateral* tecto-bulbar pathway is involved only in forward movement and snapping, whereas orienting is controlled by the monosynaptic *crossed* tecto-bulbar pathway, guiding turns contralateral to the active tectum (and away from contralateral threat), and a polysynaptic, *double-crossing*, tecto-tegmento-bulbar pathway, guiding turns ipsilateral to the active tectum (and escape from a threat crossing in front of the subject).

Both models become essentially the same if one considers Ingle's double-crossing tract to be an equivalent of Grobstein's ipsilateral (uncrossed) tract involved in ipsiversive turns. The major difference, however, is that in Grobstein's model both tectal hemispheres are involved in turns to each side, whereas Ingle assumed that one tectal half controls turns to the ipsilateral side. However, one has to bear in mind that the tegmental relay region, which Ingle assumed, receives afferents from the other tectal hemisphere.

For anurans as well as for salamanders it is reasonable to assume a crossed monosynaptic (tecto-bulbar/spinal) plus a polysynaptic (tecto-tegmento-bulbar/spinal) pathway both descending medially and controlling head and neck orientation (and maybe tongue versatility via first spinal nerve), and a laterally descending (uncrossed) monosynaptic pathway involved in the control of forward movement and in the release of snapping. This does not mean that the tractus tecto-bulbaris rectus alone is sufficient to release tongue projection. It could either give a preparatory signal, followed by a final activation of the appropriate motor nuclei through tegmental or reticular centers, or, the other way around, it could give the final release signal after the tegmental and reticular centers have "warmed up" or preset the tongue projection centers. In any case, given the shortness of time at which a feeding response can occur in tongue-projecting salamanders, we have to postulate a very fast release mechanism. It may well be that under conditions of high feeding motivation and fast feeding reaction the tectum feeds only the information "prey within snapping distance at an angle of such and such degrees from the midline" into the preactivated reticular motor systems.

Roth and Wake (in prep.) assume that in tongue-projecting salamanders the activation of the motor nuclei involved in the whole feeding action proceeds in the following way:

1. Slow contraction of the SAR by means of activation of nerves IX and X that presses the tongue against the frontal part of the jaws; expansion of the tongue pad and activation of mucous glands within the tongue pad.
2. Sudden elevation of the tongue apparatus above the rim of the lower jaw, by means of the m. intermandibularis innervated by r. intermandibularis of nerve V and of the m. interhyoideus innervated by r. hyomandibularis of nerve VII.
3. Opening of the mouth by means of the m. depressor mandibulae innervated by r. hyomandibularis of nerve VII, and simultaneous projection of the tongue out of the mouth by means of the SAR muscle innervated by nerves IX and X.
4. Simultaneous tightening of the firing cylinder (formed by the anterior part of SAR and m. suprapeduncularis, innervated by r. hypoglossus of the first spinal nerve, and control of the shooting direction of the tongue by m. geniohyoideus lateralis innervated by r. hypoglossus of the first spinal nerve, and control of head/neck movement by m. cucullaris innervated by nerve XI and partly by the first and the second spinal nerves.
5. Retraction of the tongue by RCP innervated by the first and the second spinal nerves.

As the anatomical studies show, the motor nuclei involved in the feeding reaction are arranged along a rostrocaudal line within the medulla oblongata and the cervical spinal cord according to the temporal sequence of their activity, and their entire activation occurs within a few milliseconds. As shown in Chapter 2, the temporal sequence of tongue protraction and retraction is achieved by the differing length-tension curves of the SAR and RCP muscles. Further evidence for the assumption that all mentioned motor nuclei are activated more or less synchronously, consists in the fact that the first spinal nerve is involved in the control of the firing cylinder and the tongue-shooting direction, as well as in tongue retraction, and that the XI/second spinal motor nuclei (which strongly intermingle) are involved in both guidance of neck movement during tongue protraction and in tongue retraction.

6.2 Enemy and Barrier Avoidance

Another apparent task of the tecto-pretectal system, besides that of guidance of feeding behavior, is the control of enemy-avoidance behavior: on the one hand, after destruction of the praetectum, enemy avoidance fails to occur, probably due a general "disinhibition" effect; on the other hand, frogs with an incorrectly "rewired" retinotectal map jumped toward a large threatening object instead of escaping it (Ingle 1983). This shows that the retinotectal map is also used for localization of enemies and for escape behavior. Ingle assumed that at least in frogs the same "channel" as for snapping behavior, i.e., the uncrossed tecto-bulbar tract, is used for enemy escape behavior.

In frogs, Ingle (in litt.) also found that besides optomotor behavior (see below) the praetectum is involved in the detection of wide structured objects such as large barriers, a behavior that is undisturbed after ablation of the tectum. Detection of *small* barriers can be mediated, in addition to the praetectum, by the dorsal anterior thalamus. However, the dorsal thalamus seems to require that both edges of the barrier are within the binocular field. Furthermore, this brain region seems to possess no true stereopsis, i.e., it cannot distinguish a small barrier on a similarly structured background without brightness cues. This has to be tested in salamanders. As described in Chapter 5, Finkenstädt and Ewert (1983b) found that in *Salamandra* destruction of the rostral thalamus also leads to a disinhibition of feeding behavior in the binocular part of the visual field.

6.3 Optomotor Behavior

Lesion experiments in different primary visual centers of the salamander brain have shown that the praetectum and the accessory optic system are the main first relay centers of the optokinetic response (OKR, Manteuffel 1979, Manteuffel et al. 1983). Destruction of these centers leads to a consistent loss of the horizontal OKR with the slow phase to the ipsilateral side. Neurons in the praetectum and in the nucleus Darkschewitsch are sensitive to horizontal optokinetic stimulation and differ mostly with regard to their input properties. Many neurons in the nucleus praetectalis are binocularly driven, whereas neurons in the nucleus Darkschewitsch receive input only from the contralateral eye (Manteuffel 1984a). The property of pretectal neurons being much more activated by binocular than by monocular stimulation may cause differences in the degree of compensation between monocular and binocular optomotor stimulation as described in Chapter 2.

Neurons in the praetectum receive excitatory input from the contralateral eye, like those of the nucleus Darkschewitsch, but they often receive an additional inhibitory input from the ipsilateral eye (Manteuffel, in press). The functional role of this connectivity may be to make possible a comparison of the shift of the visual environment perceived by the right and the left eye. The binocular pretectal neurons are well activated if the optomotor stimulus is moved temporonasally with respect to the eye contralateral to the neurons and nasotemporally with respect to the ipsilateral eye. However, the visual afferents to the praetectum mostly show a directional selectivity for temporonasal movements (Sperl 1984). Pretectal neurons, receiving an excitatory input from the contralateral eye and an (direct and indirect) inhibitory input from the ipsilateral eye, will, therefore, be silenced if both eyes are activated in the same (temporonasal or nasotemporal) direction, which always occurs when the animals are moving forward or backward. The pretectal system, thus, will be activated only by rotatory (i.e. turning) movements of the head. The nucleus Darkschewitsch, in contrast, also seems to be involved in the perception of translatory (i.e., forward-backward) movements (Manteuffel 1986).

The efferent fibers from the praectectal and accessory optic system are connected with the ventral and rostral tegmentum mesencephali (Naujoks-Manteuffel and Manteuffel 1986). The connection to the n. ventralis tegmenti is reciprocal, thus giving the possibility for a recurrent loop. In the rostral tegmental white matter, afferents from the vestibular nuclei are found as well as dendritic afferents from the n. fasciculi longitudinalis medialis and from the nucleus Darkschewitsch. Therefore, at the level of the tegmentum mesencephali vestibular inputs to optokinetic centers are probably present.

Manteuffel et al. (1986) showed that bilateral vestibular lesions abolish the optokinetic afternystagmus. Therefore, vestibular inputs to the optokinetic system apparently exist. In amniotes it is assumed that an interaction occurs at the level of the vestibular nuclei. However, in amphibians a clear optokinetic modulation of vestibular cells was not found, although in a recent study Caston and Bricoût-Berthout (1985) could show that optokinetic signals modulate vestibular neurons at a subthreshold level. Most probably, in salamanders vestibular and optokinetic signals converge at the level of the premotor units of the tegmentum.

Furthermore, motor neurons controlling neck musculature might be directly reached by fibers originating from cells of the n. praetectalis, the nucleus Darkschewitsch and the n. fasciculi longitudinalis medialis (Manteuffel 1984a, Naujoks-Manteuffel and Manteuffel 1986). However, it seems unlikely that these direct inputs are substantially involved in the control of the rather complex optokinetic reflex. This assumption seems to be corroborated by the fact that most fibers coming from the n. praetectalis terminate at the level of the bulbar reticular formation and only a few reach the motor neurons innervating the neck musculature.

6.4 Conclusion: What is Known and What is not yet Known About Neural Guidance of Visual Behavior in Salamanders?

Visually guided behavior of salamanders is rather well known, and little seems to remain to be studied, as far as experimental investigations are concerned. However, more detailed studies are needed on the behavior of the animals within their natural habitat as well as on the exact physical and ecological conditions of their life.

The neuroanatomy of the visual system of salamanders is known to a degree which is more or less comparable now to the situation in other lower vertebrates. Furthermore, within the last few years the function of the different visual centers such as the tectum, the praetectum, the thalamus, and the midbrain tegmentum has become clearer. It became evident that the differences between urodeles and anurans are much smaller than had been assumed before. Also, the structure and function of the bulbar and spinal motor nuclei involved in the guidance of visual behavior, especially in feeding, as well as the peripheral innervation of the related motor apparatus have become rather well known.

The role of the telencephalon in the guidance of visual behavior remains somewhat unclear. On the basis of preliminary studies, it seems that the telencephalon is involved in motivation and modulation of the activity of both feeding and avoidance behavior by influencing the thalamic and pretectal visual centers directly and the tectum indirectly. However, the exact function of telencephalic visual centers, for example, the striatum, has to be studied in more detail.

Despite a relatively large amount of electrophysiological data and an understanding at a more general level, the exact tectal, thalamic, and pretectal mechanisms of prey, enemy, and barrier recognition and distance estimation remain to be investigated in more detail. Here, intracellular recordings combined with anatomical characterization of the anatomy and connectivity of the recorded cell will provide us with important information. The same is true for processing of visual information within the retina as well as the exact nature of the retinal input to the thalamic, pretectal, and mesencephalic visual centers. Finally, the transformation of visual activity of the tectal, thalamic and diencephalic centers into motor activity as well as the integration of sensory information from different modalities influencing visually guided behavior remain to be studied more thoroughly.

Because, in addition to neuroethology and neuroanatomy, so much is known about ecology, life history, evolution, and functional morphology of salamanders, this group of vertebrates seems to represent an ideal subject for an integrated approach to test some major questions of organismal biology.

Abbreviations Used in Figures

A	auricula cerebelli
Amg	amygdala
Amgpl, APL, AL	amygdala pars lateralis
Amgpm	amygdala pars medialis
AN	anastomosis between nerves VII and IX
AOl	area octavo-lateralis
AOT	axial optic tract
acot	accessory olfactory tract
AS	nucleus accumbens
BB, bb	basibranchiale
BO	bulbus olfactorius
BOa	bulbus olfactorius accessorius
BON	basal optic neuropil
BOT	basal optic tract
BR, br	m. basiradialis
C	corpus geniculatum thalamicum
CA, COMA, ComA	commissura anterior
CB, Cb	cerebellum
CB 1, 2; cb I, II	first, second ceratobranchial
CC	corpus cerebelli
cg	m. circumglossus
CH, ch	ceratohyal
CHa, ComHab, COMHAB, ComHB	commissura habenulae
CHOP, CHO, CO	chiasma opticum
cNO	nervus opticus contralateralis
ComC	commissura cerebelli
ComHi	commissura hippocampi
COMPAL	commissura pallii
COMPOP, ComPO, CPO	commissura postoptica
CP, ComP, COMPO	commissura posterior
CTub, COMTUBPO, CTP	commissura tuberculi posterioris
CU	m. cucullaris
DI, D	diencephalon
dLOT	tractus olfactorius lateralis pars dorsalis

DM	m. depressor mandibulae
dP, dPal	dorsal pallium
dT	dorsal tegmentum
dTh, DTH	dorsal thalamus
EB	epibranchial
EMTH, eTh	eminentia thalami
EP	epiphysis
EPL	external plexiform layer
flm	fasciculus longitudinalis medialis
FlTel	fasciculus longitudinalis telencephali
Fpo	fasciculus postolfactorius
FO	fila olfactoria
FRM	formatio reticularis medialis
GC	layer of granule cells
GG, gg	m. genioglossus
GHL	m. geniohyoideus lateralis
GHM	m. geniohyoideus medialis
GL	glomeruli
Ha	habenula
HABDORS	habenula dorsalis
HABVENT	habenula ventralis
HG	m. hyoglossus
hga	m. hyoglossus anterior
hgp	m. hyoglossus posterior
Hy, Hyth	hypothalamus
Hyp	hypophysis
ig	m. intraglossus
IH	m. interhyoideus
IM	m. intermandibularis
INF	infundibulum
Inj	area of injection
iNO	nervus opticus ipsilateralis
IR, ir	m. interradialis
l	glossal ligament
LLC	lobus lateralis cerebelli
lm sp	lemniscus spinalis
lP, lPal	lateral pallium
MP, mP, mPal	medial pallium
MaOT	tractus opticus marginalis, marginal optic tract
MeOT	tractus opticus medialis, medial optic tract
MES	mesencephalon
ML	layer of mitral cells
MO, ME	medulla oblongata

MN V	motor nucleus of the fifth nerve
MN VII, IX, X	motor nuclei of the seventh, ninth, and tenth nerve
MN XI	motor nucleus of the eleventh nerve
MN 1SP	motor nucleus of the first spinal nerve
MN 2SP	motor nucleus of the second spinal nerve
MREU	massa reuniens
MS	medulla spinalis, spinal cord
NBl	nucleus Bellonci pars lateralis
NBm	nucleus Bellonci pars medialis
nBON, NBON	nucleus neuropili optici basalis, nucleus of BON
nD, ND	nucleus Darkschewitsch
NDT	nucleus dorsalis tegmenti
nDTa	nucleus dorsalis tegmenti pars anterior
nDTp	nucleus dorsalis tegmenti pars posterior
nflm, NFLM	nucleus fasciculi longitudinalis medialis
NI	nucleus isthmi
nPT	nucleus praetectalis
NPTp	nucleus praetectalis profundus
NPTs	nucleus praetectalis superficialis
nTP, NTP	nucleus tuberculi posterioris
NUCB, NB	nucleus Bellonci
NUCPOA	nucleus praeopticus pars anterior
NUCPOP	nucleus praeopticus pars posterior
NVS	nucleus visceralis secundarius
NVT	nucleus ventralis tegmenti
nVTa	nucleus ventralis tegmenti pars anterior
nVTp	nucleus ventralis tegmenti pars posterior
nIII	nucleus nervi oculomotorii
nIV	nucleus nervi trochlearis
OT	tractus opticus, optic tract
PAThDORS	thalamus dorsalis pars anterior
Ped	peduncle, tuberculum posterius
PITH	pars intercalaris thalami
PO	nucleus praeopticus
PPThDORS	thalamus dorsalis pars posterior
PT	area praetectalis, pretectal neuropil
R, r	radii
RCP, rcp	m. rectus cervicis profundus
RCPANT	m. rectus cervicis profundus anterior
rcsl	lateral slip of m. rectus cervicis
RH	ramus hyomandibularis of the facialis nerve
S	septum

SAR 1	m. subarcualis rectus 1
sd	sulcus dorsalis
se	"Sehnenplatte"
serh	sulcus entorhinalis
SH	m. subhyoideus
shab	sulcus habenulae
si	sulcus isthmi
sl	sulcus limitans
slm	sulcus lateralis mesencephali
Sm	stria medullaris
sm	sulcus medialis
spo	sulcus praeopticus
srh	sulcus rhinalis
ST	striatum
Str	corpus striatum
sv	sulcus ventralis
SYNENC	synencephalon
TEG	tegmentum
TEL, Tel	telencephalon
TI	tegmentum isthmi
TO	tectum opticum
TrBc	tractus tecto-bulbaris cruciatus
TrBr	tractus tecto-bulbaris rectus
TrIT	tractus isthmo-tectalis
TrTT	tractus tecto-tectalis
TV	tegmentum trigemini
U, UF	area uncinata, uncinate field
UH	urohyal
vLOT	tractus olfactorius lateralis pars ventralis
VOT	tractus olfactorius ventralis, ventral olfactory tract
vT	tegmentum ventrale, ventral tegmentum
vTh, VTH	thalamus ventralis, ventral thalamus
I	nervus olfactorius
II	nervus opticus
III	nervus oculomotorius
IV	nervus trochlearis
V	nervus trigeminus
VI	nervus abducens
VII	nervus facialis
VII RH	ramus hyomandibularis of the seventh nerve
VIII	nervus stato-acusticus
IX	nervus glossopharyngeus

X	nervus vagus
X SAR	branch of the vagus innervating the SAR muscle
XI	nervus accessorius
XII/1 SP	nervus hypoglossus/first spinal nerve
2 SP	second spinal nerve

References

Alberch P (1981) Convergence and parallelism in foot morphology in the neotropical salamander genus *Bolitoglossa*. I. Function. Evolution 35:84–100

an der Heiden U, Roth G (1983) Cooperative neural processes in amphibian visual prey recognition. In: Basar E, Flohr H, Haken H, Mandell AJ (eds) Synergetics of the brain. Springer, Berlin Heidelberg New York Tokyo, pp 299–310

an der Heiden U, Roth G (1986) Mathematical model and simulation of retina and tectum opticum of lower vertebrates. Biol Cybern (submitted)

Anderson JD (1968) A comparison of the food habits of *Ambystoma macrodactylum sigillatum*, *Ambystoma macrodactylum croceum* and *Ambystoma tigrinum californiense*. Herpetologica 24:273–284

Apfelbach R, Döhl J (1976) Verhaltensforschung. Fischer, Stuttgart New York

Ariens-Kappers CU, Huber CG, Crosby EC (1936) The comparative anatomy of the nervous system of vertebrates, including man. MacMillan, New York; Hafner, New York (Reprint 1967)

Arnold SJ (1976) Sexual behavior, sexual interference and sexual defense in the salamanders *Ambystoma maculatum*, *Ambystoma tigrinum* and *Plethodon jordani*. Z Tierpsychol 42:247–300

Attwell D, Wilson M (1980) Behaviour of the rod network in the tiger salamander retina mediated by membrane properties of individual rods. J Physiol (London) 309:287–315

Attwell D, Wilson M, Wu SM (1984) A quantitative analysis of interactions between photoreceptors in the salamander (*Ambystoma*) retina. J Physiol (London) 352:703–737

Autrum H (1959) Das Fehlen unwillkürlicher Augenbewegungen beim Frosch. Naturwissenschaften 46:435

Ball AK, Dickson DH (1983) Displaced amacrine and ganglion cells in the newt retina. Exp Eye Res 36:199–214

Barlow HB (1953) Summation and inhibition in the frog's retina. J Physiol (London) 119:69–88

Bas Lopez S, Rivera JG, Castro-Lorenzo A, Canal JLS (1979) Datos sobre la alimentacion de la salamandra (*Salamandra salamandra* L.) en Galicia. Bol Est Cent Ecol 8:73–78

Bechterev W (1884) Über die Funktion der Vierhügel. Pflüger's Arch Ges Physiol Menschen Tiere 33:413–439.

Becker R (1980) Verhaltensbiologische Untersuchung zum Einfluß der Beuteerfahrung auf das olfaktorisch und visuell gesteuerte Beutefangverhalten juveniler Feuersalamander (*Salamandra salamandra* L.). Diplomarbeit, Univ Bremen

Becker R (1985) Biochemische, verhaltensbiologische und elektrophysiologische Untersuchungen zum olfaktorisch gesteuertem Beutefang der Urodelen. Diss, Univ Bremen

Beer T (1899) Die Accomodation des Auges bei den Amphibien. Pflüger's Arch Ges Physiol 73:501–534

Behler JL, King FW (1979) Field Guide to North American Reptiles and Amphibians. Audubon Soc, Knopf, New York

Besharse JC, Brandon RA (1972) Optomotor response and eye structure of the troglobitic salamander *Gyrinophilus palleucus*. Am Midl Natur 89:463–467

Besharse JC, Brandon RA (1974a) Postembryonic eye degeneration in the troglobitic salamander *Typhlotriton spelaeus*. J Morphol 144:381–406

Besharse JC, Brandon RA (1974b) Size and growth of the eyes of the troglobitic salamander *Typhlotriton spelaeus*. Int J Speleol 6:255–264

Besharse JC, Brandon RA (1976) Effects of continuous light and darkness on the eyes of the troglobitic salamander *Typhlotriton spelaeus*. J Morphol 149:527–546

Birukow G (1937) Untersuchungen über den optischen Drehnystagmus und über die Sehschärfe des Grasfrosches (*Rana temporaria* L.). Z Vergl Physiol 25:92–142
Bishop SC (1941) The salamanders of New York. New York State Mus Bull No 324
Boycott BB, Dowling JE (1969) Organization of the primate retina: Light microscopy. Philos Trans R Soc London Ser B 255:109–184
Boycott BB, Wässle H (1974) The morphological types of ganglion cells in the domestic cat's retina. J Physiol (London) 240:397–419
Brandon RA (1971) North American troglobitic salamanders: Some aspects of modification in cave habitats, with special reference to *Gyrinophilus palleucus*. Nat Speleol Soc Bull 33:1–21
Bridges CDB (1972) The rhodopsin-porphyropsin visual system. In: Dartnall HJA (ed) Handbook of sensory physiology, vol VII/1. Photochemistry of vision. Springer, Berlin Heidelberg New York, pp 417–480
Brown PK, Gibbons IR, Wald G (1963) The visual cells and visual pigment of the mudpuppy *Necturus*. J Cell Biol 19:79–106
Brower LP, Zandt-Brower J van (1962) Experimental studies of mimicry. VI. The reaction of toads (*Bufo terrestris*) to honeybees (*Apis mellifica*) and their dronefly mimics (*Eristalis vinetorum*). Am Nat 96:297–307
Brunner G (1934) Über die Sehschärfe der Elritze (*Phoxinus laevis*) bei verschiedenen Helligkeiten. Z Vergl Physiol 21:296–316
Brzoska J, Schneider H (1978) Modification of prey-catching behavior by learning in the common toad (*Bufo b. bufo* (L.), Anura, Amphibia): Changes in responses to visual objects and effects of auditory stimuli. Behav Processes 3:125–136
Bucci-Innocenti S, Ragghianti M, Mancino G (1983) Investigations of karyology and hybrids in *Triturus boscai* and *T. cristatus*, with a reinterpretation of the species groups within *Triturus* (Caudata, Salamandridae). Copeia 1983:662–672
Burkhardt L (1931) Über Bau und Leistung des Auges einiger amerikanischer Urodelen. Z Vergl Physiol 15:637–651
Burton TM, Likens GE (1975) Salamander population and biomass in the Hubbard Brook Experimental Forest, New Hampshire. Copeia 1975: 541–546
Butler AB, Ebbesson SOE (1975) A Golgi study of the optic tectum of the tegu lizard, *Tupinambis nigropunctatus*. J Morphol 146:215–228
Buytendijk FJJ (1918) Instinct de la recherche du nid et expérience chez les crapauds (*Bufo vulgaris* et *Bufo calamita*). Arch Neerl Physiol Homme Anim 2:1–50
Caldwell JH, Berman N (1977) The central projections in the retina in *Necturus maculosus*. J Comp Neurol 171:455–464
Caston J, Bricout-Berthout A (1985) Influence of stimulation of the visual system on the activity of vestibular nuclear neurons in the frog. Brain Behav Evol 26:49–57
Collett T (1977) Stereopsis in toads. Nature (London) 267:349–351
Collett T, Udin SB (1982) The nucleus isthmi and the correspondence problem. Abstract from Mexico Workshop on Visuomotor Coordination 1982
Collewijn H (1981) The oculomotor system of the rabbit and its plasticity. In: Braitenberg V (ed) Studies of brain functions, vol 5. Springer, Berlin Heidelberg New York Tokyo
Collins JP, Holomuzki JR (1984) Intraspecific variation in diet within and between trophic morphs in larval tiger salamander (*Ambystoma tigrium nebulosum*). Can J Zool 62:168–174
Compoint C, Clairambault P (1986) Anatomie et developpement du système visuel de *Pleurodeles poiretti*. J Hirnforsch 27:37–43
Conant R (1975) A field guide to reptiles and amphibians of Eastern and Central North America, 2nd edn. Houghton Mifflin, Boston
Cott HB (1936) The effectiveness of protective adaptions in the hive-bee, illustrated by experiments on the feeding reactions, habit formation and memory of the common toad (*Bufo bufo bufo*). Proc Zool Soc Lond 1:113–133
Cowie RJ, Baisden RH (1982) Retinofugal projections of the terrestial salamander *Plethodon glutinosus*: An autoradiographic and peroxidase study. Manuscript
Crescitelli F (1958) The natural history of visual pigments. Proc 19th Ann Biol Coll Oregon State College 1958:30–51
Crescitelli F (1972) The visual cells and visual pigments of the vertebrate eye. In: Dartnall HJA (ed) Handbook of sensory physiology, vol VII/1. Photochemistry of vision. Springer, Berlin Heidelberg New York, pp 245–363

Cronly-Dillon JR (1964) Units sensitive to direction of movement in the goldfish optic tectum. Nature (London) 203:214–215
Cronly-Dillon JR (1968) Pattern of retinotectal connections after retinal regeneration. J Neurophysiol 31:410–418
Cronly-Dillon JR, Galand G (1966) Analyse des résponses visuelles unitaires dans le nerf optique et le tectum du triton, *Triturus vulgaris*. J Physiol (London) 58:502–503
Currie J, Cowan WM (1974a) Some observations on the early development of the optic tectum in the frog (*Rana pipiens*), with special reference to the effects of early eye removal on mitotic activity in the larval tectum. J Comp Neurol 156:123–142
Currie J, Cowan WM (1974b) Evidence for the late development of the uncrossed retinothalamic projections in the frog, *Rana pipiens*. Brain Res 71:133–139
Custer NV (1973) Structurally specialized contacts between the photoreceptors of the retina of the axolotl. J Comp Neurol 151:35–56
Dann JF, Beazley LD (1982) The development of connections between the isthmic nucleus and the tectum in *Xenopus* and *Limnodynastes* tadpoles. Neurosci Lett 33:107–113
Dartnall HJA (1953) The interpretation of spectral sensitivity curves. Br Med Bull 9:24–30
David RS, Jaeger RG (1981) Prey location through chemical cues by a terrestrial salamander. Copeia 1981: 435–440
Dickson DH, Hollenberg MJ (1971) The fine structure of the pigment epithelium and photoreceptor cells of the newt, *Triturus viridescens dorsalis* (Rafinesque). J Morphol 135:389–432
Dieringer N, Cochran SL, Precht W (1982) Differences in the central organization of gaze stabilizing reflexes between frog and turtle. J Comp Physiol 153:495–508
Donner KO, Reuter T (1976) Visual pigments and photoreceptor function. In: Llinas R, Precht W (eds) Frog neurobiology. Springer, Berlin Heidelberg New York, pp 251–277
Douglas RH, Collett TS, Wagner HJ (1986) Accomodation in anuran amphibia and its role in depth vision. J Comp Physiol 158A:133–143
Dowling JE (1970) Organization of vertebrate retinas. The Jonas M. Friedenwald Memorial Lecture. Invest Ophthalmol 9:655–680
Dowling JE, Werblin FS (1969) Organization of retina of the mudpuppy, *Necturus maculosus*. I. Synaptic structure. J Neurophysiol 32:315–338
Dowling JE, Ehinger B, Hedden WL (1976) The interplexiform cell: A new type of retinal neuron. Invest Ophthalmol 15:916–926
Dunn ER (1926) The salamanders of the family Plethodontidae. Smith College, Northampton, Massachusetts
Dunn-Meynell AA, Sharma SC (1986) The visual system of the channel catfish (*Ictalurus punctatus*). I. Retinal ganglion cell morphology. J Comp Neurol 247:32–55
Durand JP (1971) Récherches sur l'appareil visuel du proteé *Proteus anguinus* Laurenti urodèle hypoge. Ann Speleol 26:497–824
Ebbesson SOE, Schroeder DM (1971) Connections of the nurse shark's telencephalon. Science 173:254–256
Edwards JL (1976) Spinal nerves and their bearing on salamander phylogeny. J Morphol 148:305–328
Ehrenhardt H (1937) Formensehen und Sehschärfenbestimmung bei Eidechsen. Z Vergl Physiol 24:248–304
Eibl-Eibesfeld I (1951) Nahrungserwerb und Beuteschema der Erdkröte (*Bufo bufo* L.). Behaviour 4:1–35
Elias P, Wake DB (1983) *Nyctanolis pernix*, a new genus and species of plethodontid salamander from northwestern Guatemala and Chiapas, Mexico. In: Rhodin AGJ, Mitaya K (eds) Advances in herpetology and evolutionary biology: Essays in honor of Ernest E. Williams. Cambridge Mass Mus Comp Zool Harvard Univ, pp 1–12
Estes R (1981) Gymnophiona, Caudata. In: Wellnhofer P (ed) Handbuch der Paläoherpetologie, Teil 2. Fischer, Stuttgart New York, pp 1–115
Ewert J-P (1967) Aktivierung der Verhaltensfolge beim Beutefang der Erdkröte (*Bufo bufo* L.) durch elektrische Mittelhirnreizung. Z Vergl Physiol 54:455–481
Ewert J-P (1968) Der Einfluß von Zwischenhirndefekten auf die Visuomotorik im Beute- und Fluchtverhalten der Erdkröte (*Bufo bufo* L.). Z Vergl Physiol 61:41–70
Ewert J-P (1974) The neural basis of visually guided behavior. Sci Am 230:34–42

Ewert J-P (1976) The visual system of the toad: Behavioral and physiological studies on a pattern recognition system. In: Fite KV (ed) The amphibian visual system. A multidisciplinary approach. Academic Press, London New York, pp 142–202
Ewert J-P (1984) Tectal mechanisms that underlie prey-catching and avoidance behaviors in toads. In: Vanegas H (ed) Comparative neurology of the optic tectum. Plenum Press, New York London pp 247–416
Ewert J-P, Gebauer I (1973) Größenkonstanzphänomen im Beutefangverhalten der Erdkröte (*Bufo bufo* L.). J Comp Physiol 85:303–315
Ewert J-P, Hock FJ (1972) Movement-sensitive neurones in the toad's retina. Exp Brain Res 16:41–59
Ewert J-P, Seelen W von (1974) Neurobiologie und System-Theorie eines visuellen Muster-Erkennungsmechanismus bei Kröten. Kybernetik 14:167–183
Ewert J-P, Wietersheim A von (1974) Der Einfluß von Thalamus/Prätectum-Defekten auf die Antwort von Tectum-Neuronen gegenüber bewegten visuellen Mustern bei der Kröte (*Bufo bufo* L.). J Comp Physiol 92:149–160
Ewert J-P, Burghagen H, Schürg-Pfeiffer E (1983) Neuroethological analysis of the innate releasing mechanism for prey catching in toads. In: Ewert J-P, Capranica RR, Ingle D (eds) Advances in vertebrate neuroethology. NATO ASI Ser vol 56, Plenum Press, New York London, pp 413–475
Fiebig E, Ebbesson SOE, Meyer DL (1983) Afferent connections of the optic tectum of the piranha (*Serrasalmus nattereri*). Cell Tiss Res 231:55–72
Finch DJ, Collett TS (1983) Small-field, binocular neurons in the superficial layers of the frog optic tectum. Proc R Soc London Ser B 217:491–497
Finkenstädt T, Ewert J-P (1983a) Processing of area dimensions of visual key stimuli by tectal neurons in *Salamandra salamandra*. J Comp Physiol 153:85–98
Finkenstädt T, Ewert J-P (1983b) Visual pattern discrimination through interactions of neural networks: A combined electrical brain stimulation, brain lesion, and extracellular recording study in *Salamandra salamandra*. J Comp Physiol 153:99–110
Finkenstädt T, Ebbesson SOE, Ewert J-P (1983) Projections to the midbrain tectum in *Salamandra salamandra* L. Cell Tiss Res 234:39–55
Fite KV, Scalia F (1976) Central visual pathways in the frog. In: Fite KV (ed) The amphibian visual system. A multidisciplinary approach. Academic Press, London New York, pp 87–118
Flower SS (1927) Loss of memory accompanying metamorphosis in amphibians. Proc Zool Soc Lond 1:155–156
Francis EBT (1934) The anatomy of the salamander. Clarendon Press, Oxford
Fraser DF (1976) Empirical evaluation of the hypothesis of food competition in salamanders of the genus *Plethodon*. Ecology 57:459–471
Freda J (1983) Diet of larval *Ambystoma maculatum* of New Jersey. J Herpetol 17:177–179
Freed AN (1982) A tree frog's menu: Selection for an evening's meal. Oecologia (Berl) 53:20–26
Freytag GE (1970) Die Lurche. In: Freytag GE, Grzimek B, Kuhn O, Thenius E (Hrsg) Grzimeks Tierleben, Bd V. Fische 2/Lurche. Kindler, Zürich, pp 289–358
Fritzsch B (1980) Retinal projections in european Salamandridae. Cell Tiss Res 213:325–341
Fritzsch B (1981a) The pattern of lateral line afferents in urodeles. A horseradish peroxidase study. Cell Tiss Res 218:581–594
Fritzsch B (1981b) Efferent neurons to the labyrinth of *Salamandra salamandra* as revealed by retrograde transport of horseradish peroxidase. Neurosci Lett 26:191–196
Fritzsch B, Himstedt W (1980) Anatomy of visual afferents in salamander brain. Naturwissenschaften 67:203
Fritzsch B, Will U, Nikundiwe A (1983) The area octavolateralis of amphibians: A reinterpretation. In: Duncker HR, Fleischer DG (eds) Functional morphology of vertebrates (Fortschritte der Zoologie Bd 30). Fischer, Stuttgart, pp 603–606
Frontera JG (1952) A study of the anuran diencephalon. J Comp Neurol 96:1–69
Frost DO, So K-F, Schneider GE (1979) Postnatal development of retinal projections in Syrian hamster: A study using autoradiographic and anterograde degeneration techniques. Neuroscience 4:16–49
Frost DR (ed) (1985) Amphibian species of the world. A taxonomic and geographic reference. Allen Press and Association of Systematic Collections, Lawrence, Kansas

Gaillard J (1985) Binocularly driven neurons in the rostral part of the frog optic tectum. J Comp Physiol 157:47–55

Gallego A (1983) Organization of the outer plexiform layer of the tetrapoda retina. Progr Sens Physiol 4:83–114

Gauss CH (1961) Ein Beitrag zur Kenntnis des Balzverhaltens einheimischer Molche. Z Tierpsychol 18:60–66

Glasser S, Ingle D (1978) The nucleus isthmi as a relay station in the ipsilateral visual projection to the frog's optic tectum. Brain Res 159:214–218

Glickstein M, Millodot M (1970) Retinoscopy and eye size. Science 168:605–606

Granit R (1942) Colour receptors of the frog retina. Acta Physiol Scand 3:137–151

Grobstein P, Comer C (1983) The nucleus isthmi as an intertectal relay for the ipsilateral oculotectal projection in the frog *Rana pipiens*. J Comp Neurol 217:54–74

Grobstein P, Comer C, Kostyk SK (1983) Frog prey capture behavior: between sensory maps and directed motor output. In: Ewert J-P, Capranica RR, Ingle D (eds.) Advances in vertebrate neuroethology. NATO ASI Ser, vol 56, Plenum Press, New York London, pp 331–347

Grover BG, Sharma SC (1979) Tectal projections in the goldfish (*Carassius auratus*): A degeneration study. J Comp Neurol 184:435–454

Grover BG, Sharma SC (1981) Organization of extrinsic tectal connections in goldfish (*Carassius auratus*). J Comp Neurol 196:471–488

Gruberg ER (1972) Optic fiber projections of the tiger salamander *Ambystoma tigrinum*. J Hirnforsch 14:399–411

Gruberg ER, Harris WA (1981) The serotonergic somatosensory projection to the tectum of normal and eyeless salamanders. J Morphol 170:55–69

Gruberg ER, Lettvin JY (1980) Anatomy and physiology of a binocular system in the frog *Rana pipiens*. Brain Res 192:313–325

Gruberg ER, Udin SB (1978) Topographic projections between the nucleus isthmi and the tectum of the frog *Rana pipiens*. J Comp Neurol 179:487–500

Grüsser O-J, Grüsser-Cornehls U (1968) Neurophysiologische Grundlagen visueller angeborener Auslösemechanismen beim Frosch. Z Vergl Physiol 59:1–24

Grüsser O-J, Grüsser-Cornehls U (1970) Die Neurophysiologie visuell gesteuerter Verhaltensweisen bei Anuren. Verh Dtsch Zool Ges 64:201–218

Grüsser O-J, Grüsser-Cornehls U (1976) Neurophysiology of the anuran visual system. In: Llinas R, Precht W (eds) Frog neurobiology. Springer, Berlin Heidelberg New York, pp 297–385

Grüsser-Cornehls U (1984) The neurophysiology of the amphibian optic tectum. In: Vanegas H (ed) Comparative neurology of the optic tectum. Plenum Press, New York London, pp 211–245

Grüsser-Cornehls U, Himstedt W (1973) Responses of retinal and tectal neurons of the salamander (*Salamandra salamandra* L.) to moving visual stimuli. Brain Behav Evol 7:145–168

Grüsser-Cornehls U, Himstedt W (1976) The urodele visual system. In: Fite KV (ed) The amphibian visual system. A multidisciplinary approach. Academic Press, London New York, pp 203–266

Guillery RW, Updyke BV (1976) Retinofugal pathways in normal and albino axolotls. Brain Res 109:235–244

Hairston NG (1949) The local distribution and ecology of the plethodontid salamanders of the southern Appalachians. Ecol Monogr 19:47–73

Hairston NG (1980) Species packing in the salamander genus *Desmognathus*: What are the interspecific interactions involved? Am Nat 115:354–366

Halliday TR (1974) Sexual behavior of the smooth newt *Triturus vulgaris* (Urodela, Salamandridae). J Herpetol 8:277–292

Hanken J (1980) Morphological and genetic investigations of miniaturisation in salamanders (genus *Thorius*). Doctoral Thesis, Univ California

Hanken J (1982) Appendicular skeletal morphology in minute salamanders genus *Thorius* (Amphibia, Plethodontidae). Growth regulation, adult size determination and natural variation. J Morphol 174:57–78

Hanken J (1983) Miniaturization and its effects on cranial morphology in plethodontid sala-

manders, genus *Thorius* (Amphibia, Plethodontidae). II. The fate of the brain and sense organs and their role in skull morphogenesis and evolution. J Morphol 177:255–268
Harnischfeger G (1979) An improved method for extracellular marking of electrode tip positions in nervous tissue. J Neurosci Meth 1:195–200
Hartline HK (1938) The response of single optic nerve fibers of the vertebrate eye to illumination of the retina. Am J Physiol 121:400–415
Hecht M, Edwards J (1977) The methodology of phylogenetic inference above the species level. In: Hecht M, Goody P, Hecht B (eds) Major patterns in vertebrate evolution. Plenum Press, New York London, pp 3–51
Hendrickson A (1966) Landolt's club in the amphibian retina: A Golgi and electron microscope study. Invest Ophthalmol 5:484–496
Henning J (1985) Untersuchung der funktionellen Cytoarchitektur von Neuronen im Tectum opticum des Bergmolches (*Triturus alpestris*). Diplomarbeit, Technische Hochschule Darmstadt
Herrick CJ (1933) The amphibian forebrain. VI. *Necturus*. J Comp Neurol 58:1–288
Herrick CJ (1941) Optic and postoptic systems of fibers in the brain of *Necturus*. J Comp Neurol 75:487–544
Herrick CJ (1942) Optic and postoptic systems in the brain of *Amblystoma tigrinum*. J Comp Neurol 77:191–353
Herrick CJ (1948) The brain of the tiger salamander *Ambystoma tigrinum*. Univ Chicago Press, Chicago, Illinois
Hershkowitz M, Samuel D (1973) The retention of learning during metamorphosis of the crested newt (*Triturus cristatus*). Anim Behav 21:83–85
Himstedt W (1967) Experimentelle Analyse der optischen Sinnesleistungen im Beutefangverhalten der einheimischen Urodelen. Zool Jahrb Physiol 73:281–320
Himstedt W (1971) Die Tagesperiodik von Salamandriden. Oecologia (Berl) 8:194–208
Himstedt W (1972) Untersuchungen zum Farbensehen von Urodelen. J Comp Physiol 81:229–238
Himstedt W (1973a) Die spektrale Empfindlichkeit von Urodelen in Abhängigkeit von Metamorphose, Jahreszeit und Lebensraum. Zool Jahrb Physiol 77:246–274
Himstedt W (1973b) Die spektrale Empfindlichkeit von *Triturus alpestris* (Amphibia, Urodela) während des Wasser- und Landlebens. Pflüger's Arch 341:7–14
Himstedt W (1979) The significance of color signals in partner recognition of the newt *Triturus alpestris*. Copeia 1979:40–43
Himstedt W (1982) Prey selection in salamanders. In: Ingle D, Goodale MA, Mansfield RJW (eds) Analysis of Visual Behavior. MIT Press, Cambridge Mass London, pp 47–66
Himstedt W, Fischerleitner E (1975) Die Antworten von Retinaneuronen auf Farbreize bei Urodelen. Zool Jahrb Physiol 79:128–147
Himstedt W, Roth G (1980) Neuronal responses in the tectum opticum of *Salamandra* to visual prey stimuli. J Comp Physiol 135:251–257
Himstedt W, Freidank U, Singer E (1976) Die Veränderung eines Auslösemechanismus im Beutefangverhalten während der Entwicklung von *Salamandra salamandra* L. Z Tierpsychol 41:235–243
Himstedt W, Tempel P, Weiler J (1978) Responses of salamanders to stationary visual patterns. J Comp Physiol 124:49–52
Himstedt W, Helas A, Sommer TJ (1981) Projection of color coding retinal neurons in urodele amphibians. Brain Behav Evol 18:19–32
Himstedt W, Heller K, Manteuffel G (1986) Neuronal responses to moving visual stimuli in different thalamic and midbrain centers of *Salamandra salamandra (L.)*. Zool Jahrb Physiol (in press)
Hinsche G (1935) Ein Schnappreflex nach „Nichts" bei Anuren. Zool Anz 111:113–122
Hutchins JB, Polans AS, Werblin FS (1984) Localization of cholinesterase activity in the outer plexiform layer of the larval tiger salamander retina. Brain Res 292:303–315
Immelmann K (1975) Ecological significance of imprinting and early learning. Annu Rev Ecol Syst 6:15–37
Ingle DJ (1968) Visual releasers of prey-catching behavior in frog and toads. Brain Behav Evol 1:500–518

Ingle DJ (1972) Depth vision in monocular frogs. Psychon Sci 29:37–38
Ingle DJ (1973) Two visual systems in the frog. Science 181:1053–1055
Ingle DJ (1976) Spatial vision in anurans. In: Fite KV (ed) The amphibian visual system. A multidisciplinary approach. Academic Press, London New York, pp 119–140
Ingle DJ (1980) Some effects of pretectum lesions on the frog's detection of stationary objects. Behav Brain Res 1:139–163
Ingle DJ (1983) Brain mechanisms of visual localization by frogs and toads. In: Ewert J-P, Capranica RR, Ingle DJ (eds) Advances in vertebrate neuroethology. NATO ASI Series, vol 56, Plenum Press, New York London, pp 177–226
Ingle DJ, Cook J (1977) The effect of viewing distance upon size preference of frogs for prey. Vision Res 17:1009–1013
Jacobson M (1978) Developmental Neurobiology. Plenum Press, New York London
Jacobson M, Gaze RM (1964) Types of visual response from single units in the optic tectum and optic nerve of the goldfish. Q J Exp Physiol 49:199–209
Jaeger RG (1972) Food as a limited resource in competition between two species of terrestial salamanders. Ecology 53:535–546
Jaeger RG (1978) Plant climbing by salamanders: Periodic availability of plant-dwelling prey. Copeia 1978: 686–691
Jaeger RG, Barnard DF (1981) Foraging tactics of a terrestrial salamander: Choice of diet in structurally simple environments. Am Nat 117:639–664
Jaeger RG, Gergits WF (1979) Intra- and interspecific communication in salamanders through chemical signals on the substrate. Anim Behav 27:150–156
Jaeger RG, Rubin AM (1982) Foraging tactics of a terrestrial salamander: Judging prey profitability. J Anim Ecol 51:167–176
Jaeger RG, Barnard DE, Joseph RG (1982) Foraging tactics of a terrestial salamander: Assessing prey density. Am Nat 119:885–890
Jakway JS, Riss W (1972) Retinal projections in the tiger salamander, *Ambystoma tigrinum*. Brain Behav Evol 5:401–442
Jordan M, Luthardt G, Meyer-Naujoks C, Roth G (1980) The role of eye accommodation in the depth perception of common toads. Z Naturforsch 35C:851–852
Kalinina AV (1974) Classification of frog retina neurons by their quantitative characteristics. Vision Res 14:1305–1316
Källén B (1951) Some remarks on the ontogeny of the telencephalon in some lower vertebrates. Acta Anat 11:537–548
Kaneko A (1979) Physiology of the retina. Ann Rev Neurosci 2:169–191
Kaneko A (1983) Retinal bipolar cells: Their function and morphology. Trends Neurosci 6:219–223
Keefe JR (1971) The fine structure of the retina in the newt, *Triturus viridescens*. J Exp Zool 177:263–294
Keen WH (1975) Foraging strategies in two species of plethodontid salamanders. Diss, Kent State Univ
Kennedy MC, Rubinson K (1984) Development and structure of the lamprey optic tectum. In: Vanegas H (ed) Comparative neurology of the optic tectum. Plenum Press, New York London, pp 1–13
Kirschfeld K (1976) The resolution of lens and compound eyes. In: Zettler F, Weiler R (eds) Neural principles in vision. Springer, Berlin Heidelberg New York, pp 354–370
Kirschfeld K (1984) Linsen- und Komplexaugen: Grenzen ihrer Leistung. Naturwiss Rundschau 37:352–362
Kokoros JJ, Northcutt RG (1977) Telencephalic efferents of the tiger salamander *Ambystoma tigrinum* (Green). J Comp Neurol 173:613–628
Kopp J, Manteuffel G (1984) Quantitative analysis of salamander horizontal head nystagmus. Brain Behav Evol 25:187–196
Kostyk SK, Grobstein P (1980) Visual prey aquisition behavior in the frog: Effects of various unilateral lesions. Soc Neurosci Abstr 6:75
Kostyk SK, Grobstein P (1983) Visual orienting deficits in frogs with various unilateral lesions. Behav Brain Res 6:379–388
Kühlhorn F (1959) Beitrag zur Kenntnis der Ernährungsbiologie unserer heimischen Amphibien. Veröff Zool Staatssamml München 5:145–188

Kuhlenbeck H (1975) The central nervous system of vertebrates, vol. IV. Spinal cord and deuterencephalon. Karger, München Paris London New York Sydney
Kuhlenbeck H (1977) The central nervous system of vertebrates, vol. V/1. Derivatives of the prosencephalon: Diencephalon and telencephalon. Karger, München Paris London New York Sydney
Kupferman I, Weiss KR (1978) The command neuron concept. Behav Brain Sci 1:3–39
Larsell O (1967) The comparative anatomy and histology of the cerebellum from myxinoids through birds. Jansen J (ed). Univ Minnesota Press, Minneapolis
Larsen JH, Guthrie DJ (1975) The feeding system of terrestial tiger salamanders (*Ambystoma tigrinum melanostictum* Baird). J Morphol 147:137–154
Lasansky A (1973) Organization of the outer synaptic layer in the retina of the larval tiger salamander. Philos Trans R Soc London Ser B 265:471–489
Lázár G (1971) The projections of the retinal quadrants of the optic centres in the frog. A terminal degeneration study. Acta Morphol Acad Sci Hung 19:325–334
Lázár G (1973) Role of the accessory optic system in the optokinetic nystagmus of the frog. Brain Behav Evol 5:443–460
Lázár G (1978) Application of the cobalt filling technique to show retinal projections in the frog. Neuroscience 3:725–737
Lázár G (1984) Structure and connections of the frog optic tectum. In: Vanegas H (ed) Comparative neurology of the optic tectum. Plenum Press, New York London, pp 185–210
Lázár G, Székely G (1967) Golgi studies on the optic center of the frog. J Hirnforsch 9:329–344
Lázár G, Székely G (1969) Distribution of optic terminals in the different optic centers in the frog. Brain Res 16:1–14
Lázár G, Tóth P, Csank E, Kicliter E (1983) Morphology and location of tectal projection neurons in frogs: A study with HRP and cobalt filling. J Comp Neurol 215:108–120
Leghissa S (1962) L'evoluzione del tetto ottico nei bassi vertebrati (I). Arch Ital Anat Embriol 67:343–413
Lettvin JY, Maturana HR, McCulloch WS, Pitts WH (1959) What the frog's eye tells the frog's brain. Proc Inst Radio Eng NY 47:1940–1951
Lettvin JY, Maturana HR, McCulloch WS, Pitts WH (1961) Two remarks on the visual system of the frog. In: Rosenblith W (ed) Sensory communication. MIT Press, Cambridge Mass, pp 757–776
Levine RL (1980) An autoradiographic study of retinal projection in *Xenopus laevis* with comparison to *Rana*. J Comp Neurol 189:1–29
Liebman PA (1972) Microspectrophotometry of photoreceptors. In: Dartnall HJA (ed) Handbook of sensory physiology, vol VII/1. Springer, Berlin Heidelberg New York, pp 481–528
Liebman PA, Entine G (1968) Visual pigments of frog and tadpole (*Rana pipiens*). Vision Res 8:761–775
Lindquist SB, Bachmann MD (1982) The role of visual and olfactory cues in the prey catching behavior of the tiger salamander, *Ambystoma tigrinum*. Copeia 1982:81–90
Linke R, Roth G, Rottluff B (1986) Comparative studies on the eye morphology of lungless salamanders, family Plethodontidae, and the effect of miniaturization. J Morphol 189:131–143
Lock A, Collett T (1979) A toad's devious approach to its prey: A study of some complex uses of depth vision. J Comp Physiol 131:179–189
Lombard RE, Wake DB (1976) Tongue evolution in the lungless salamanders, family Plethodontidae. I. Introduction, theory and a general modell of dynamics. J Morphol 148:265–286
Lombard RE, Wake DB (1977) Tongue evolution in the lungless salamanders, family Plethodontidae. II. Function and evolutionary diversity. J Morphol 153:39–80
Luiten PGM (1981) Afferent and efferent connections of the optic tectum in the carp (*Cyprinus carpio* L.). Brain Res 220:51–65
Luthardt G (1981) Verhaltensbiologische Untersuchungen zum visuell gesteuerten Beutefangverhalten von *Salamandra salamandra* (L.). Minerva, München
Luthardt G, Roth G (1979a) The relationship between stimulus orientation and stimulus movement pattern in the prey catching behavior of *Salamandra salamandra*. Copeia 1979: 442–447
Luthardt G, Roth G (1979b) The influence of prey experience on movement pattern preference in *Salamandra salamandra* L. Z Tierpsychol 51:252–259

Luthardt G, Roth G (1983) The interaction of the visual and the olfactory systems in guiding prey catching behavior in *Salamandra salamandra* (L.). Behavior 83:69–79
Luthardt-Laimer G (1983) Distance estimation in binocular and monocular salamanders. Z Tierpsychol 63:233–240
Luthardt-Laimer G, Roth G (1983) Reduction of visual inhibition to stationary prey by early experience in *Salamandra salamandra* (L.). Z Tierpsychol 63:294–302
Lynch JF (1985) The feeding ecology of *Aneides flavipunctatus* and sympatric plethodontid salamanders in Northwestern California. J Herpetol 19:328–352
Madison DM (1972) Homing orientation in salamanders: A mechanism involving chemical cues. In: Galler SR, Schmidt-Koenig K, Jacobs GJ, Belleville RE (eds) Animal orientation and navigation. Nat Aeronautics and Space Admin, Washington DC, pp 485–498
Madison DM (1975) Intraspecific odor preferences between salamanders of the same sex: Dependence of season and proximity of residence. J Can Zool 53:1356–1361
Maiorana VC (1978a) Difference in diet as an epiphenomenon: Space regulates salamanders. Can J Zool 56:1017–1025
Maiorana VC (1978b) Behavior of an unobservable species. Diet selection by a salamander. Copeia 1978: 664–672
Manteuffel C (1985) Verbindungen des Prätectum und des basalen optischen Systems beim Feuersalamander (*Salamandra salamandra*). Eine Meerrettichperoxidase Studie. Diss, Technische Hochschule Darmstadt
Manteuffel G (1979) Übergang vom optomotorischen Verhalten zur Fixation beim Kammolch (*Triturus cristatus*). Zool Jahrb Physiol 83:526–539
Manteuffel G (1982) The accessory optic system in the newt, *Triturus cristatus*: Unitary response properties from the basal optic neuropil. Brain Behav Evol 21:175–184
Manteuffel G (1984a) Electrophysiology and anatomy of direction-specific pretectal units in *Salamandra salamandra*. Exp Brain Res 54:415–425
Manteuffel G (1984b) A 'physiological' model for the salamander horizontal optokinetic reflex. Brain Behav Evol 25:197–205
Manteuffel G (1985) Monocular and binocular optic inputs to salamander pretectal neurons: An intracellular recording and HRP-labelling study. Brain Behav Evol 27:1–10
Manteuffel G (1986) Der Beitrag synencephaler Nuclei zum Problem der Objekt-Hintergrund-diskriminierung bei Salamandern. Verh Dtsch Zool Ges 79:226
Manteuffel G, Wess O, Himstedt W (1977a) Messungen am dioptrischen Apparat von Amphibienaugen und Berechnung der Sehschärfe in Wasser und Luft. Zool Jahrb Physiol 81:395–406
Manteuffel G, Plasa L, Sommer TJ, Wess O (1977b) Involuntary eye movements in salamanders. Naturwissenschaften 64:533
Manteuffel G, Petersen J, Himstedt W (1983) Optic nystagmus and nystagmogen centers in the european fire salamander (*Salamandra salamandra*). Zool Jahrb Physiol 87:113–125
Manteuffel G, Kopp J, Himstedt W (1986) The amphibian optokinetic afternystagmus: properties and comparative analysis in various species. Behav Brain Res 28:186–197
Mariani AP (1986) Photoreceptors of the larval tiger salamander retina. Proc R Soc London Ser B 227:483–492
Mariani AP, Lasansky A (1984) Chemical synapses between turtle photoreceptors. Brain Res 310:351–354
Marshal LM, Werblin FS (1978) Synaptic transmission of the horizontal cells in the retina of the larval tiger salamander. J Physiol (London) 279:321–346
Matthes E (1924a) Das Geruchsvermögen von Triton beim Aufenthalt unter Wasser. Z Vergl Physiol 1:57–83
Matthes E (1924b) Das Geruchsvermögen von Triton beim Aufenthalt an Land. Z Vergl Physiol 1:590–606
Matthes E (1924c) Die Rolle des Gesichts-, Geruchs- und Erschütterungssinnes für den Nahrungserwerb von Triton. Biol Zentralbl 44:72–87
Matthews G (1983) Physiological characteristics of single green rod photoreceptors from toad retina. J Physiol (London) 342:347–359
Maturana HR (1959) Number of fibers in the optic nerve and the number of ganglion cells in the retina of anurans. Nature (London) 183:1406–1407

Maturana HR, Lettvin JY, McCulloch WS, Pitts WH (1960) Anatomy and physiology of vision in the frog (*Rana pipiens*). J Gen Physiol (Suppl 2) 43:129–175
McCormick CA (1982) The organization of the octavolateralis area in actinopterygian fishes: A new interpretation. J Morphol 171:159–181
McGavin M (1978) Recognition of conspecific odors by the salamander *Plethodon cinereus*. Copeia 1978: 356–358
McKibben PS (1911) The nervus terminalis in urodele amphibia. J Comp Neurol 21:249–289
Meek J (1983) Functional anatomy of the tectum mesencephali of the goldfish. An explorative analysis of the functional implications of the laminar structural organization of the tectum. Brain Res Rev 6:247–297
Menner E (1929) Untersuchungen über die Retina mit besonderer Berücksichtigung der äußeren Körnerschicht. Ein Beitrag zur Duplizitätstheorie. Z Vergl Physiol 8:761–826
Merchant H (1972) Estimated population size and home range of the salamanders *Plethodon jordani* and *Plethodon glutinosus*. J Wash Acad Sci 62:248–257
Meyer P (1976) Taschenlexikon der Verhaltenskunde. UTB, Schöningh, Paderborn
Meyer-Naujoks C (1982) Vergleichende Untersuchung der olfaktorischen und vomeronasalen Projektionen bei Urodelen. Diplomarbeit, Univ Bremen
Möller A (1951) Die Struktur des Auges bei Urodelen verschiedener Körpergröße. Zool Jahrb Physiol 62:138–182
Montgomery N, Fite KV, Bengston L (1981) The accessory optic system of *Rana pipiens*: Neuroanatomical connections and intrinsic organisation. J Comp Neurol 203:595–612
Montgomery N, Fite KV, Taylor M, Bengston L (1982) Neural correlates of optokinetic nystagmus in the mesencephalon of *Rana pipiens*: A functional analysis. Brain Behav Evol 21:137–150
Montgomery N, Fite KV, Grigonis AM (1985) The pretectal nucleus lentiformis mesencephali of *Rana pipiens*. J Comp Neurol 234:264–275
Muntz WRA (1963) Phototaxis and green rods in urodeles. Nature (London) 109:620
Naujoks-Manteuffel C, Manteuffel G (1986) Internuclear connections between the pretectum and the accessory optic system in *Salamandra salamandra*. Cell Tiss Res 243:595–602
Naujoks-Manteuffel C, Manteuffel G, Himstedt W (1986a) Deszendierende Bahnen aus dem Di- und Mesencephalon des Feuersalamanders (*Salamandra salamandra* L.). Verh Dtsch Zool Ges 79:373
Naujoks-Manteuffel C, Manteuffel G, Himstedt W (1986b) The nucleus-ruber problem: comparison between a quadruped and a limbless amphibian (*Salamandra salamandra* versus *Ichthyophis kohtaoensis*). Neurosci Lett Suppl. 26:S448
Neary TJ (1975) Architectonics of the thalamus of the bullfrog (*Rana catesbeiana*). A histochemical analysis. Anat Rec 181:434–435
Neary TJ (1984) Anterior thalamic nucleus projections to the dorsal pallium in ranid frogs. Neurosci Lett 51:213–218
Neary TJ, Northcutt RG (1983) Nuclear organisation of the bullfrog diencephalon. J Comp Neurol 213:262–278
Nicholas JS (1922) The reaction of *Amblystoma tigrinum* to olfactory stimuli. J Exp Zool 35:257–281
Norris HW (1908) The cranial nerves of *Amphiuma means*. J Comp Neurol Psychol 18:527–568
Northcutt RG (1977) Retinofugal projections in the lepidosirenid lungfishes. J Comp Neurol 174:553–574
Northcutt RG (1980) Retinal projections in the australian lungfish. Brain Res 185:85–90
Northcutt RG (1982) Localization of neurons afferent to the optic tectum in longnose gars. J Comp Neurol 204:325–335
Northcutt RG, Kicliter E (1980) Organization of the amphibian telencephalon. In: Ebbesson SOE (ed) Comparative neurology of the telencephalon. Plenum Press, New York London, pp 203–255
Ogden TE, Mascetti GG, Pierantoni R (1984) The internal horizontal cell of the frog: Analysis of receptor input. Invest Ophthalmol Vis Sci 25:1382–1394
Ogden TE, Mascetti GG, Pierantoni R (1985) The outer horizontal cell of the frog retina: Morphology, receptor input and function. Invest Ophthalmol Vis Sci 26:643–656
Olmo E (1983) Nucleotype and cell size in vertebrates: A review. Bas Appl Histochem 27:227–256

Opdam P, Nieuwenhuys R (1976) Topological analysis of the brain stem of the axolotl *Ambystoma mexicanum*. J Comp Neurol 165:285–306
Özeti N, Wake DB (1969) The morphology and evolution of the tongue and associated structures in salamanders and newts (family Salamandridae). Copeia 1969: 91–123
Pettigrew JD (1972) The neurophysiology of binocular vision. Sci Am 227:84–95
Pettigrew JD, Konishi M (1976) Neurons selective for orientation and binocular disparity in the visual wulst of the barn owl (*Tyto alba*). Science 193:675–678
Prechtl HFR (1951) Zur Paarungsbiologie einiger Molcharten. Z Tierpsychol 8:337–348
Pope CH (1947) Amphibians and Reptiles of the Chicago Area. Nat Hist Mus Chicago 14
Ramon y Cajal S (1892) The Structure of the Retina. Transl. by Thorpe SA and Glickstein M. Thomas, Springfield 1972
Ranvier L (1875) Traité technique d'histologie. Paris
Rensch B (1948) Histological changes correlated with evolutionary changes of body size. Evolution 2:218–230
Rettig G (1984) Neuroanatomische Untersuchungen der visuellen Projektionen bei Salamandern (Ordnung Caudata). Diss, Universität Bremen
Rettig G, Roth G (1982) Afferent visual projections in three species of lungless salamanders (Family Plethodontidae). Neurosci Lett 31:221–224
Rettig G, Roth G (1986) Retinofugal projections in salamanders of the family Plethodontidae. Cell Tiss Res 243:385–396
Rettig G, Fritzsch B, Himstedt W (1981) Development of retinofugal neuropil areas in the brain of the alpine newt, *Triturus alpestris*. Anat Embryol 162:163–171
Rettig G, Roth G, Wake DB (1987) Precocial development of retinofugal projections in a direct developing plethodontid salamander (submitted for publication)
Reuter T (1969) Visual pigments and ganglion cell activity in the retinae of tadpoles and adult frogs (*Rana temporaria* L.). Acta Zool Fennica 122:1–64
Reuter T, White RH, Wald G (1971) Rhodopsin and porphyropsin fields in the adult bullfrog retina. J Gen Physiol 58:351–371
Riss W, Knapp H, Scalia F (1963) Optic pathways in *Cryptobranchus allegheniensis* as revealed by the Nauta technique. J Comp Neurol 121:31–43
Roth G (1974) Experimentelle Untersuchungen zum Beutefang von *Hydromantes italicus* Dunn (Amphibia, Plethodontidae). Diss, Univ Münster
Roth G (1976) Experimental analysis of the prey catching behavior of *Hydromantes italicus* Dunn (Amphibia, Plethodontidae). J Comp Physiol 109:47–58
Roth G (1978) The role of stimulus movement patterns in the prey catching behavior of *Hydromantes genei* (Amphibia, Plethodontidae). J Comp Physiol 123:261–264
Roth G (1982a) Responses in the optic tectum of the salamander *Hydromantes italicus* to moving prey stimuli. Exp Brain Res 45:386–392
Roth G (1982b) Beuteerkennungsmechanismen im Tectum opticum von Amphibien – eine vergleichende Untersuchung. Funkt Biol Med 1:90–98
Roth G (1986) Neural mechanisms of prey recognition: An example in amphibians. In Feder ME, Lauder GV (eds) Predator-prey relationships. Univ Chicago Press, Chicago London, pp 42–68
Roth G, Himstedt W (1978) Response characteristics of neurons in the tectum opticum of *Salamandra*. Naturwissenschaften 65:657
Roth G, Jordan M (1982) Response characteristics and stratification of tectal neurons in the toad *Bufo bufo* L. Exp Brain Res 45:393–398
Roth G, Luthardt G (1980) The role of early sensory experience in the prey catching responses of *Salamandra salamandra* to stationary prey. Z Tierpsychol 52:141–148
Roth G, Wake DB (1985a) Trends in the functional morphology and sensorimotor control of feeding behavior in salamanders: An example of the role of internal dynamics in evolution. Acta Biotheoret 34:175–192
Roth G, Wake DB (1985b) The structure of the brainstem and cervical spinal cord in lungless salamanders (family Plethodontidae) and its relation to feeding. J Comp Neurol 241:99–110
Roth G, Wiggers W (1983) Responses of the toad *Bufo bufo* to stationary prey stimuli. Z Tierpsychol 61:225–234
Roth G, Grunwald W, Linke R, Rettig G, Rottluff B (1983) Evolutionary patterns in the visual system of lungless salamanders (Family Plethodontidae). Arch Biol Med Exp 16:329–341

Roth G, Wake DB, Wake MH, Rettig G (1984) Distribution of accessory and hypoglossal nerves in the hindbrain and spinal cord of lungless salamanders, family Plethodontidae. Neurosci Lett 44:53–57

Royce GJ, Northcutt RG (1969) Olfactory bulb projection in the tiger salamander (*Ambystoma tigrinum*) and the bullfrog (*Rana catesbeiana*). Anat Rec 163:254

Rubinson K (1968) Projections of the tectum opticum of the frog. Brain Behav Evol 1:529–561

Satou M, Ewert J-P (1985) The antidromic activation of tectal neurons by electrical stimuli applied to the caudal medulla oblongata in the toad *Bufo bufo* L. J Comp Physiol A 157: 739–748

Satou M, Matsushima T, Ueda K (1984) Neuronal pathways for the tectal "snapping-evoking area" to the tongue-muscle-controlling motoneurons in the japanese toad: Evidence of the intervention of excitatory interneurons. Zool Sci 1:829–832

Satou M, Matsushima T, Takeuchi H, Ueda K (1985) Tongue-muscle-controlling motoneurons in the japanese toad: Topography, morphology and neuronal pathways from the 'snapping-evoking area' in the optic tectum. J Comp Physiol 157A:717–737

Scalia F (1976) The optic pathway of the frog: Nuclear organisation and connections. In: Llinas R, Precht W (eds) Frog neurobiology. Springer, Berlin Heidelberg New York, pp 386–406

Scalia F, Fite KV (1974) A retinotopic analysis of the central connections of the optic tectum in the frog. J Comp Neurol 158:455–478

Schaeffer AP (1911) Habit formation in frog. J Anim Behav 1:309–335

Schipperheyn JJ (1965) Contrast detection in frog's retina. Acta Physiol Pharmacol Néerl 13:231–277

Schmidt A, Manteuffel C (1984) Olfaktorische Projektionen beim Feuersalamander *Salamandra salamandra* L. Verh Dtsch Zool Ges 77:325

Schneider D (1954) Beitrag zu einer Analyse des Beute- und Fluchtverhaltens einheimischer Anuren. Biol Zentralbl 73:225–282

Schürg-Pfeiffer E, Ewert J-P (1981) Investigations of neurons involved in the analysis of gestalt prey features in the frog *Rana temporaria*. J Comp Physiol 141:139–152

Sites JW (1978) The foraging strategy of the dusky salamander, *Desmognathus fuscus* (Amphibia, Urodela, Plethodontidae): An empirical approach to predation theory. J Herpetol 12:373–383

Sivak JG, Warburg MR (1980) Optical metamophosis of the eye of *Salamandra salamandra*. Can J Zool 58:2059–2064

Skrzypek J (1984) Electrical coupling between horizontal cell bodies in the tiger salamander retina. Vision Res 24:701–711

Slaughter MM, Miller RF (1983) The role of excitatory amino acid transmitters in the mudpuppy retina: An analysis with kainic acid and N-methyl aspartate. J Neurosci 3:1701–1711

Smeets WJAJ (1982) The afferent connections of the tectum mesencephali in two chondrichthyans, the shark *Scyliorhinus canicula* and the ray *Raja clavata*. J Comp Neurol 205:139–152

Smith HM (1978) Amphibians of North America. Golden Press, New York

Sperl M (1984) Neuronale Antworten aus dem prätectalen Neuropil des Feuersalamanders (*Salamandra salamandra* L.) auf verschiedene optisch relevante Reizmuster. Diplomarbeit, Technische Hochschule Darmstadt

Sperry RW (1945) Restauration of vision after crossing of optic nerves and after contralateral transplantation of eye. J Neurophysiol 8:15–28

Stebbins RC (1951) A Field Guide to Western Reptiles and Amphibians. Houghton Mifflin, Boston

Steedman JG, Stirling RV, Gaze RM (1979) The central pathways of optic fibers in *Xenopus* tadpoles. J Embryol Exp Morphol 50:199–215

Stell WK (1972) The morphological organization of the vertebrate retina. In: Fuortes MGF (ed) Handbook of sensory physiology, vol VII/2. Physiology of photoreceptor organs. Springer, Berlin Heidelberg New York, pp 111–213

Stell WK, Witkovsky P (1973) Retinal structure in the smooth dogfish, *Mustelus canis*. General description and light microscopy of giant ganglion cells. J Comp Neurol 148:1–32

Stenner G (1976) Untersuchung über die Funktion von Reizfiltermechanismen im Beutefangverhalten des Feuersalamanders nach der Metamorphose. Staatsexamensarbeit, Technische Hochschule Darmstadt

Stephan P, Weiler R (1981) Morphology of horizontal cells in the frog retina. Cell Tiss Res 221:443–449

Sternthal DE (1974) Olfactory and visual cues in the feeding behavior of the leopard frog (*Rana pipiens*). Z Tierpsychol 34:239–335

Stirling RV, Brändle K (1982) Expansion of the visual projection to the tectum of axolotls during metamorphosis. Dev Brain Res 5:343–345

Stone LS (1964) The structure and visual function of the eye of larval and adult cave salamanders *Typhlotriton spelaeus*. J Exp Zool 156:201–218

Ströer WFH (1940) Das optische System beim Wassermolch (*Triturus taeniatus*). Acta Neerl Morphol 3:178–195

Szabó I (1962) Nahrungswahl und Nahrung des gefleckten Feuersalamanders (*Salamandra salamandra* L.). Acta Zool Acad Sci Hung 8:459–477

Székely G, Lázár G (1976) Cellular and synaptic architecture of the optic tectum. In: Llinas R, Precht W (eds) Frog neurobiology. Springer, Berlin Heidelberg New York, pp 407–434

Tembrock G (1978) Verhaltensbiologie. UTB, Fischer, Stuttgart New York

Thexton AJ, Wake DB, Wake MH (1977) Tongue function in the salamander *Bolitoglossa occidentalis*. Arch Oral Biol 22:361–366

Thiesmeier B (1982) Beitrag zur Nahrungsbiologie der Larven des Feuersalamanders, *Salamandra salamandra* (L.) (Amphibia, Caudata, Salamandridae). Salamandra 18:86–88

Thorn R (1968) Les salamandres d'Europe, d'Asie et d'Afrique du Nord. Lechevalier, Paris

Tinbergen N (1951) The study of instinct. Oxford Univ Press, New York

Tóth P, Csank G, Lázár G (1985) Morphology of the cells of origin of descending pathways to the spinal cord in *Rana esculenta*. A tracing study using cobaltic-lysine complex. J Hirnforsch 26:365–383

Trachtenberg MC, Ingle D (1974) Thalamo-tectal projections in the frog. Brain Res 79:419–430

Tretjakoff D (1906) Der Musculus protractor lentis im Urodelenauge. Anat Anz 28:25–32

Tristram DA (1977) Intraspecific olfactory communication in the terrestial salamander *Plethodon cinereus*. Copeia 1977: 597–600

Udin SB (1978) Permanent disorganization of the regenerating optic tract in frog. Exp Neurol 58:455–470

Udin SB (1985) The role of visual experience in the formation of binocular projections in the frog. Cell Mol Neurobiol 5:85–102

Vallerga S (1981) Physiological and morphological identification of amacrine cells in the tiger salamander retina. Vision Res. 21:1307–1317

Vanegas H, Ebbesson SOE (1973) Retinal projections in the perch-like teleost *Eugerres plumieri*. J Comp Neurol 151:331–357

Vanegas H, Ebbesson SOE, Laufer M (1984) Morphological aspects of the teleostean optic tectum. In: Vanegas H (ed) Neurology of the Optic Tectum. Plenum Press, New York London, pp 93–120

Van Sluyters RC, Stewart DL (1974) Binocular neurons of the rabbit's visual cortex: Receptive field characteristics. Exp Brain Res 19:166–195

Vial JL (1968) The ecology of the tropical salamander *Bolitoglossa subpalmata* in Costa Rica. Rev Biol Trop 15:13–115

Wake DB (1966) Comparative osteology and evolution of the lungless salamanders, family Plethodontidae. Mem S Cal Acad Sci 4:1–111

Wake DB (1982) Functional and developmental constraints and opportunities in the evolution of feeding systems in urodeles. In: Mossakowski D, Roth G (eds) Environmental adaption and evolution. Fischer, Stuttgart, pp 51–66

Wake DB (1986) Adaptive radiation of salamanders in the Middle American cloud forests. Ann Missouri Bot Gardens, (in press)

Wake DB, Brame A (1969) Systematics and evolution of neotropical salamanders of the *Bolitoglossa helmrichi* group. Contr Sci Nat Hist Mus Los Angeles Co 175:1–40

Wake DB, Elias P (1983) New genera and a new species of Central America salamanders, with a review of the tropical genera (Amphibia, Caudata, Plethodontidae). Nat Hist Mus Los Angeles Co Contrib Sci 345:1–19

Wake DB, Lynch JF (1976) The distribution, ecology and evolutionary history of plethodontid salamanders in tropical America. Nat Hist Mus Los Angeles Co Sci Bull 25:1–65

Wake DB, Özeti N (1969) Evolutionary relationships in the family salamandridae. Copeia 1969: 124–137
Wake DB, Mason LR, Wurst GZ (1978) Genetic differentiation, albumin evolution and their biogeographic implications in plethodontid salamanders of California and southern Europe. Evol 32:529–539
Wake DB, Roth G, Wake MH (1983) Tongue evolution in lungless salamanders, family Plethodontidae. III. Patterns of peripheral innervation. J Morphol 178:207–224
Wald G (1947) The chemical evolution of vision. Harvey Lect 41:117–160
Wald G (1952) Biochemical evolution. In: Guzman Barrón ES (ed) Modern trends in physiology and biochemistry. Academic Press, New York, pp 337–376
Walls GL (1942) The vertebrate eye and its adaptive radiation. Cranbook Institute of Science, Bloomfield Hill, Michigan
Wang SR, Yan K, Wang YT, Jiang SY, Wang XS (1981) Neuroanatomy and electrophysiology of the lacertilian nucleus isthmi. Brain Res 243:215–224
Werner C (1983) Zielorientierung im Beutefangverhalten des Feuersalamanders *Salamandra salamandra* L. Diss, Technische Hochschule Darmstadt
Werner C, Himstedt W (1985) Mechanism of head orientation during prey capture in salamander (*Salamandra salamandra* L.). Zool Jahrb Physiol 89:359–368
Westheimer G (1972) Optical properties of vertebrate eyes. In: Fuortes MFG (ed) Handbook of sensory physiology, vol VII/2. Springer, Berlin Heidelberg New York, pp 449–482
Wicht H (1984) Retinofugale projektionen im Thalamus von *Triturus alpestris* (Laur.). Versuche zur Darstellung der postsynaptischen Strukturen. Diplomarbeit, Technische Hochschule Darmstadt
Wicht H, Himstedt W (1986) Two thalamo-telencephalic pathways in a urodele, *Triturus alpestris*. Neurosci Lett 68:90–94
Wilczynski W, Northcutt RG (1977) Afferents to the optic tectum of the leopard frog: An HRP study. J Comp Neurol 173:219–230
Wilder ILW, Dunn ER (1920) The correlation of lungless salamanders with a mountain brook habitat. Copeia 84:63–68
Wong-Riley MTT (1974) Synaptic organization of the inner plexiform layer in the retina of the tiger salamander. J Neurocytol 3:1–33
Zippelius HM (1948) Untersuchungen über das Balzverhalten heimischer Molche. Verh Dtsch Zool Ges 1948: 127–130

Subject Index

Accomodation 121–124, 268, 269
Amacrine cells 103, 104
Amygdala
- pars lateralis 135
- pars medialis 136
Axial optic tract 153

Barrier avoidance 275
Basal optic neuropil
- in plethodontids 156, 161
- in salamandrids 153
Basal optic tract 153, 156, 161
Bipolar cells 102, 103
Brainstem (see Medulla oblongata)

Cerebellum 143, 144
Cervical spinal cord 148–151
- role in feeding 274, 275
Cones 96–98
Corpus geniculatum thalamicum
- in plethodontids 156
- in salamandrids 153, 155

Depth perception 52–58, 263–269
- binocular 54–58, 263–268
- monocular 54–58, 268, 269
- neural mechanisms 263–269
Diencephalon
- connectivity 190–197
- connectivity in anurans 196–198
- cytoarchitecture 180–182
- morphology 136–138
- morphology in anurans 139, 140
Diet (see Food)
Distance estimation (see Depth perception)
Dorsal pallium 135
Dorsal tegmentum 142, 143
Dorsal thalamus
- morphology 137, 138
- neurophysiology 232–234

Enemy avoidance 275
Epithalamus 137
Experience
- larval, premetamorphic 81–82

Eye
- aquatic 89, 90
- morphology 90–94
- refraction 119–124
- terrestrial 89, 90
Eye convergence 53
Eye degeneration 127, 128
Eye movement
- respiratory 51
- undirected 51

Feeding
- aquatic 15, 20, 21
- ecology 27–34
- in darkness 59
- larval (see aquatic)
- olfactory guidance 60–65
- ontogeny 65–79
- terrestrial 15, 20, 22
Feeding apparatus 14–27
Feeding sequence 34–36
- neural control 269–275
Feeding strategy
- change in 79–81
Food
- larval 29
- natural 27–34
- seasonal variation 29, 30
Forebrain
- effect of electric stimulation 240
- effect of lesions 241
- morphology 132–136

Horizontal cells 102
Hyobranchial apparatus (see Feeding apparatus)
Hypothalamus 138

Illumination level 45–47
Imprinting 76–79

Lateral pallium 135

Marginal optic tract 153–160, 163, 164, 166
Medial optic tract 153–160, 163, 164, 167

Medial pallium 134, 135
Medulla oblongata 144–150
– role in feeding 274
Motion parallax 53

Neuropil Bellonci pars lateralis
– in plethodontids 156
– in salamandrids 153–155
– Neuropil Bellonci parts medialis
– in plethodontids 156
– in salamandrids 154, 155
Nucleus Bellonci
– connectivity 183, 192, 194, 197
– morphology 137
Nucleus Darkschewitsch 139, 190, 276, 277
Nucleus isthmi
– function 267, 268
– morphology 143, 185, 194
Nucleus mesencephalicus nervi trigemini 174, 175
Nucleus praeopticus 138
Nucleus praetectalis
– function 276, 277
– morphology and connectivity 139, 183, 190, 194, 197
– neurophysiology 234–239

Olfactory bulbs 132, 133
Optic nerve 152, 153
Optokinetic afternystagmus 84, 85
Optokinetic nystagmus 82
Optomotor behavior 82–85
– neural mechanisms 235–239, 275–277

Photoreceptors
– density 115
– number 115
– ontogenetic development 114
– size 115
Posterior thalamic neuropil
– in plethodontids 160
– in salamandrids 154, 155
Praetectum
– afferents 190–192
– effect of electric stimulation 240
– effect of lesion 241
– efferents 190–192
– morphoogy 139
Pretectal neuropil (see Posterior thalamic neuropil)
Pretectal neurons
– response properties 234–239
Prey (see also Food)
– correlation with predator size 30–33
– motion "postulate" 61, 65
– natural size 30–33
– odor 60
– spectrum 28
Prey catching behavior (see Feeding)
Prey localization
– in darkness 59
– in light 59, 62–64
Prey object identification 243, 244
Prey preferences
– experience dependence 68–82
– ontogeny 65–79
Prey recognition
– neural mechanisms 244–248, 255–257

Retina
– connectivity 111–113
– morphology 95–113
– thickness 91–94
Retina ganglion cells
– color coding 210–214
– displaced 110
– functional classification 201–208
– morphological classification 106–110
– number 117, 118
– response properties 201–210
– velocity function 208
Retinal disparity 263, 266, 267
Retinofugal afferents/projections
– development 166–169
– to praetectum 154–156, 160–162
– to tectum opticum 155, 156, 160–162
– to thalamus 153, 156
– topic organization 162–166
Rod-cone relationship 115–117
Rods 96–98

Sensitive period 76–78
Sensori-motor integration 269–275
Septal nuclei 135, 136
Size constancy 257–263
Snapping 35
– distance 52
– success/accuracy 54
Stationary prey
– detection 47–51
– stimulus preferences 48
Stimulus background contrast 45–47
Stimulus configuration/orientation 39–43
Stimulus movement pattern 43–45
Stimulus preferences 36–45
Stimulus size 36–38, 257–263
Stimulus size recognition 257–263
Stimulus velocity 38, 39
Striatum 136

Tectal neurons
– classification in anurans 177, 178
– constitution of response types 250–252
– functional classification 175–177, 248–250

- morphology 171–177
- response properties in anurans 227–231
- response properties in *Hydromantes italicus* 222–227
- response properties in *Salamandra salamandra* 214–222
- responses to monochromatic light 241, 242

Tectum mesencephali/opticum
- afferents 182–188
- cytoarchitecture 170–177
- cytoarchitecture in anurans 177, 178
- effects of ablation/less 240, 246
- effects of electric stimulation 239, 240
- efferents 188–190
- morphology 140–142, 170, 171
- retinofugal afferents 155, 160, 161

Tegmentum isthmi 143
Telencephalon (see Forebrain)
Thalamic neurons
- response properties 232–234

Thalamus
- afferents 192
- effects of electric stimulation 239, 240
- effects of lesion 240, 241
- efferents 192

Tongue projection 22–24, 26, 27
- velocity 25

Tongue skeleton 16, 17 (see also Feeding apparatus)
TP-Phenomenon 252–255

Uncinate field
- in plethodontids 156, 160
- in salamandrids 154

Ventral tegmentum 143
Ventral thalamus 138
Vision
- aquatic 89, 90
- interaction with olfaction 59–65
- terrestrial 89, 90

Visual acuity 124–127
Visual inhibition 61, 62, 75, 76
Visual mate recognition 85–88
Visual pigments 98–101
- change in 100, 101

"Worm-antiworm" discrimination 39–43, 252–255

MIX
Papier aus verantwortungsvollen Quellen
Paper from responsible sources
FSC® C105338

If you have any concerns about our products,
you can contact us on
ProductSafety@springernature.com

In case Publisher is established outside the EU,
the EU authorized representative is:
Springer Nature Customer Service Center GmbH
Europaplatz 3, 69115 Heidelberg, Germany

Printed by Libri Plureos GmbH
in Hamburg, Germany